BARONS
OF THE
SKY

FROM EARLY FLIGHT
TO STRATEGIC WARFARE

THE STORY
OF THE
AMERICAN AEROSPACE
INDUSTRY

BARONS
OF THE
SKY

WAYNE BIDDLE

The Johns Hopkins University Press
Baltimore and London

For Sam

Originally published in a hardcover edition by Simon & Schuster, Inc., 1991
Johns Hopkins Paperbacks edition, 2001
2 4 6 8 9 7 5 3 1

Designed by Irving Perkins Assoc.
Picture Editor: Vincent Virga

The Johns Hopkins University Press
2715 North Charles Street
Baltimore, Maryland 21218-4363
www.press.jhu.edu

Library of Congress Cataloging-in-Publication Data

Biddle, Wayne.
Barons of the sky : from early flight to strategic warfare : the story of the
American aerospace industry / Wayne Biddle.
p.   cm.
Originally published: New York : Simon & Schuster, 1991.
Includes bibliographical references and index.
ISBN 0-8018-6828-9 (pbk. : alk. paper)
1. Aerospace industries—Military aspects—United States—History.
2. Aircraft industry—Military aspects—United States—History.
3. Airplanes, Military—United States—History. 4. Aeronautics, Military—
United States—History. I. Title.

HD9711.5.U6 B54 2001
338.4′762913334′0973—dc21
2001038606

A catalog record for this book is available from the British Library.

CONTENTS

PREFACE TO THE
JOHNS HOPKINS EDITION

WHEN HENRY ADAMS ENCOUNTERED the mammoth dynamos displayed at the Paris Exposition of 1900, their power struck him "as a moral force, much as the early Christians felt the Cross." His instinct, he wrote, was to pray to them, as perhaps one might worship any "symbol of infinity" and "ultimate energy."

A century later, the power of technology is so commonplace that unless it actually injures us it has mostly lost such personal impact, but the moral force that Adams sensed has never gone away. High technology is still an imperative equated with progress, well-being, and truth, despite the occasional disaster. Respectful critics of this god risk being thrown out of church by believers or scorned by atheists. So it is today, as it was a decade ago, when *Barons of the Sky* was first published.

It seemed then that with the dissolution of the Soviet Union might come a retreat from the mountainous Cold War defense budgets that had rescued the aviation business after World War II and fueled its growth into the modern aerospace empire. But after a few years of consolidation, the weapons builders settled into the old cushion again, supported by military spending that remains enormous by any standard. The lack of an enemy state of covalent stature made no difference. Now an elusive terrorist may seem to justify almost any expenditure. One could say that there is a moral force behind buying the newest, most advanced weaponry, regardless of how suitable it is to the world at hand.

Readers coming anew to this book should therefore regard it not just as a history of technology with requisite economic and political context, but also as the record of an especially American sense of national identity. The military-industrial partnership thrives because Americans believe in it (and presumably in the threats to national security it advertises), not because its goods are always wanted or even work well. Perhaps the contemporary chimera of anti-missile defense best demonstrates this cultural trait, whereby decades pass with little or no technical advancement while billions are poured into a problem whose very definition shifts beneath the analytical foundation engineers usually require.

Maybe this rubric is too high-minded, however. Congressional investigations, court records, and the Pentagon's own internal housekeeping reports would lavishly support an alternative explanation for the perennial durability of the aerospace business, namely that the defense establishment long ago became so vast as to be self-perpetuating and unmanageable. It is certainly not the pet of any one president or party. Voters (even the small minority who follow defense issues) have no control over a wasteful system corrupted by pork-barrel politics and old-fashioned fraud. Americans don't believe in it—they're just stuck with it.

The early-twentieth-century founders of the industry would surely recognize the latter explanation for what their baby grew up to be. They were hard-nosed businessmen not given to philosophical introspection. They built airplanes, an expensive product with no mass market. Like most entrepreneurs, they never much liked the government in Washington, but the military turned out to be their most reliable customer in an epoch of one war after another. If their product somehow became part of the American psyche, then so it goes.

Moral force or fiscal quagmire, the technology that the Wright brothers pioneered and then lost to the weapon-makers biographied here is a fundament of our age. The names of the original barons are now strung together in mergers like Lockheed Martin or dropped completely like Douglas, but the process of designing, assembling, and selling their singular merchandise depends on the same troublesome symbiosis between state and corporate purpose that coalesced in the days of linen-covered spruce.

W.B.

Baltimore

September 2001

BARONS
OF THE
SKY

FOREWORD

CORPORATIONS DESTROY HISTORY. Those few that have any interest at all in their past—beyond meeting mundane legal requirements—tend to preserve records that make them look good and throw out the rest. Even the most flattering material may be lost after a generation or so, unless it carries some advertising value. That the largest companies have played central roles in American history, especially during this century, as their power reached the level of international sovereignty, is undeniable. Yet critical accounts of any commercial enterprise are among the rarest of scholarly or journalistic works. There is no National Archives for the Fortune 500.

The problem is compounded when a business has been associated for many years with "national security," a tenebrous realm that hides failures and embarrassments. In the aerospace industry, this has contributed to a morass of petty secrecy, public relations ploys, and facile patriotism that can stupefy even serious students of its technical achievements. Much writing about individual companies thus seems to wind up in territory populated by hardware buffs, whether or not technology was the author's intended focus. But hardware is always trivial when isolated from the political and economic forces that spawned it.

My ambition for this book was to trace the rise of some of the great American aerospace weapons companies—not just by examining what remains of their paper trails, but by following the lives of their founders, who were a rather eccentric lot. Flying was considered craziness in the early days, and anyone who chose to build flying machines for a living dwelt on the outskirts of polite society. It was war, a condoned form of insanity, that brought the backyard "birdmen" into the mainstream. None of them was diabolical enough to contrive this singular path to success, but none ever turned away from it.

The seminal companies that are the focus here—Martin, Douglas, Lockheed, and Northrop—were bright reflections of their founders' strengths and weaknesses. The men who launched them occupied the heart of the business for five decades, beginning with the cottage-industry era before World War I. Although they were sometimes outdone by relatively faceless conglomerates, they were widely recognized as flagships of a new technological, political, and economic power. (Two other companies that might claim a place in this elite—Boeing and Grumman—have not been treated as deeply, for separate reasons: Boeing was absorbed by a large holding company rather early in its history, diffusing the influence of its pioneer; the family of Grumman's founder did not save his personal papers, making probative biography virtually impossible.) Like Ford, they became synonymous with what they built, for better or worse.

In an ironic twist, by the end of the Second World War their products were so vital to the nation's sense of itself as a great power that the patriarchs' oddities could no longer be tolerated. Some were squeezed out; others stumbled badly in postwar competition. The most illustrious enjoyed a certain succès d'estime, while their namesake companies carried on without them in what must have felt like an icy new order.

There is a pattern here that can be seen in other big businesses—autos and computers come immediately to mind. Brash pioneers, often downright oddballs, start a shaky commerce that begins as a personal obsession and turns gradually into a cultural bulwark. The business grows so important that the qualities apparently needed to initiate it are soon perceived as counterproductive. The pioneers become legends, perhaps, but their likes are never allowed again in the boardroom.

Where the aerospace industry differs is that none of the legendary founders wanted their commerce to turn into the kind of bulwark it became—a "Gargantua Behemoth," a military giant dependent on the federal government for sustenance. Glenn Martin, Donald Douglas, the Loughead (Lockheed) brothers, and Jack Northrop each started out as teenagers awestruck by the airplane's promise as a means of transportation. They were infatuated with the idea of flying. Robert Gross had not the slightest interest in weapons production when he plucked the Lockheed company out of bankruptcy court during the Great Depression. He wanted to expand upon the aviation business success he had tasted as an investment banker before the Crash. Yet they all embraced the military sooner or later, because there was no steadier source of income. It appeared to be their duty, as well.

Without doubt, their youthful ideals were terribly compromised (in some cases more willingly than others), though they tried to find non-military markets whenever possible (in some cases far more successfully than others). But unlike automobile manufacturing, which could be sustained by a middle-class consumer economy, aviation was always a super-luxury endeavor that could not thrive without massive government aid. And throughout this bellicose century, government aid for technological enterprises has carried a military imperative. A central imponderable is what the great burgeoning of aerospace science and technology would have been like without two world wars and myriad smaller conflicts—or whether there would have been any burgeoning at all. How darksome it is, especially now after yet another foreign war celebrating new technology, to consider the extent to which the most fertile fields of American engineering have been shaped by wartime values. Aerospace industry founders recognized this quandary at various stages of their careers, but neither the men nor the times were inclined toward regret.

In attempting to weave together institutional history and individual biography, I was fortunate to find two extraordinary collections of personal papers in the Library of Congress—those of Glenn Martin and Robert Gross—which had not been explored previously in fullest detail. They contained a wellspring of information not just about Martin and Gross, but about nearly every notable figure in the military-industrial world of aviation during their lifetimes.

The Martin papers were especially poignant. A lifelong bachelor who always lived with his mother, he had no one to leave the minutiae of his life to, and thus seems to have emptied his file cabinets into the Library's outstretched hands. There were old Christmas cards and country-club bills jumbled among the most momentous corporate documents, but it was not a haphazard collection. Martin was careful to present a refined personality to the public from the first days of his career. Yet by the account of every contemporary who cared to remark, he was by nature a very peculiar fellow.

Was he gay? Almost certainly yes, in style if not by act (there is no common evidence of a sexual relationship of any kind during his entire life). This sensitive question is of importance only in that it points to what must have been an excruciating aspect of his success. He started out as a circus performer, where provincial audiences paid to see departures from what they thought was normal. But he wound up in a politically ultraconservative business where any perceived social deviation was anathema. After he became famous, it was often said that he had no life other than his business. But there was probably no other life that he would have been permitted to have, given his apparent disposition.

The Gross papers revealed a writer of considerable elegance. In his vast business correspondence he maintained a remarkable ingenuousness, clarity of purpose, and sense of integrity that more than made up for his complete lack of technical acumen in aeronautics. In his personal letters he expressed the self-effacement that always helps make an individual's wealth and power more palatable. Some of these attractive traits might have stemmed from the security of his upper-class New England roots, though the same background has certainly produced its share of monsters. More than other industry leaders he was prone to doubt about weapons production, yet under his guidance Lockheed became one of the biggest armorers of all. In short, he was a splendid synthesis of contradictions, and his papers provided a broad window into the industry's culture.

It is a pity that so much less documentary evidence remains of the life of Donald Douglas. He was in many ways the most appealing figure of his coterie, the first spectacular success in Southern California's garden of high technology. Having watched his business grow literally from a storefront to a sprawling kingdom, he harbored few illusions about the glory of it all. Among engineers, who are not known for their emotional color, he was a complex personality, with a stormy romantic life behind the drab veneer of airplane builder. He had a certain talent for writing, too, which, though undeveloped, might have helped him leave a useful record of his experience. Unfortunately, many of his personal papers have been lost, leaving a void to be filled by public relations officials far less sophisticated than he deserved.

As for Jack Northrop, who rose from abject poverty with a provincial high school education to design several of this century's technological masterpieces, few direct remnants of his life exist besides images of Vegas, Alphas, Gammas, and flying wings that in their heyday made all surrounding objects look like artifacts. This is as it should be, perhaps, for the only technical genius among the founding fathers. Northrop still has a loyal following of buffs who defend his most radical and beautiful creation, the 1940s flying-wing bomber, but anyone who seeks deeper clues to his legacy must be content with rather dry utterances from a man who described himself as "unsocial."

All in all, the rise of these men and their companies was neither as wholesome as the annual reports would have us believe nor as evil as the "merchants of death" theorists would sometimes contend. At bottom, it was business as usual, and Americans should understand by now that business never makes a pretty picture, on either a personal or a cultural level. Sooner or later, dog eats dog, a fact of life borne out by the extraordinary careers of a few individuals who loved flying, loved their country, and made a great deal of money when the two loves combined in war.

THE RESEARCHING AND WRITING of this book would have been impossible without the support of my wife, Mimi. She never questioned the value of a project that made little sense according to most measures of everyday life. My deepest gratitude is expressed here by giving her first mention among these acknowledgments.

For a fellowship that defrayed research and living expenses, and also provided a circle of colleagues at an early stage of work that might otherwise have brought isolation, I thank the Alicia Patterson Foundation. Hoc Noble, Gilbert Rogin, David Sanford, and Pierre Sprey wrote recommendations that helped secure this godsend. Leonard Downie, Linda Greenhouse, William Greider, Stephen Hess, Anthony Marro, Peggy Simpson, and Betty Anne Williams judged my proposal to be worthwhile. Peggy Engel met every administrative need. Gaylord Shaw graciously extended the resources of *Newsday*'s Washington bureau as an additional benefit.

The Library of Congress, especially the manuscript division that holds the various collections of papers cited here, gave space and professional assistance. For the latter I especially thank Fred Bowman, Charles Kelly, and Mary Wolfskill for guidance through the Martin and Gross collections. I am also grateful to Mary Ison and Beverly Brannan of the prints and photographs division for patient supervision while I studied the unprocessed Glenn Martin photo collection.

At the National Air and Space Museum library, Larry Wilson, Tim Cronen, and Bob Dreesen were always helpful in locating document files and photographs pertaining to aviation companies, founders, and other historical figures.

In Columbia University's oral history research office, I thank Ron Grele and Mary Marshall Clark for their help in using transcripts of interviews with Donald Douglas and other aviation industry notables. I am also grateful to David Russell of the University of California at Santa Barbara library for the availability of a 1972 interview with Jack Northrop, a copy of which was provided by Northrop Corporation historian Ira Chart.

I am especially indebted to W. Bruce Arnold for allowing access to the H. H. Arnold collection at Columbia. Barbara Douglas Arnold, Lee Atwood, Donald Douglas Jr., Charles Ducommun, Rose Grumman, and John Lockheed graciously provided personal recollections and memorabilia.

For her faith in this project from the start, I thank Harriet Wasserman. At Simon & Schuster, Allen Peacock provided the best editorial guidance I have ever known. Alice Mayhew took responsibility for the manuscript at a critical juncture, doing the solid job for which she is widely respected. Ari Hoogenboom saw it through the production phase, answering all of my anxious questions. Vincent Virga handled the photographic section with a close interest for which I am grateful.

Finally, I would like to thank those editors of *The New York Times* who first permitted me to enter the Pentagon labyrinth on a regular basis. Credentials come and go, but the experience of enormous clout is indelible.

W.B.
Washington D.C.
March 1991

The right is more precious than peace, and we shall fight for the things which we have always carried nearest our hearts—for democracy, for the right of those people who submit to authority to have a voice in their own governments, for the rights and liberties of small nations, for a universal dominion of right by such a concert of free peoples as shall bring peace and safety to all nations and make the world itself at last free.

WOODROW WILSON, address to Congress
April 2, 1917

Is there any man here or any woman—let me say, is there any child—who does not know that the seed of war in the modern world is industrial and commercial rivalry?

WOODROW WILSON, speech in St. Louis
September 5, 1919

PROLOGUE

FANTASIES

IT WAS A SPRING MORNING in Mexico, 1913. The whole world had not yet been changed forever by total war. Somewhere there were still kings and queens of actual consequence, real ladies and gentlemen, unregulated capitalists, feudal masses, and smartly tailored legionnaires. All would be swept away by the coming bloodbath, leaving a red border between the mannered past and the technological future. In these last months of the old ways, there were moments of fantastic juxtaposition, even in places where people had never heard of Serbia, Croatia, or the Archduke Ferdinand.

In Sonora Province, illiterate revolutionaries who controlled the region fought Victoriano Huerta's autocratic regime in Mexico City much as their fathers had battled against the dictator Porfirio Díaz— wearing leggings and sombreros, on foot or horseback, with Remington carbines smuggled down from the Coloso del Norte. Yaqui and Maya Indians siding with the rebels carried bows and arrows, still quite effective in the bush.

But at Maylorena, a railroad junction forty miles north of Guaymas on the Gulf of California, a twenty-seven-year-old Frenchman named Didier Masson stepped from his customized Pullman car parked on a desert spur. He enjoyed the rank of captain in the army of Alvaro Obregón, a young, innovative Constitutionalist who became one of the revolution's greatest generals. The desperate Obregón hired the obscure Masson for the luxurious wage of $300 a month to perform a military mission that was as far outside anyone's experience as gas masks were from sombreros. Masson would on that day steer a machine that could fly over Federal ships in Guaymas Bay and attempt to drop bombs onto their decks. It was as simple as that, but it had never been done before.

The queer contraption that faced him in the dust was not much advanced from the Wright brothers' invention. It might remind someone who was mechanically inclined of a self-propelled kite, but there was nothing obvious about its shape or construction that would build confidence in an average bystander. It looked more like motorized beach furniture, perhaps, or an abstruse combination of English bicycle and Dutch windmill parts. It was, to many normal adults of the time, an expression of psychosis. Ten years after the Wrights' epochal success, flying was still an activity for visionaries and fools, with far more of the latter willing to climb on board such a rickety deathtrap.

Much less reliable than a burro, Masson's machine was nonetheless capable with plentiful luck of carrying a passenger and 150 pounds of explosives for a hundred miles. The pilot sat on a plank braced between two fifty-foot-long wings made of cotton fabric stretched across wooden spreaders, with an unmuffled seventy-five-horsepower gasoline engine directly behind his head. There was no protective fuselage. There was nothing to hold on to except an unsteady steering wheel. Obregón had acquired this device in Los Angeles for $5,000 through the good offices of the Aero Club of Southern California, a gentlemen's sporting group not well known for kindness to Mexicans but eager to promote the business of flying in any way possible. To them, the transaction was perhaps like selling Jack some magic beans.

The plane was shipped dismantled in five crates by rail to Tucson, then on a truck to a ranch outside Nogales. There it was seized as contraband by a U.S. marshal acting on a rumor, just as Masson and his Australian mechanic were heading off toward the border in a horse-drawn wagon. They insisted that they merely wished to set the machine up and test it in flat country to avoid accidents, an almost plausible excuse. Two young men with a few boxes of staves, wire, cloth panels, and engine parts could still assemble the world's most advanced weapon in a few hours, with most passersby having not the slightest idea of what was in store.

Without proof that the aeroplane was indeed bound for war in Mexico—a prospect that even many military experts would have found difficult to believe—the marshal left a one-legged deputy to guard the camp. It seems likely that some rebel cash helped fix this arrangement, given the rollicking arms market of the border region in those days. Or perhaps a *mordida* is what the marshal had in mind all along. In any case, the deputy was soon persuaded to accompany the expedition across the border to Hermosillo, where he was made a major in the Obregón forces.

Masson had by now already made a number of reconnaissance flights over Federal positions, scattering propaganda sheets along the way. So when he strapped himself into the *Sonora,* as the little biplane was called, and revved up its motor, he was fairly certain of his invulnerability as long as he managed to stay aloft. But Obregón's thousands had not been squandered with leaflets in mind; the colonel badly needed to neutralize the power of the gunboat *Guerrero* in Guaymas harbor.

At such distance, both chronological and technological, from the early years of flight, it is impossible to conjure the shock that accompanied the appearance of aircraft. Civilized people at exhibitions screamed and cried and rushed the pilots as though they were archangels. As late as 1910, seven years after the first powered flight, a mob watching a demonstration in Denver ripped souvenirs from a crashed plane and stole the gloves from the dead pilot's hands before police cleared the site. Seeing a man in the air aboard one of these machines was evidently a paroxysmal experience, a raw confrontation with the power of technology that we can no longer feel.

Knowing that the birdman might also kill with impunity was an immediate terror that has never gone away. What Obregón sensed was that it made little difference whether the Frenchman and his absurd machine caused real damage—the psychological trauma alone would be worth the expense.

Masson was joined by a Mexican captain who would actually drop the eight homemade bombs strung beneath the plane. If the mission succeeded, it would not do to have a foreigner take all the honor. Each bomb consisted of a fifteen-inch piece of iron gas pipe packed with sticks of dynamite and scrap metal. A rod ran from the bomb's nose through the center of the pipe to a detonator screwed into the tail end. When the rod struck the ground, it jammed into the detonator, which was made from a dynamite cap and match heads. In a later epoch, delinquent teenagers might create similar ordnance from intuition alone. The bombs were released by yanking a string tied to a wire clip on each one. A crude aiming sight was fitted next to the bomber's seat, though it accomplished little except adding a bit of confidence-building pseudoscience to the escapade. In just thirty-two years, Hiroshima would be leveled by the same concept.

Thus the terror machine was ready. Masson taxied away from his Pullman car and its refrigerator stocked with real ice as part of the deal, clattering across the desert floor with harebrained optimism that the collection of garage hardware would stay together. The machine got airborne—an event that must have astonished him and surely his passenger every time. Airplanes were just barely workable, with inefficient wings, dull resistance to the wind, and strained engines that stank of leaking castor oil. He nonetheless climbed the *Sonora* to about 2,000 feet and headed for Guaymas. If we try hard, we can imagine how much like a god this foolish young man felt, with all of poor Mexico rolling past below at his mercy.

Masson flew five sorties for Obregón between May and August 1913. On the first, bombs fell around the *Guerrero* but did nothing besides splash water on its deck. This was enough to send terrified sailors diving overboard when he returned the next day. He then flew some experimental runs back at Maylorena with his mechanic as bombardier, trying to figure out how to make the pipes land where they should.

On his last attack, under hapless fire from the gunboat as usual, the *Sonora*'s engine died over the bay and he glided to a landing almost inside friendly territory while cannon shells dug up the earth around him, just like in a real war. All the way down he trailed a bomb at the end of a fouled release string, but luckily it was a dud. As enemy troops fought toward the plane, he tore some loose components from the engine and ran for his life, turning to notice that the Federals were not interested in the pilot as much as the terror machine, which they quickly surrounded. Useless with a blown engine, however, the *Sonora* was soon recaptured by those who had spent so much to bring her there.

The next day Masson handed his resignation to Obregón—sick, disgusted, and unpaid for the past month. The last factor was probably the deciding one, with dozens of different kinds of Mexican money in circulation, some printed on toilet paper.

Before leaving Obregón's band, Masson was persuaded to teach the nephew of Venustiano Carranza, the leader of the Constitutionalist Army, how to fly. This young man took the *Sonora* to Mazatlan, where he generated mortal fear among the population before dashing the machine to bits. Masson maintained near the end of his life that he had refused Obregón's demand to bomb cities, though there is at least one account, from May 1913, that reported Masson circling over Guaymas, tossing bombs which "fell in the principal business street, causing some loss of life and doing great damage to property." The alleged attack "of this dragon of war created a reign of terror such as the city has never seen." Guernica was still two decades away.

It is possible that this story was concocted or embellished by Federals who hoped to enrage foreign sensibilities against the barbarous new weapon, which they would soon be using themselves. After the nephew's Mazatlan bombings, the commander of the United States Pacific Fleet sent two aides to the Constitutionalist camp to protest the use of airplanes. (The U.S. Navy had been experimenting with flying machines since 1911, and an American admiral patented the first aerial torpedo in 1912. This diplomatic mission was a bald attempt to maintain primacy, at least in the hemisphere.) But the world was about to leave such gentilities in the dust.

Masson drifted around Sonora until the start of the Great War, "doing business," as he would say. But there was little honest business for him or anyone else. He returned to France and eventually joined the Escadrille Lafayette, winning the Croix de Guerre in 1916, which was proof of courage and extraordinary luck. After the war he went back to Latin America and tried to launch various export ventures over the years, including chewing gum and canned crayfish, but none caught on.

He died in 1950 as manager of the Hotel Iris in Chetumal, Quintana Roo.

LURING THIS MERCENARY, whose most notable accomplishment in the New World prior to the Guaymas raids had been flying a load of *Los Angeles Times* newspapers to San Bernardino in 1911, was not difficult. Smuggling an airplane across the Arizona line called for some youthful bravura, though the border barely existed as a barrier for illegal arms. It was all quite easy, actually, which is the lasting significance of an episode that otherwise would slip from memory as the century turned infinitely more brutal.

At the other end of the Mexican arms pipeline was an oddball young businessman named Glenn Luther Martin. He lived with his

mother in Los Angeles and built airplanes on a shoestring, supporting the work by stunt-flying at state fairs and rural carnivals. Among
his threadbare employees was Masson, who taught local rich boys
how to fly, and the Australian mechanic, who liked to play with
explosives. Martin looked so foppish in his circus costume of riding
breeches and black leather jacket that other exhibition pilots mocked
him as the "Flying Dude," about the same in those days as being
called a faggot. But he was a nervy self-promoter who foresaw that
war could be a golden realm for these new machines. With a few
other men who were infatuated with airplanes and needed to make
a living, he would go on to found one of the most powerful and
vexatious industries in American history. Within his lifetime, it
would absorb vast resources of the U.S. Treasury—sometimes saving the nation, often bleeding it.

Didier Masson's *Sonora* was his first export sale.

CHAPTER ONE

INVENTION BEFORE INDUSTRY

SINCE PEARL HARBOR, the United States has been in a constant state of either fighting or preparing for war, a strange fate for a liberal democracy that has allowed the military to gain enormous influence on everyday life. By trying to keep its armies in perpetual readiness, the nation cultivated an armaments industry that now seems as natural to many Americans as television and computers.

Most accounts of the modern weapons business use the late 1940s as a starting point. Competition between capitalism and communism, lethalized by the spread of atomic bombs, caused World War II armorers to continue thriving instead of withering as was the usual case after mass calamities. Forty years later, when the Soviet Union allowed a renewal of nationalism along its border with Europe, the rationale for keeping arsenals stocked to their brims became as pervious as the Berlin Wall. But the taproots of the greatest American weapons-building fortunes—the ones with family names almost as famous as Ford—run much deeper than any modern political tangle, stretching all the way back to the first years of the century when machines that could fly became foremost instruments of war.

While 1913 marked the first time an American-built airplane was used for killing, and the first such naval battle anywhere, the event was by no means a sudden quirk. The invention and early development of flying machines may seem like a folksy age of innocent fascination—at some moments perhaps it was. But the military held an immediate and soon decisive influence on the founders of the aviation business, which steadily evolved into today's aerospace weapons industry.

The entrepreneurs who survived those early years to establish great corporations did so by becoming weapons salesmen, not by putting a flying machine in every garage or between every city. For the most part self-motivated and ingenuous inventors, they were hardly "merchants of death," as the rhetorical charge would first be made during the 1930s—at least not any more or less so than the Manhattan Project physicists were sinners. Their goal in life was to build airplanes and make money. The only reliable way to achieve this turned out to be producing weapons for the federal government, a process that mixed patriotism with cutthroat business tactics. This was a potent combination, a double-edged sword that has obviously proved difficult for society to manage.

Tracing the industry's long roots means entering a time that is almost gone from living memory and altogether ancient as measured by the products or organization of advanced technology. That Didier Masson's *Sonora* and the B-2 Stealth bomber could both exist within an average human life span is dizzying evidence of the pace of change. Consequently, the industry often seems disconnected from any historical interest other than that of buffs who dote upon famous airplanes. But this is an illusion, as shall be seen, because the imbroglios that perennially darken the business—all of which stem from an inability to draw clear lines between public and private institutions—were there at the beginning.

THE SOLUTION TO THE PROBLEM of human flight did not come from a scientific or engineering elite, as it invariably would today. Attacking the subject successfully required a combination of naiveté and shop-floor common sense possessed mainly by provincial mechanics working outside professional circles. For a while, this social factor had a determining effect on who entered the business and how they made their way. If some of the future magnates of the aerospace empire were to come back again as young men in the late twentieth century, they might be found pumping gas or doing tune-ups out on the Interstate.

Biographers of Wilbur and Orville Wright have tended to insist somewhat loftily that what distinguished the brothers was their step-by-step pursuit of technical details, a process that transcended the wasteful cut-and-try approach of other experimenters. But they were finical Ohio shopkeepers whose orderly approach to building the first flying machine can be interpreted either as scientific method or a natural manifestation of straitlaced fussiness. Glenn Curtiss—a contemporary of the Wrights who occupies a lower rung in history but had a stronger sense of how to make flying practical—was a roughneck who raced motorcycles in the farming village of Hammondsport, New York, which is better known for viticulture than as an early hotbed of technology. He was a gifted technician with a far more pragmatic intuition for aeronautics than Alexander Graham Bell, who became his mentor in 1907 after purchasing a Curtiss-built engine to power large kites. Yet he was still enough of a rube to be conned out of his entire manufacturing facility in 1909 by a charlatan who claimed to hold patent rights predating the Wrights'.

The Wrights and Curtiss, whose names have faded from the modern weapons industry, sprang from a network of cranks and semi-serious researchers that gained respectability very slowly even after having brought forth a successful machine. Their first task was to make flying believable, then to demonstrate its usefulness as something more than a carnival stunt. This process consumed the ten years after the Wrights' 1903 maiden flight, when technical progress in aeronautics was otherwise slight, at least in the United States. That it paralleled the start of the First World War is one of the century's darksome coincidences, like the discovery of nuclear fission just before World War II.

Until 1903, one major concern of all flying machine seekers was to avoid being pilloried not just by a skeptical public but especially by educated observers. "Man's desire to fly like a bird is inborn in our race," remarked the eminent astronomer and mathematician at Johns Hopkins University, Simon Newcomb, in an often quoted *McClure's* magazine article of 1901. "We can no more be expected to abandon the idea than the ancient mathematicians could have been expected to give up the problems of squaring the circle." Even H. G. Wells, the best-selling futurist, would only venture to predict a successful first flight "long before the year 2000 A.D., probably before 1950." Those who entered the field either had little to lose socially or were so obsessed by the notion of heavier-than-air flight that they swallowed the risks of censure.

On a more abstract level, these researchers were also brought together by a new pecking order in the technical professions wherein academically trained engineers shared some of the territory once held exclusively by craftsmen. There was still a broad area, especially in fields such as aeronautics that lacked a mature theoretical basis, where happy-thought and rule-of-thumb could be as powerful as higher mathematics. This was the opening through which a few bright young garage hands eventually pushed their incredible invention.

By the 1880s, an unruly mix of minds and methods had produced enough preliminary information about flight to embolden several reputable American technologists to take their obsession public. In 1886, Octave Chanute, a vice president of the American Association for the Advancement of Science who had become a highly successful civil engineer after apprenticing in the railroad business, called for an aeronautical program to be part of the AAAS's annual meeting in Buffalo, New York. Chanute was a solid figure, the builder of the Chicago stockyards and the first bridge across the Missouri River at Kansas City. As an engineer whose training had been non-academic but who was firmly ensconced in his profession's establishment, he straddled the two worlds of flying machine research. He was joined by Robert Thurston, dean of Cornell University's engineering college, an elite institution nonetheless founded on the ideal of universal education. Thurston had surveyed aeronautics for *Science* magazine in 1884, reaching the audacious conclusion that "the art of aerostation is much nearer a practical state than scientific men generally suppose."

Yet neither Chanute nor Thurston felt comfortable about revealing his personal fascination with flying at the AAAS conclave. Instead, Chanute invited an Illinois farmer named Israel Lancaster, an amateur ornithologist who claimed to have flown small models of birds for more than five hundred yards, to speak about his experiments with "soaring effigies." Lancaster was a reputable bird watcher, but his thoughts about "attaining atmospheric dominion" were strictly homemade. Not surprisingly, the lecture turned into a farce as the learned audience hooted at Lancaster's untutored theories. Chanute skipped a scheduled demonstration of the models, where they proved unworkable and a few raucous scientists shouted joke offers of cash for a good flight.

But the public sacrifice of poor Lancaster served to attract to the strange coterie of aeronautical enthusiasts an equally peculiar figure in nineteenth-century American science, Samuel Pierpont Langley. A pompous, idiosyncratic man, he had published a few notable papers during the 1870s based on observations of the sun, and his public reputation had risen to heights far above his contribution to science. The connection of his prominent titles to semi-scientific speculations about flying would help propel the field upward, socially if not technically. His career deserves a close look here because it was in many ways typical of those of the self-concocted individuals who would later start the aviation weapons industry.

LANGLEY, a handsome fifty-one-year-old bachelor and an internationally prominent figure in science at the time of the AAAS meeting, was from a middle-class Boston family that had become well to do in the wholesale produce business. After graduating from high school, he worked in the Midwest as a civil engineer and draftsman—loosely defined occupations which in those days did not require college training. Little else is known about his youth, because of his secretiveness and the destruction of personal records. Apparently discouraged by his prospects after a dozen years, he came home to Boston in 1864 and decided to be an astronomer, another occupation that could still be pursued without much formal training, though career openings were scarcer than for apprentice engineers.

With the help of his Harvard-educated younger brother, John, he tried building telescopes, grinding lenses, and doing other fabrication at his grandfather's farm outside Boston. He and his brother then took a grand tour of Europe, visiting some observatories along the way. When they returned in 1865, the director of the Harvard Observatory was somehow persuaded to hire him as an assistant, a rather lowly job that nonetheless offered opportunities to look through a large telescope and be exposed to the daily routines of legitimate research.

On the basis of the director's recommendation, Langley parlayed this position into an assistant professorship of mathematics and director of the observatory at the United States Naval Academy, which was evidently so desperate for staff after the Civil War that it would hire someone with only a classical high school diploma and virtually no experience. In 1866, he leaped again to the Western University of Pennsylvania (now the University of Pittsburgh) as professor of physics and director of the Allegheny Observatory, a decrepit back-water from where he would sculpt a gigantic persona over the next twenty years.

The observatory had been born several years earlier as a venture of Pittsburgh citizens who hoped to bring some culture to the raw city through amateur astronomy, then a widespread fad. They had purchased an expensive German telescope from a dealer in New York, but none of the auxiliary equipment needed to point it. One of the group later called the project a "well-meant blunder of a few gentlemen." As the private association floundered, the telescope and an observatory building were deeded to the local college with enough money to endow a meager professorship. Only Langley and one other candidate, apparently a local man about whom nothing is known, applied for the job.

When Langley arrived, he found a dilapidated observatory structure, a useless telescope that had been damaged under its leaky roof dome, and no funding beyond his salary. But he also discovered that one of the richest men in the region, railroad executive William Thaw, was the observatory's most influential trustee. Langley soon cultivated a relationship with Thaw that outfitted the observatory and freed him from teaching duties that he was neither inclined nor educated to fulfill. This relationship became so personal that by 1871 Thaw stipulated in a gift pledge that $100,000 would go to the the university only if Professor Langley was not forced to teach any students.

Thaw also helped Langley set up a system of providing the exact time to railroad companies, bringing in money to supplement Thaw's own largesse. Until the mid-1880s, this was enough to hire technicians who not only ran the time service but made up for Langley's shortcomings in basic science and math. Langley used the income to add to his salary, too, which while modest was still the faculty's highest. The combination of self-serving intimacy with a powerful trustee and the profit-making use of a public institution irked Langley's academic colleagues, leading to his virtual isolation at the observatory. The situation was not helped when Thaw orchestrated the hiring of Langley's brother as a chemistry professor in 1872.

Nonetheless, Langley was able during the 1870s to carry out original research in solar astronomy that attracted international attention. Other than the brief stints in Cambridge and Annapolis, his only exposure to professional astronomy had been during two solar eclipse expeditions in 1869 and 1870. But knowledge of the sun was still so minimal in the 1870s that almost any serious questions offered open territory. He shrewdly launched his career with a visual study of sunspots, a project which did not require much mathematics or advanced physics and chemistry.

Langley was by nature and necessity an empiricist, not a theorist—a trait he shared with many of the early flying machine inventors. As it turned out, Allegheny was the only well-equipped observatory in the country devoted primarily to solar research, a happenstance which helped raise its rank in a relatively short time. Working before photography was adapted for such studies, Langley produced beautiful drawings of sunspots and lucid descriptions of the sun's surface. Prodigious patience and an attractive writing style made up for lack of scientific background. When he tried to go beyond empirical work, however, he often stumbled badly. In 1874, he announced that sunspots directly affect climates on earth by decreasing the planet's temperature, though he did not have adequate technical or theoretical support.

In the late 1870s, Langley developed an instrument he called the "bolometer" to measure the sun's heat. It was based on the known principle of the Wheatstone Bridge, where the electrical resistance of a wire varies inversely with temperature. Langley exposed the tiny wires of his instrument to various portions of the sun's spectrum and deduced with simple equations the related heat energy. The bolometer was maddeningly delicate and difficult to use—few other experimenters had any luck with it—but it brought Langley more acclaim.

The success of the bolometer pushed Langley to conceive an elaborate 1884 expedition, financed substantially by Thaw, to take the bolometer and other instruments to Mount Whitney in order to measure the heat-absorbing action of the atmosphere. Hastily and poorly organized—involving tons of equipment, a private railway car (provided by Thaw), a twenty-mule train for the wilderness, and a fourteen-member retinue that included a military contingent from the U.S. Army Signal Corps—the project was a tortuous fiasco. Langley nonetheless produced two years later a dramatic 239-page report that was widely read. That it contained bogus data and erroneous conclusions did not surface before lavish praise from prominent scientists further secured his place in the top echelon of international science. It was not until the early 1900s that Langley's value for the solar constant (a measure of the amount of energy reaching the earth's atmosphere)—which was too great by about 50 percent—was seriously questioned. The error was not discussed in print until 1914, eight years after his death.

As one of America's most venerated scientists, with numerous honorary degrees and memberships in scientific academies, Langley became secretary of the Smithsonian Institution in 1887. That summer, a year after attending the AAAS lectures on flying where Israel Lancaster was vilified, he ordered the construction at Allegheny of a large whirling arm device to test some of the existing assumptions about aerodynamic lift and drag. He referred to these experiments, which were as usual conducted by an assistant, as his "work in pneumatics" in order to hide his sudden interest in flying from university officials. Thaw again picked up the bill for equipment and staff. After an intricate series of heavily instrumented but technically specious tests of stuffed birds and small model airplanes that lasted until 1894, he concluded willy-nilly that "mechanical flight is possible with engines we now possess." He also declared that the faster a surface moved through the air, the less power was needed to keep it up.

As had often been the case for his solar physics research, in the absence of comparable studies elsewhere Langley's experiments were widely accepted and the speed-power relationship became sanctified—at least for a while—as "Langley's Law." Even a scientist of the stature of Lord Kelvin, who pointed out mistakes in Langley's calculations in 1894, could not shake the apparently widespread inclination to view the Allegheny experiments as certifying that flying machines were a valid prospect. Langley's stature was the most important imprimatur on American aeronautics during the fertile last decade of the century, despite what Orville Wright himself would later call "unfortunate lapses in scientific accuracy."

Langley also oversaw until 1903 a lavish series of trials, under Smithsonian auspices, devoted to flying a large self-propelled model airplane. When he is remembered today, it is for these experiments. He hired various experts to do the work, since he lacked the basic engineering skills for building such a machine. For funds, he raided an 1891 bequest to the Smithsonian intended to support "investigation of the properties of atmospheric air," meaning topics such as tuberculosis and air pollution. He even consulted a philologist to coin a name for his new technology, but mistakenly picked the term "aerodrome," which in Greek is closer to "airport" than "airplane."

The experiments were plagued by the same expensive overcomplexity that had undermined his expedition to Mount Whitney. Langley's imperious habits, such as an insistence that subordinates walk behind him in a formal procession, created the same level of resentment that had soured his years in Pittsburgh. Nonetheless, he succeeded by 1896 in flying an unmanned steam-powered machine over the Potomac River for several thousand feet. Unquestionably a milestone achievement, it would have won him a clear niche in aeronautical history had he quit then and there.

In June 1897, Langley wrote to Octave Chanute asking if he knew anyone willing to put $50,000 "at my disposal" for building a man-carrying airplane. No such private donor existed, so Langley turned to his own employer, the federal government. With the Spanish-American war looming, he played up the potential military value of a manned airplane. The Navy wisely rejected him, despite Assistant Secretary Theodore Roosevelt's recommendation that "it seems to me worthwhile for this government to try whether it will not work on a large enough scale to be of use in the event of war." After the conflict with Spain was over, Langley finally obtained the funds from the War Department's Board of Ordnance and Fortification, which in those days was the body that financed Army research and development. In a move that presaged the modern way of obtaining advanced weapons, Langley insisted on total control of the project unfettered by congressional oversight, plus absolute secrecy.

By the fall of 1901, with a small fortune already exhausted and no flying machine in hand, Langley took an additional $23,000 from several discretionary research funds under his control as head of the Smithsonian. Despite his 1894 pronouncement, a great deal of time and money was spent on developing a small gasoline engine to power the aerodrome, which remained structurally and aerodynamically unsound. In fact, the 200-pound forty-five-horsepower five-cylinder radial engine—built by a Cornell graduate, Charles Manly, who was sent to him by Robert Thurston—was to be the only solid achievement of the whole venture, though it played no further role in aeronautics. The inevitable disaster came on December 8, 1903, when the 800-pound machine pitched through the ice of the Potomac immediately after launching. The resulting furor in Congress and the press brought Langley's involvement with aeronautics to an ignominious end, just nine days before the Wright brothers' elegant success on the Outer Banks of North Carolina.

"We hope that Prof. Langley will not put his substantial greatness as a scientist to further peril by continuing to waste his time, and the

money involved, in further airship experiments," the *New York Times*
intoned. He evidently took this advice to heart. In 1906—the same
year that William Thaw's name was blackened when his son, Harry
K. Thaw, murdered architect Stanford White—Langley died of a
stroke hastened by severe depression.

Many years later, Orville Wright would estimate the entire cost of
his and Wilbur's research at about a thousand dollars. In 1901, they
had refused money from Langley and in 1902 declined his invitation
to Washington to discuss their aerodynamic control system. Even
Octave Chanute, who championed the sharing of knowledge about
aviation, found Langley's duplicitous solicitations to be "cheeky."
Today his image graces postage stamps as an aviation pioneer.

The importance of Samuel P. Langley does not derive from his
technical contributions to the future aerospace weapons industry.
They turned out to be irrelevant when not simply wrong. But the
fact that the secretary of the Smithsonian Institution was serious
about flying made it easier for more vulnerable men to get involved.
American science was still somewhat hokey compared to European.
When the Wright brothers decided to begin their work, the first
thing Wilbur did was write to the Smithsonian for reference materials. He received four free pamphlets, all reprints of articles that had
appeared in the institution's annual reports under Langley's authority. Langley's dependence on the military for financial support was
a harbinger, but his public humiliation set such relationships back
by a decade in the United States, leaving the Wrights to fend for
themselves. This turned out to be a blessing.

AFTER ALMOST AN ENTIRE CENTURY filled with technological miracles
since the two wallflowers from Dayton accomplished theirs, the
familiar story of Kitty Hawk has acquired the quaint distance of
myth. At first glance, there is nothing—not in their education, which
ended unnotably at high school; not in their family, which was
colorlessly ministerial; not even in their names, which sounded as
hickish in 1900 as today—to help locate the Wright brothers' real
genius. The chief obstacle to late-twentieth-century appreciation of
the Wrights lies in imagining a time when high technology could
spring from workshop odds and ends, when the frontiers of engineering still intersected the neighborhood garage. Brilliant discoveries were just as difficult to make as today, of course, but they could
occur in what now seem like very lowly circumstances.

Wilbur and Orville were neither idiot-savants nor prodigies. They were mildly eccentric, prosperous, and ingenious small-town businessmen who worked exhaustively toward the solution of a carefully defined problem. Along the way they engaged in all of the activities nowadays ascribed to professional engineering—consultation of advanced literature, creation of models and instrumented tests, experimentation with full-scale prototypes. Their native parsimony and social insularity were as crucial to their success as their mechanical aptitude, saving them from grandiose schemes that ruined well-situated men like Langley who wanted to fly as much as they did.

After writing to the Smithsonian in May 1899 and immersing themselves in reference works, the brothers first built a small kite biplane with a wingspan of five feet. They figured out an easy way to turn the kite in midair by twisting its wings, and became confident enough to try a man-carrying glider. Many other experimenters around the world had already reached this point, but the Wrights were distinguished by their fine economical craftsmanship and ability to simplify the problems they encountered. Steady winds strong enough to launch a large glider were to be found along the North Carolina coast, according to Weather Bureau charts they read, so in September 1900 they made their first trip to the remote fishing village there called Kitty Hawk.

Though Nags Head, an old summer resort, lay just twelve miles south, living on the dunes near Kitty Hawk was no vacation. The Outer Banks were a marshy wilderness, a place of great natural beauty made horrendously inhospitable by insects, storms, drifting sand, and arduous access. Even today, the Wrights' restored campsite—a tourist attraction strangled by highways and shopping malls—still faintly echoes the strength of their commitment. There has never been a more serious research and development team.

During the first three weeks of October, the brothers managed several hours of practice with the seventeen-foot-wingspan glider, mostly unmanned. They learned about its dynamic equilibrium, wind resistance, and control properties. Unlike the gliders of other designers, where the rider usually dangled from horizontal bars with his feet toward the ground, the Wright craft confidently carried its passenger prone between the wings, where his arms were free to work the wing-twisting (they called it "warping") controls. "Setting out as we did, with almost revolutionary theories on many points, and an entirely untried form of machine, we considered it quite a point to be able to return without having our pet theories completely knocked in the head by the hard logic of experience, and our brains dashed out in the bargain," Wilbur summarized the first trip. When they left, they gave the glider's French sateen wing fabric to a local family, where it soon became new dresses for two little girls.

In the summer of 1901, the brothers returned to Kitty Hawk with an improved glider and materials to build a shed to house it near the bottom of a steep dune. They were joined briefly by the illustrious Octave Chanute, who had recognized the quality of their effort in letters they wrote to him over the preceding winter. The tests this time were not encouraging, however, and they were worn down physically by mosquitoes and gruesome weather. Chanute nonetheless invited Wilbur to give a lecture in September before a professional engineering society in Chicago, which was then published in various journals, including *Scientific American* and the Smithsonian's annual report. They continued to experiment at home with model airfoils, using a homemade wind tunnel and other instruments of marvelous simplicity. By now they had reached a level of technical sophistication unmatched anywhere in the world.

Again they returned to Kitty Hawk in the summer of 1902, with yet another full-sized glider gleaned from the results of their laboratory work. To avoid the deprivations of past visits, they built a tiny cabin next to the existing storage shed, outfitting it with loft beds and a decent kitchen. The new glider had a wingspan of thirty-two feet, plus all the refinements of detail construction that flowed from two years of experience. A lively craft, it soared easily in wind speeds of just twelve miles per hour. It was also powerful enough to be dangerous, leading to several crashes that luckily broke no bones. They made several record glides of over 500 feet and twenty to twenty-five seconds in duration. In the process, they brought to maturity (and patentability) a system of controlling—through wing-twisting and a vertical tail rudder—an airplane in three-dimensional space. Smelling success, Chanute urged them to get a motor and propeller.

The machine that took form in their Dayton bicycle shop beginning in February 1903 was called, with what sounds today like staggering laconism, the *Flyer*, indicating the brothers' absolute faith in what was to come. They could not find an engine from commercial manufacturers that suited their needs, so they built one themselves in six weeks. It developed twelve horsepower, 50 percent more than they thought they required. When existing information about marine propellers proved unsuitable, they started from zero and designed two airplane propeller blades, eight feet in diameter, each made from three laminations of fine spruce. Connecting the propellers to the motor posed no problem for experienced bicycle mechanics—sprockets, gears, and chains were their stock in trade.

In September 1903, they shipped a heavy collection of tools and supplies to North Carolina, including camera gear; bicycle accessories; instruments such as a tachometer, anemometer, and stopwatch; and, of course, mosquito netting. A photograph of the campsite, now a full-fledged experimental station, shows two rough pine-plank sheds on a sandy plain, which could have been mistaken for fishing shanties if not for something parked close by that might have been from another planet. Like many of the pictures snapped by the brothers of the momentous scene, it conveys pure guilelessness, rock-steady nerve, an almost ascetic absence of clutter or ornamentation—the Wright stuff, as it were.

They proceeded to carry out gliding trials that demonstrated the *Flyer*'s airworthiness. By the first week of November, they were ready to start testing the homemade motor. Then a twisted propeller shaft brought two weeks of delay while new ones were shipped from Dayton. During this time, they fine-tuned the *Flyer*'s trussing and double-checked the transmission system between engine and propellers. Octave Chanute, with a lifetime of engineering experience to back him up, had predicted that 20 percent of the engine's power would be lost to friction as the transmission chains passed over sprocket wheels—disastrously more than their own estimate of 5 percent. To find out the correct answer, they slung a drive chain across a sprocket and tied bags of sand on each end to approximate the tension during actual operation. Then they gradually added weight to one end until the sprocket wheel just started to rotate. This additional pull was equivalent to the wheel's friction, which translated into a power loss of about 5 percent. It was the kind of bone-simple analysis that would have made any science professor in the world swell with pride.

New shafts were mounted on the *Flyer* on November 19. Further tests revealed the need for solid shafts rather than the tubular ones they were using, so on November 30 Orville left for Dayton to fetch them himself. Nine days later, on a train returning to North Carolina with the prized parts, he read a newspaper account of Samuel Langley's costly debacle on the Potomac.

On Saturday, December 12, the complete 600-pound *Flyer* was rolled out of its shed, only to find that the fabled Outer Banks winds were too light for a flight. On Sunday, as always, they rested, despite the fact that they were standing on the brink of their destiny. On Monday the fourteenth, their formidable patience exhausted, they attempted a trial run even though the wind was still weak. They were rewarded with a splintered framework after an erratic hop of about one hundred feet, which they did not consider a true flight. But now they knew everything would work.

On December 17, wearing their usual conservative business attire among the dunes, Orville climbed aboard the repaired aeroplane, with Wilbur holding up its right wing tip for balance. The engine fired, the propellers whirled, and when the machine attained a ground speed of about seven miles per hour it left the sand, rising to an altitude of about ten feet before landing 120 feet away. A photograph taken fortuitously by a local helper just as the *Flyer* was lifting off shows Wilbur, thirty-six years old, running alongside in his suit and cap, the thirty-two-year-old Orville prostrate on the lower wing between the propellers' aureoles, a shovel in the sand, and an overcast sky. Here was the matter-of-fact culmination of thousands of years of daydreaming.

They made four flights that day, the last measured at 825 feet in fifty-nine seconds. Then it seemed like the gods decided to vent their anger, flipping the parked *Flyer* over with a gust of wind that cracked its wing ribs, bent its chain guides, and snapped off its engine frame legs. It never flew again. But it did not have to.

THOUGH THE WRIGHTS had not begun their work with mercantile ambitions, they soon recognized that the *Flyer* could be the foundation for great fortune. They also thought that their research had left potential competitors far behind. "We do not believe there is one chance in a hundred that anyone will have a machine of the least practical usefulness within five years," Wilbur wrote in 1906, sounding exactly like all the skeptics who had doubted the success of any flying machine before December 17, 1903. A man very much like Wilbur would make him eat his words, spawning a legal battle over patent rights that would paralyze American aeronautical development.

Glenn Curtiss, the motorcycle racer and mechanic from upstate New York, was drawn into the circle of flying machine experimenters when he sold a secondhand motorcycle engine to a carnival balloonist in the summer of 1904. The engine was mounted on a fifty-three-foot dirigible called the *California Arrow,* which then made the first closed-circuit flight by an aircraft of any type in the United States (over San Francisco Bay) and performed at the St. Louis world's fair that autumn. Sensing a new market, in October 1905 he displayed his engines at an aeronautical exhibition in New York City, where Alexander Graham Bell purchased a fifteen-horsepower 125-pound unit to power man-carrying tetrahedral kites. Bell, a longtime aviation enthusiast who had been part of Samuel Langley's social circle, organized a team of five men at his Nova Scotia estate to start developing these kites into flying machines during the summer of 1907, with Curtiss as the resident engine expert.

By March 1908, the Aerial Experiment Association, as the group called itself, had abandoned Bell's ungainly kites to try a powered biplane. Like the Wright *Flyer,* it carried elevator surfaces in front of its wings and a vertical rudder behind. It was not fully controllable, however, and was followed in May by an improved model that used hinged wing flaps, or ailerons, for turning. Though such mechanisms had already appeared in France, Bell seems to have suggested them independently, spurred by the group's feeling that the Wright method of wing-twisting was inherently dangerous.

In June 1908, the AEA's work crystallized in a third machine designed by Glenn Curtiss, which Bell dubbed the *June Bug.* The pilot sat upright on a bench just ahead of the wings, a more commanding position than lying prone between them. Rather than using two motors and propellers connected by a vulnerable chain transmission system, Curtiss took advantage of just one of his superior engines mounted at the center of the wingspan. Strong and relatively easy to fly, it put the thirty-year-old Curtiss on the path to acclaim when he won a trophy that month sponsored by *Scientific American* for flying a distance of one kilometer. Though the Wrights had amply exceeded this mark in 1905, they refused to compete because the contest rules stipulated that the airplane must take off on its own set of wheels, rather than from a single-rail track as they favored. They thus handed Curtiss a widely publicized victory, which became the first stepping-stone in his career as an airplane manufacturer.

Immediately after the *Scientific American* contest, Bell sent a patent lawyer to examine the *June Bug.* His past legal battles over the invention of the telephone had made him wary about theft of lucrative design secrets. Several elements—including its ailerons, tricycle landing gear, and steering system—were declared to be patentable. When *Scientific American* mistakenly reported that the AEA was ready to start selling June Bugs to the public for $5,000, the Wrights promptly warned Curtiss about infringing their patents, which they thought covered any adjustment of wings for control—including ailerons.

Curtiss replied that the AEA was considering the patent problem and that in any case he was not going into business. But in 1909, he split from Bell's association and started his own company, which soon delivered its first airplane to the Aeronautic Society of New York—the first such commercial order in the United States. The society had considered buying a Wright model, but decided in favor of Curtiss's more practical machine. The Wrights, by this time hotly defensive and almost paranoid about what they considered to be their invention, responded by suing Curtiss, thus starting a courtroom battle that would stymie domestic aviation developments until the nation went to war in Europe.

THE SHIFTING, rough-hewn state of American engineering during these years just before and after the turn of the century meant that powered flight, an almost theoryless subject, would be mastered by self-educated mechanics like the Wrights and Curtiss, not by the traditional elites represented, however poorly, by Langley. His failure was an early warning of the corrupting influence of power and money that would later haunt the industry. Given the bourgeois perception of flying as a rather lunatic thing to do, aviation turned out to be a gathering place first for introverted radicals, not for eminent authorities. America had lots of Wilburs and Orvilles in the provinces, and they would flock to the new invention like charmed children.

Their eccentricity, which would show itself in the personalities of the founders and in their instincts about aeronautics, would be tolerated for an extraordinarily long time by governmental and financial powers that soon allied with the industry. Only when the business of building airplanes, and much later missiles, became a foundation for American strategic power would the pioneers be pushed rudely aside. Flying simply became too important to be left in the hands of those who had first risked figuring it out.

*   *   *

FROM 1903 TO THE BEGINNING of World War I, an American aircraft "industry" did not exist in any usual sense of the word. It is a stark comment on the pace of technological innovation in those days, as well as the apparent outlandishness of flying, that a decade would pass with little advancement or investment in the field. But here and there around the country there were cottage enthusiasts, a few of whom would persevere to create a rich business in weapons based on the new technology. This would, in turn, support the gradual development of airplanes for commercial transportation.

Data about the number of airplanes produced before 1914 are unreliable, indicating the relative unimportance of what little activity there was. In August 1907, due mostly to the aviation interests of President Theodore Roosevelt, the U.S. Army established an Aeronautical Division in the Army Signal Corps, staffed by one officer and two enlisted men. In 1908, the corps had three officers and ten enlisted men operating one Wright airplane under test, plus three balloons. In 1909, the Army allotted $30,000 to pay the Wrights for their latest machine, using special presidential funds provided by Roosevelt. That same year, attracted by the scent of federal dollars and the possibility of a patent monopoly, a group of financiers led by Cornelius Vanderbilt, Howard Gould, and August Belmont capitalized the Wright Company at a million dollars.

In 1911, two second lieutenants who would become World War II generals—Thomas Milling and Henry "Hap" Arnold—were sent to Dayton for training by Orville Wright. The Navy detailed a few officers to a flying camp run by Glenn Curtiss on North Island near San Diego, which would become the primary prewar center for evolution of military aircraft. By 1913, though dozens of airplane companies had already sprung up and disappeared, there were probably no more than a few score airplanes in military and civilian hands in the whole United States, with a much smaller number of competent pilots. Total expenditures of the War Department for aircraft between 1908 and 1913, when the first air combat squadron was formed, came to just $250,000.

Where Vanderbilt and his fellow financial barons evidently saw a promising investment, the U.S. military saw mostly a strange contraption. The invention of the airplane had happened at an awkward time for the Army, which was just beginning to emerge from a long period of social isolation. The Spanish-American War was a brief interruption of more than thirty years of peace, discounting various missions against the Indians. In military circles the most interesting innovation of 1903 was the Springfield rifle. West Point cadets still aspired to be cavalry officers. But a technological revolution in weaponry was already under way that would result in the appalling confrontations of Ypres and Verdun. The civilian and military worlds were being slowly drawn together, if only improvisationally.

In Europe, the situation was far different, but this did not mean much to American airplane builders until well after the war began in earnest. Shoulder-to-shoulder proximity of rival powers on the Continent no doubt made the threat—and usefulness—of aerial warfare more palpable there. When a Frenchman, Louis Blériot, flew across the English Channel in July 1909, for example, the feat had clear implications for England's Splendid Isolation. (The Hague Conference of 1907 tried to ban "bombardment by whatever means" of "undefended" cities, but only the United States and Great Britain were willing to ratify such a prohibition.) Seeing Blériot's monoplane displayed above a Paris street soon after the flight, the young Hap Arnold wondered with childlike awe: "If one man could do it once, what if a lot of men did it together at the same time?"

European aviation had been energized by a series of public exhibitions flown in France by Wilbur Wright between September 1908 and March 1909, which were watched by tens of thousands of people. Designers quickly absorbed the Wrights' emphasis on controllability, then introduced improvements such as ailerons for turning. In 1912, the French Deperdussin company built a hundred-mile-per-hour racing aircraft that featured a "monocoque" (literally, single shell) wooden fuselage and a rotary piston engine. It made the Wright planes look like toys. European technology accelerated even faster when regional conflicts in Libya and the Balkans attracted Italian and French airplanes and pilots just before the First World War.

By 1913, France was spending $7.4 million a year on aviation, Germany and Russia $5 million each, England $3 million, and Italy $2.1 million. Each had active aeronautical laboratories with government support. Even poor Mexico spent some $400,000 in the course of its civil wars, not including the cost of rebel operations like Didier Masson's.

The Wrights first approached the War Department with their invention, which they thought might be useful for "scouting and carrying messages," in 1905 after having been visited the previous year by a British officer who urged them to sell it to his government. But Washington was so distrustful following Langley's debacle—and so put off by the Wrights' initial asking price of $100,000—that it took another three years to approve a contract.

This reluctance was understandable, given the revolutionary nature of the machine. Some hint of the proportion of cranks to serious researchers can be gleaned from the fact that of forty-one bids received by February 1, 1908, by the new Aeronautical Division of the Signal Corps, just two led to signed contracts. Of these, only the Wrights produced a successful aircraft. Even then, Congress refused to appropriate funds specifically for aviation until 1911, despite personal appeals from the Secretary of War that the airplane would "profoundly affect modern warfare."

Meanwhile, the pieces for a supremely American industry were falling randomly into place, waiting to be exploited. Light and powerful internal combustion engines were increasingly available, along with abundant supplies of oil to fuel them. Automobile shops spawned tools and techniques to fabricate the kinds of parts needed in airplanes, though there was little direct interplay between the two technologies. The Navy's ongoing acquisition of a new fleet of armored steam-and-propeller-driven ships carrying modern ordnance was creating a prototypical production team of politicians, military officers, and businessmen.

But political instability and militarism would be the most important long-term stimuli to aviation in America as well as Europe, making most other factors seem like window dressing. Drawn inexorably into these dark crosscurrents were a few rather earnest young men who thought flying was fun.

CHAPTER TWO

MONEY-MINDED ECCENTRICS

GLENN L. MARTIN was born in Macksburg, Iowa, population 300, on January 17, 1886, the same year that Samuel Langley was inspired by a farmer's talk about stuffed birds to delve into what had always been considered laughable. Martin and Langley never met, but they would have either loved or hated each other out of mutual recognition.

Martin was, of course, too young to figure in the birth of flying machines. It was his fortune to come of age during the years between 1903 and World War I, when aviation was turning from a dream come true into workaday reality. Rather than create something wholly new, he adapted what already existed, especially the inventions of Glenn Curtiss. But unlike the Wrights and Curtiss, his name survives in the modern aerospace weapons industry as a progenitor—Martin Marietta. It is therefore with Glenn Martin that an unbroken historical thread begins, not of a technology but of a technological business based on advanced weapons.

Martin's arrival as a serious experimenter came about six years after the Wrights' first powered flight. He had been raised by strict Presbyterian parents in central Kansas, where his father, Clarence, sold hardware and farm implements. The frontier there was still fresh, with Indian massacres a proximate memory and formal education a crude template. By high school, he had begun to work in Salina carriage and bicycle shops—typical jobs for bright country boys not tied to farm chores, especially if Father was a tool salesman. After his sophomore year he quit to take bookkeeping courses at Kansas Wesleyan Business College, a somewhat unusual step in his milieu, undoubtedly pushed by his ambitious mother, Araminta, who had briefly been a teacher before her marriage.

Minta, as she was called, was the orphaned daughter of a Union soldier. She had met Clarence Martin, a transplanted Pennsylvanian, as he followed the wheat growers and the railroad across Iowa. At various times he had managed his own store, operated a freight wagon service between towns, and even tried his hand at the wheat itself. But bad weather or inattentiveness had always derailed him, and Minta's hopes for good fortune had been gradually transferred to her only son.

Something of a mystic in the middle-brow fashion of post–Civil War years, she was fixated by a dream she had had just before his birth. She and her baby were flying in a warm blue sky above prim Macksburg, with all the townspeople below pointing up in amazement. It was a premonition, she would always insist—though perhaps it was just a strong wish, since the boy weighed twelve pounds at birth on a morning when the temperature was twenty-six below zero.

After his business courses, Glenn Martin tried selling Queen automobiles by driving from town to town in a demonstration model, but customers must have been rare. In the vanguard of a great migration off the grim Plains, the family moved west to Santa Ana, California, in 1905, ostensibly for Minta's health but probably more to placate her restlessness with their bottom-rung economic status. They settled on a small apricot ranch, where the fruit business provided marginal income. Clarence Martin again worked in a hardware store while his son found a job servicing automobiles. Being more inclined toward ledgers than shopwork, at the end of the first year Glenn presented an estimate of what he needed to start his own garage to a local bank and carried off an unsecured loan of $800. He acquired a Ford and Maxwell car distributorship, rented a garage, hired a mechanic named Roy Beall, and proceeded to run the business with his father as sales manager. Whatever else could be said about the young man, who by all accounts was girlish and annoyingly prim, he possessed a steely sense of commerce, underpinned by his mother's driving will. When he was no more than six, she had helped make ends meet by selling kites he built to his playmates for the exorbitant price of a quarter each.

From this point, Glenn Martin could conceivably have grown old as a prosperous car dealer, perhaps someday running for local office—a not uncommon American life in this century. But he was painfully introverted, ridiculed as a dreamer by his father, who found him repulsive, yet prized by his mother as a way to escape her disappointments. Despite an outwardly shy and formal manner, which must have made it difficult to get along in the coarse world of machine shops—Santa Ana rakes called him a "mollycoddle" and a "milksop"—he had a reputation for occasional theatrics, once gunning a Ford up the Santa Ana courthouse steps to demonstrate its horsepower. There was clearly something shackled inside Glenn Martin. In 1907, he began to think about building airplanes.

It is not possible to pinpoint the source of Martin's inspiration—like Samuel Langley he was tight-lipped or elusive about his youth after he became famous—other than his own or possibly his mother's imagination. The year 1907 brought a Wall Street panic and nationwide economic downturn, so perhaps he was casting idly about for ways to widen his base. Though he was aware of the Wrights' invention, published details about their machines were still very sketchy. The average person and the press were largely indifferent to or ignorant of the distinction between motor-driven dirigible balloons, which were old hat, and new heavier-than-air flying machines. The insular Wrights did not make public flights until 1908. It was in July of that year that Glenn Curtiss won the trophy sponsored by *Scientific American* for the first public flight of one kilometer.

Plans and specifications for a hang glider designed by Octave Chanute were widely available, however, and many were built by amateur sportsmen. Essentially small biplane kites—weighing perhaps twenty-five pounds with a wingspan of fifteen feet—they were only as safe as the handiwork of their maker, but they were the first step in any serious experimentation. In 1905 and 1906, manned gliders designed by John Montgomery, who was one of many early experimenters who drifted back and forth across the line between fakery and legitimate research, had made highly publicized drops from hot-air circus balloons in several California towns. The gliders were tandem-wing monoplanes similar to Langley's aerodromes.

After trying out a glider, Martin's first home-brew attempt at powered airplane construction was simply to bolt on a twelve-horsepower Model N Ford automobile engine. Again like Langley, Martin depended on the manual skills of others, mostly Beall and a few local mechanics at his garage. And like so many other half-blind efforts of the era, their creation was only vaguely airworthy.

On its first test, Martin was nearly decapitated when the machine spun around him on the ground with the engine's throttle wide open as he clung to the framework. A year's labor and garage profits were torn to pieces. But his car business was solvent, and he was at least in a position to wave off his father's disgust at the waste of money. Mother Minta, who was as enamored with the notion of flight as her son, prodded him to try again.

This time, instead of working behind the garage in broad daylight, he rented an abandoned Methodist church for twelve dollars a month, blacking out its tall windows so he could carry on his project away from ogling neighbors. Folklore says that he worked alone at night with Minta holding a lantern, but this story—which he encouraged in later years—rings more like a metaphor for the lifelong partnership he had with his mother than an accurate description of how he obtained his next airplane. A half-dozen local mechanics again did most if not all of the construction. He would one day claim to have gained insight into wing construction from studying library books about bridge truss design, but he lacked the mathematics to get much beyond looking at the pictures.

Whatever they started out with in mind, Martin's crew ended up by the latter half of 1909 with their own rendition of Glenn Curtiss's *June Bug* biplane, powered by a lightened version of the Ford engine. Details and pictures of the Curtiss machine had not been in circulation prior to May of that year, so one may assume that Martin's later attempts to date his aircraft in 1908 were disingenuous. He would declare that this effort had taken thirteen months and $3,000—figures that would certainly explain his father's anger. "You were so interested in the project in the little old church, it was hard for you to stay at the automobile place," a woman who kept the garage's books would remind him many years later. "Your dad used to get so disgusted when he came in and found you gone." Clarence Martin had led a hardscrabble life, and here was his boy throwing money to the wind.

In August 1909, Martin claimed one successful hop with the new plane across a grassy field on the 80,000-acre Irvine ranch near Santa Ana. If true, he thus became the first person to fly in Califor-

nia. But whether he got airborne or not, he realized that this machine was dangerously underpowered, because only a year later he
had obtained another with a thirty-horsepower engine. Instead of
risking his own neck to test it, he offered fifty dollars to two other
local enthusiasts, one of whom drove it into an irrigation ditch on
the first time out. It was with this machine, which weighed about
750 pounds unmanned, that he then taught himself to fly—away
from his father, away from the garage, even away from his milksop
self.

IF THERE WAS a brief age of relative innocence for flying machines,
1910 was its apex. European and American aviators were earning
tens of thousands of dollars at gaudy air shows, breaking altitude
and speed records for prize money. The sensationalistic yellow press
was at the beck and call of promoters. Glamorous professional exhibition teams trained by the Wrights and Curtiss, as well as half-
cocked freelancers, played to the crowds' "savage desire to look
upon mangled bodies and hear the sob of expiring life," as *Scientific
American* lamented. "Come Josephine in My Flying Machine" was
the whimsical hit song of the day, but it was gory sport. The spectacle had the power to turn Edwardian audiences into mobs that
would rip apart crashed machines, so fantastic was the act of flight.
To Glenn Martin, the supposedly humorless mama's boy who was
not known to drink, smoke, or date, this raucous world offered a
cathartic release. He ran away and joined the circus, as it were,
though he turned the experience to his own very serious counting-
house purpose.

The first elaborate air show in the United States was held on
ranchland outside Los Angeles in January 1910. Grandstands were
built to seat 25,000 people, and circus tents served as hangars.
Among the promoters were Harry Chandler, notorious publisher of
the *Los Angeles Times,* and Frank Garbutt, a wealthy oilman and
member of the Aero Club of California (the parent Aero Club of
America had been founded in 1907 as an offshoot of the Automobile Club of America, reflecting the social overlap of drivers and
pilots). It was there that Martin saw how to perfect his 1909 Curtiss-
type machine, because Glenn Curtiss himself was among the con-
testants. (A few contemporaneous newspaper articles suggest that
Martin's interest in aviation was triggered by the January 1910

show, raising the question of whether he built anything before that time. At the show, Martin made the acquaintance of Charles Day, a local pilot who was trying to compete in his own plane. Day was a graduate of Rensselaer Polytechnic Institute and helped Martin examine the Curtiss machine. He soon became Martin's "chief engineer," and may in fact have then built Martin's first airplane. There is no definitive record.) Martin was no doubt also impressed by the $19,000 in prize money won by Louis Paulhan, a famous French aviator whose black leather flying togs he would soon adopt, and a nearly equal purse taken by Curtiss. These were princely sums, representing many years' wages for anyone of Glenn Martin's class.

During the carnival, a Signal Corps officer detailed there as an observer went aloft with Glenn Curtiss to try out a makeshift bombsight. From an altitude of 250 feet, he dropped fifteen-pound sandbags at a circle on the ground, consistently missing but marking the first attempt by the Army to use such a sighting device. Given Martin's close attention to everything else Curtiss did, it is certain that he watched this relatively quiet event with as much interest as the breakneck stunts.

In August 1910, the *Los Angeles Times* carried an article about Martin's flying machine projects, headlined "Aeroplane Up At Santa Ana." It contained the earliest recorded utterance of the future weapons industrialist, which though perhaps rewritten according to the journalistic habits of the day, captured his dour veneer:

"I would not consider it worthwhile building this machine if it were not for the fact that I want the experience. I have had some flying machine schemes in my mind for a long while and expect to carry them out in building a new machine. I built the biplane I now have in order to get used to flying."

So said the twenty-four-year-old devotee of the era's wildest sport. By the fall of 1910, Martin was committed, at least intellectually, to earning a living from aviation—one of the era's more radical career choices. In September, his mother received a pleading letter from the family doctor: "For Heavens sake, if you have any influence with that wiled-eyed [sic], Hallucinated, Vissionary [sic] young man, call him off before he is killed. Have him devote his energies to substantial, feasable [sic] and proffitable [sic] pursuits, leaving dreaming to the professional dreamers." But she and her strange boy shrugged it off. He put his machine on display at Santa Ana's "Carnival of Progress," a promotion for local businesses, and made his first public flight across a cow pasture before several thousand people. The *Santa Ana Register* announced the exhibition on November 21, 1910, with rapt simplicity:

> Tomorrow afternoon Glenn L. Martin will fly. The exhibition will be free to the public, and will be given under the auspices of the Santa Ana Merchants & Manufacturers Association.
>
> Martin has been doing some very fine work in his practice flights. He has accomplished wonders with his aeroplane, which he built and by his own experiments learned to fly. On a number of occasions the aviator has made flights of over two miles, circling to the place of starting.
>
> Martin has planned a program for tomorrow to demonstrate the work of the aeroplane. The flights will be varied to show the rise and the dip and the circle. Weather conditions may vary the program somewhat, but if the wind is not too strong Martin will do some work that cannot be surpassed by any amateur in the country.
>
> The exhibition will be the first flying exhibition to be held in this county. That there will be a big crowd in attendance is assured by the inquiry and the talk that has been heard concerning the affair.
>
> The flights will be from James McFadden's pasture, a few rods east of the residence at the corner of South Main and McFadden streets, and will commence at 4 o'clock. No flying will be done before that hour.
>
> The Pacific Electric will run special cars to and from the field.
>
> The schools have made arrangements to dismiss all classes so that the pupils may witness the exhibition. It will be the kind of thing that may not be seen here again for a long time.
>
> Automobiles and rigs will be admitted into the field, and with the spectators will be kept on the west side of the field. The east side must be kept clear for starting and stopping purposes.

There was something awestruck about the announcement, yet also a fearful undertone, stemming no doubt from the high probability that Glenn Martin might kill himself. He made three flights that day for the incredulous Santa Anans, each several miles long at about a hundred feet off the ground and a speed of forty miles per hour. This was a remarkably consistent operation for a makeshift flying machine, showing considerable mechanical discipline—if not on Martin's behalf, then at least on the part of his employees. More remarkable still was the fact that a young man widely regarded as a sissy flew with his right arm in a sling, having torn a ligament several weeks earlier in a crash. Keeping such primitive biplanes under control was a highly physical task, requiring brute strength and unhesitant reflexes. Martin's performance thus revealed a bravura totally at odds with his outward personality.

It was all a modest prelude to the second Los Angeles air show, billed as "The Most Colossal Air Circus in the World." Martin apparently hoped to fly a monoplane of his own design at this event, perhaps to avoid a confrontation with Curtiss men over the obvious pattern of his biplane. This machine was to be outfitted with a device for "automatic lateral stability," then something of a chimera among airplane builders, which he wanted to patent himself. He brought both the new monoplane and the proven biplane up to Los Angeles in late December 1910, along with the thirty-horsepower engine that could fit either one. But like Langley, his attempt to make a sudden leap beyond empirical success was a failure—he wrecked the monoplane during a trial run when the wind tipped it after rising only a few feet off the ground. The engine was promptly bolted onto the biplane.

Though a talented pilot with a relatively dependable machine, Martin was not yet a celebrity of the caliber of Wright and Curtiss team professionals who flew at the Los Angeles show. He was relegated to events for local novices, which he easily dominated. Though his first run past the grandstands turned into a quick embarrassment when his engine stalled, he took first place and $450 several days later in a race. His businesslike presence—like the spruce Orville Wright, he went everywhere in a suit and starched collar—stood in stark contrast to the garish show, attracting the confidence of Southern California millionaires who would become his first customers. Besides, Wright and Curtiss planes had to be shipped from far away in Ohio and New York, but Martin was ready, or so he said, in Santa Ana.

Without enough up-front money to finance orders, however, he decided to use the exhibition circuit as a source of capital. While local bankers could be persuaded to underwrite a new auto garage, flying machines were out of the question. Fortunately for Martin, Glenn Curtiss never patented his early biplanes, leaving Martin free to turn professional unthreatened by the kind of legal maelstrom that the Wrights had unleashed on Curtiss. To his credit, he wrote to the Wrights saying that he was using their invention, but was not able to pay their license fee. He even asked what amount they would be willing to accept, under the circumstances. Such a fine Christian gesture must have appealed to the brothers, because they wrote back telling him not to worry about the fee.

Martin needed cash more than encouragement. Imitating the Curtiss organization, he expanded his home work force of mechanics to seven and hired a teenaged boy to travel with him on the air-show circuit. They played the carney bush leagues, from cantaloupe festivals in the Imperial Valley to Fourth of July celebrations at San Diego. It was the world of the tattooed and toothless, not a genteel existence. Yet by summer they were flying a new biplane worth $4,000 with a sixty-horsepower Hall & Scott marine engine, for which Martin had cannily obtained the Southern California agency. Still basically a Curtiss type, refined by Martin's own growing experience in the air, the plane was larger than previous versions, and its wings were covered with a new fabric introduced by Goodyear Rubber especially for airplanes. Newspaper articles began to mention a Martin factory and pilot school in Santa Ana, though this was more the result of his knack for self-promotion than an accurate description of the rag-tag outfit there.

He was now earning $1,500 for two days of exhibition flying, all of which was invested in the business. One of his favorite maneuvers was to swoop down on gate-crashers, "frightening them out of their wits" into paying for a safe seat the next time. At a San Diego show he began to shoot balloons with a rifle, showing off a skill from his Kansas boyhood. At the Pomona Valley Fair in September 1911, he masculinized his persona by tossing a bouquet of carnations onto the ground near the pageant queen. But the region was already growing blasé about birdmen, and the smaller air circuses were not breaking even. In the fall, he toured Kansas and Nebraska farm towns under contract with the Curtiss company, taking the place of a recently killed pilot called J. J. Frisbie. The experience was grueling and cash poor, however, and he cut short the trip as soon as winter weather made flying even more suicidal than usual.

Besides the weather, there may have been another reason why Martin decided to break off the tour. In the first week of November, his parents committed his sister to a state mental hospital at Patton, California. Della Martin was twenty-eight, three years older than Glenn, and unmarried. The petition stated that she was "acting in a strange and incoherent manner; imagining she can read other people's thoughts and that they can read hers; has various delusions and hallucinations of hearing; extremely religious; believes she can hear God's voice talking to her and that she has been commissioned to redeem the world"—common symptoms of schizophrenia, a disease not then understood as such. She would remain in various hospitals for the rest of her long life.

Her existence was a tightly guarded secret, expunged from all family records, unmentioned in any of the biographical notes or popular writings that would celebrate Glenn Martin's industrial achievements. To the rest of the world, he was his mother's only child. Minta herself never stated otherwise. The fact that this peculiar birdman had a mad sister was far more than the family thought prudent to reveal.

When he arrived back in California at the end of November, Martin gave a lecture before a Santa Ana businessmen's club on "The Art of Flying." Despite the weight of evidence to the contrary, he predicted that in the "near future the airplane would become a thoroughly practical means of transportation for passengers and freight." He declared that Samuel Langley deserved the real credit for discovering how to fly. And he announced that a new principle had recently been found whereby radio waves could nullify the force of gravity. These were the gushings of a well-intentioned bumpkin, no doubt taken at face value by his worshipful audience. They showed the mix of blind faith in aviation and blithe vacuity about other matters that would color his long career.

DURING THE FIRST month of 1912, Martin and three other flyers helped promote the third Los Angeles air show by joining a luridly publicized police manhunt in the San Fernando Valley. The following story appeared under his own byline in the *Los Angeles Herald* for January 15:

> It is but a question of a short time before the aeroplane will be used by the police in peace as well as by the military in war. My recent trip through the air after the San Fernando bandits convinced me of this. The country lay before me like an open book. My vision was not limited. Had I taken the trail fresh and began my flight in the early stages of the manhunt when the fugitives were in flight over the hills I could have directed the progress of land posses with ease.
>
> I believe that in time there will be no more powerful factor in the successful culmination of a manhunt than the aeroplane. The police of every city and the sheriffs of every county will soon realize this and no

well directed department of criminal investigation will be without an
aircraft and a birdman.

A police aviator should be accompanied by an assistant, who will
occupy the position of a lookout. The assistant could carry a field glass
and a gun. Probably they will some day carry dynamite, and by dropping well directed bombs into canyons drive the desperado from the
brush into the open, where the land posse, guided to the spot by the
falling shells, may surround their quarry.

I do not regret my aero manhunt last week. It proved to me the
practibility of a lighter-than-air [sic] machine. I could see little groups
of armed men hanging about the approaches to the rough country,
powerless to invade the brush and rocks because they could not see
farther ahead than the next rise of ground.

From my station I could see every moving object for miles around
me. I could tell whether a man was armed or not by the reflection of
the sunlight on the metal of his gun.

It is thus that the police must soon realize, as well as the military,
that the aeroplane will be a powerful factor in bringing to justice those
who escape from the scene of their crimes and seek safety in the
recesses of the hills and the mountains.

As one of the first and most explicit predictions by an American
aviator of the way flying machines would be used in war, the article
marked the beginning of Martin's efforts to prove how useful the
new technology could be, rather than merely how exciting it was.
His advocacy of dropping bombs on civilians, albeit "bandits," was
so brazen as to suggest that the *Herald* readership, at least, was not
appalled by such prospects. Between the lines—besides a certain
cold-bloodedness—lurked his hunch that the government would be
a crucial source of economic support.

That year for the first time, too, the air show included a "mimic
aerial battle" and a nighttime "battle from the sky spectacle," besides such typical carnival diversions as "marriage in aeroplane"
and "race contested by man, beast, motorcycle, automobile and
aeroplane." News from Europe—such as the war between Italy and
Turkey that saw Italian flyers in Libya carry out the first bombing
operations since Austrians had lofted explosive-laden balloons over
Venice in 1849—no doubt fueled such entertainment. (A similar
show had been planned in Atlantic City the previous summer, but
the Navy cancelled the participation of a warship because it feared
foreign agents might be watching.) Under the headlines "Airmen's
Bombs Hit Mark" and "Aviators in Darkness Easily Demolish For-

tress," the *Los Angeles Tribune* breathlessly described the new kind of entertainment:

> Dropping missiles with a precision that would be hard to duplicate in daylight under the most favorable conditions, aviators Lincoln Beachey [a member of the Wright team, popularly known as the "Flying Fool"] and Glenn Martin poured a rain of bombs into the "forbidden city" that entirely destroyed it and brought the spectators to their feet with wild applause.
>
> From the harbor at San Pedro could be seen the searchlights of the Paul Jones, Perry and Preble, the destroyers detailed by the Secretary of War, under command of Lieutenant Glassford, to assist in the maneuvers. At times the aviators could be seen plainly in the beams of the searchlights as they darted and turned in their maneuvers.

The *Los Angeles Times* called the show an "epoch-making event," saying that "a crowd of 3000 persons cheered the daring airmen and a squad of naval officers" who witnessed the display. "The birdmen showered bombs on an imaginary fort from an altitude of 1000 feet and destroyed the entire place." Taking an editorial voice, the paper found that "most thrilling of all the maneuvers of the evening was the bomb throwing and destruction of the fort. For the first time in history these famous birdmen demonstrated that an aerial fleet can be made the most powerful enemy in time of war. Soldiers at the fort fired volley after volley at the airmen, but were unable to reach them." The last sentence confused theater with reality, if only in an attempt to convey the dazzle of the evening.

The *Los Angeles Express* reported "such a spectacle as has never before been seen on the Pacific coast, or anywhere else. With their machines illuminated with gas and electric lights, [Beachey and Martin] took the air and there ensued the most thrilling scenes imaginable. Beachey and Martin carried bombs with which they bombarded the miniature city and fortress, their attack being repelled by the Signal Corps and a company of the state militia.

"During these maneuvers in the air, searchlights from three torpedo boats in the harbor at San Pedro were piercing the atmosphere, now and then throwing into striking relief the circling and pitching aeroplanes. Lincoln Beachey carried on his machine red fuse light, with which he signalled to those on the ground his intentions while making the attack on the fortress. Glenn Martin carried a searchlight, whose piercing gleam aided materially in the night attack."
The *Express* added that Beachey "was in communication at all

times with the war boats in the harbor, and during his flight he received from the searchlights of the ships messages and answered them with fuses" supplied by the Navy. "Beachey while in the air dropped bomb after bomb upon the enemies' warships and by actual measurements made by naval officers he dropped them accurately. Martin did equally well in three of his circles over the beleaguered port." The newspaper opined that this "proved more conclusively than any other experiments ever tried that the aeroplane is practical in war and is not only useful as a fighting part of a military force, but as a scout and signal machine."

Among the dozen or so professional birdmen listed as attending the show was the Frenchman Didier Masson, who had come to America as Louis Paulhan's mechanic in 1910.

This cooperative test put on by the Navy, Army, and air-show promoters heralded several developments in American aviation that would accelerate as war in Europe grew closer. First, though the military still did not have the wherewithal to stage such "experiments" with its own men and equipment, it felt so unfettered that participation in a public circus was quite acceptable. Second, there was some nascent awareness that flying machines could be specialized for fighting, bombing, scouting, and signaling—clearly a boon to anyone contemplating a warplane industry. Third, there was apparently no serious outrage or horror at the prospect of turning this revolutionary technology into a new weapon for destroying cities, albeit miniature "forbidden" ones. Josephine was being replaced on the flying machine by a load of explosives.

By the beginning of 1912, then, the fusion of warfare with professional aviation was well rooted. Middle-class air-show crowds were not shocked but enthusiastic. The act of aerial bombardment was greeted with cheers. For Martin, it was a pivotal experience, like the sight of Glenn Curtiss's tight little biplane only two years before. He would continue to try almost every stunt imaginable to fan public interest in aviation and further his business, while military applications became more and more prevalent.

"This show and others like it accomplished its purpose—it attracted the attention of the War Department," Martin recalled thirty years later.

Shortly after the Los Angeles show, he took his mother flying for the first time. "Aviation is a sure cure for the blues," she gaily told a reporter from the *Santa Ana Blade* upon landing. The loquacious Mrs. Martin was known to do most of the talking while Glenn listened respectfully, gaining a reputation among local newspaper-men as "the best and most enthusiastic press representative any man ever had." As her son's fame grew, so would her inclination to aggrandize their role together as aviation pioneers. While Glenn pursued the War Department, Mother Minta dished out the apple pie.

IF 1912 BROUGHT the brief age of flying machine innocence to an end, it is worth considering why this happened so easily. It may be that air-show promoters were desperate for any new way to attract spectators, who were no longer enthralled by "the rise and the dip and the circle." With war clouds gathering in Europe, why not capitalize on the public's morbid fascination, which had already been primed by sensational novels such as H. G. Wells's *The War in the Air* (1908)—in which German bombers attack New York City. Glenn Martin and other aviators were well aware of the U.S. government's growing interest in warplanes, though as yet it meant next to nothing financially.

The evident enthusiasm for the Los Angeles exhibition suggests, in any case, that the coming together of new technology with a military-commercial sponsor seemed quite natural to many Americans. Certainly the true terror of aerial warfare was too much for anyone yet to understand. Predictive literature of the day even argued that the rapid destructive force of modern weapons would make combat more humane, perhaps too difficult to wage at all. In their effort to draw crowds, the promoters and complicit journalists naturally tried to exaggerate the power of warplanes, not knowing how dreadful reality would be. By cheering on the formidable "aerial fleet," they were also expressing a common optimism that technological progress would promote peace, just as it had cured so many other human miseries. It was a naiveté that lasted until the first months of American participation in the coming world war.

"Sentiments seemingly opposed to each other—hope for peace and fear of modern weapons—coexisted comfortably because this [prewar] generation found hope in their very terror," historian Michael S. Sherry has marveled. The birdmen themselves were heroes, of course, masters of a machine that epitomized courage and free will.

Only after the slaughter of 1914–18 did it become clear that the new weapons—machine guns, asphyxiating gases, tanks, battleships, submarines, and airplanes—could cause an atrocious kind of prolonged conflict. In 1912, especially in the United States, the military and the civilian public were just beginning to grow serious about a dangerous new amusement.

IN THE SPRING of 1912, Glenn Martin's craftsmen finished a new biplane, still basically a Curtiss-type pusher, mounted on a kayak-shaped pontoon instead of wheels so that it could land on and take off from water. Curtiss himself had been experimenting with "hydro-aeroplanes" since 1909. At a time when there were no airports, hydros held great promise for the Navy and for carrying passengers and mail along coastal routes. They had particular appeal in the warm climate of Southern California, where land speculators were hungry to develop the coastline near Los Angeles. They also held a special charm for Martin, whose advocacy of huge seaplanes and "flying boats" would persist long after they had become obsolete.

It was with his new hydro that Martin first became a national name in May 1912, rising above the coterie of circus performers to embody the new technology's promise. With Harry Chandler and other Los Angeles businessmen using air shows to stir the region's commercial expansion, Martin was an obvious asset as a birdman with a hometown product. When he began to test the hydro from beaches along Newport Bay—often wearing a blue serge suit, white collar, and tie while airborne—a real estate developer who owned Balboa Island, W. S. Collins, recognized a gimmick to draw crowds to an otherwise lonely piece of property. The deal suited Martin's new bookings manager, Roy Knabenshue, a former carnival balloon pitchman in the "Professor Marvel" tradition who had left the Wright company after the brothers withdrew their teams from the death-plagued exhibition circuit in 1911. (Knabenshue had piloted the Curtiss-powered *California Arrow* dirigible in 1904.)

On May 10, Knabenshue and Collins arranged a flight from Balboa to Catalina Island and back to coincide with a Shriners convention in Avalon. The trip was the first by air to the resort and just happened to exceed Louis Blériot's 1909 trans-Channel record by a few miles (Glenn Curtiss had flown farther over water between two points along a continuous shoreline). In a way that presaged Charles Lindbergh's feat fifteen years later, it was the kind of stunt that crystallized for the public just what a flying machine might mean to everyday transportation. It transformed Martin into a leader of West Coast aviation and quickly attracted other investors to his fledgling airplane business.

Riding the crest of publicity, Glenn Martin proposed one month later the formation of a stock company to several Santa Ana businessmen. The town's Chamber of Commerce passed a resolution to give him "proper encouragement," but Martin apparently balked at the investors' stipulation that he quit the carnival life. They soon dropped this prudent demand, and on August 16 the Glenn L. Martin Company was incorporated with capitalization of $100,000 "for the purpose of manufacturing and selling aeroplanes and hydro-aeroplanes." Directors included W. S. Collins, the developer; William Loftus, an oil millionaire; W. A. Zimmerman, a Santa Ana banker who was also involved in the exploitation of Newport Bay; and Glenn and Clarence Martin. Collins and Loftus put up cash, with Zimmerman presiding over a loan secured by the Martin garage (hence the presence of Clarence on the board, who otherwise had only grudging interest in his son's exploits). Though Glenn Martin would endure myriad business transmutations over the next four decades, he was fond of tracing the lineage of his company to this date.

As both a new celebrity and a business authority, Martin began to curry the kind of associations that would make him part of the circle of economic power in Southern California. In August, he was invited to defend the prestigious Gordon Bennett Cup for international racing at Chicago. He set his factory to work on a new tractor-type biplane (where the engine is mounted in front of the pilot and the propeller thrust pulls the plane forward, instead of pushing from behind) designed by William Stevens, chief of his mechanics. But the organizers insisted that he fly a machine made in Illinois, a shoddy monoplane which turned out to be far slower than an entry from France. Nonetheless, he carried off top winnings of $4,854 from other events flown in one of his older biplanes and—though he could not know it at the time—put his company in a strategic market position with the new tractor design.

Names of the Los Angeles aristocracy now began to appear next to Glenn Martin's in newspaper headlines. Frank Garbutt, who had become Martin's William Thaw and was described as his "warm friend," accompanied him on a hazardous attempt to rescue two pilots whose hydroplane had sunk during a flight from Los Angeles to San Francisco. Martin's own plane was wrecked during the mission, which the *New York Evening Post* called "a valuable contribution to military and naval aviation." His jittery corporate partners were far less impressed and threatened abandonment of the business. Garbutt responded by buying them all out.

With a repaired machine and friendlier capital, Martin continued to a crescendo of publicity. The chief of Army Engineers for Southern California, Colonel C. H. McKinstry, was taken aboard Martin's hydroplane from a moving boat while inspecting Newport Bay for harbor improvements. Pacific Electric railroad tycoon Paul Shoup and millionaire rancher James Irvine took joy rides. And the motion picture colony discovered the magic of Martin's technology, sending cameramen and film starlets on flights over the Hollywood hills. These events, which now sound so prosaic, generated steady press coverage for all of Glenn Martin's enterprises.

The headlines indicated that Martin was especially fond of Hollywood promotional jobs, whether out of affinity for the cinema colony or hunger for easy notice. But the close proximity of voluptuous actresses sometimes brought out the contradictions in his personality. An article in the *Long Beach Press* of November 23, 1912, described one such outing with musical comedy siren Valeska Suratt:

> As the actress was getting into her seat the members of her company began to sing "Hello, Hello, Frisco Town," an aviation song of "The Kiss Waltz." While admitting some nervousness, Miss Suratt laughed good naturedly. Martin gave the passenger an excellent ride. He went up the bay to West Newport, came back down the coast side of the sandspit, over the heads of the spectators, along the rocky shore below Corona del Mar and back to the landing place. The highest altitude reached was 600 feet. Miss Suratt was met with cheers. Martin was unconcernedly extracting himself from the straps that tie him to the framework when Miss Suratt leaned over toward him and quickly planted a kiss on the aviator's cheek. Martin was the most surprised man in East Newport. The surprise was so evident that everybody shrieked with laughter.

"Mr. Martin is a businessman, a manufacturer of airplanes. He lives in a very quiet community," Roy Knabenshue later tried to spin the incident to the *Los Angeles Times*, which splashed a story across page one headlined "No Kissing for Martin."

*   *   *

BY THE END of 1912, the carney-cum-manufacturer was ensconced at the center of Southern California's fetal aviation industry. In December he opened a 16,000-square-foot assembly building in downtown Los Angeles, which joined the largest hangar at Griffith Aviation Park (now Los Angeles International Airport), a testing site on Balboa Island, and a school for hydroplane flying in East Newport, where Didier Masson worked as an instructor. Martin employed fourteen men and was making enough money to buy himself a green Stoddard-Dayton touring car with nickel trim and mohair top, a vehicle more in keeping with his Hollywood friends than his Chamber of Commerce backers. When the Japanese government—which then owned a total of four flying machines—sent a military officer to study the American aviation industry, Martin's hydroplanes impressed him more than Glenn Curtiss's.

"The United States is far behind in military aeroplanes, but one of these days we will wake up and lead the other countries," Martin said, noting that Secretary of the Navy George von L. Meyer was recommending more aircraft and that "all of the Army and Navy men to whom I have talked have advocated the increase of the aeroplane arm of the service." He must have been talking only to Army and Navy pilots, because the rest of the military establishment was still largely unimpressed with aviation.

The years 1913 and 1914 were an inexorable slide into war, the final hours before one of the great dividing lines of Western history. As such, accounts of Glenn Martin's exhibition shows take on a sad quaintness when read with hindsight of what was about to occur. Military observers became ubiquitous alongside millionaire sportsmen. The machines themselves started to look more like "hammerhead sharks" than breezy pleasure craft, a transformation that writers at the time found "queer." Without the war, it seems possible that flying might have evolved slowly into a form of transportation far more personalized than we know it today. But with war came dual-use, whereby aircraft designed for battle were then reconfigured for civilian purposes. War brought an unquestionable acceleration in aeronautical progress, but it also shaped the end product in ways that now look so basic as to be inseparable from the definition of aviation.

In May 1913, while Didier Masson was splashing water on Mexican gunboats, Martin unveiled an armored biplane built to specifications circulated privately among a number of American manufacturers by the War Department. Essentially, it was William

Stevens's tractor design upgraded for military applications. Instead of the wide-open box-kite look of a typical exhibition plane like the one funneled to Sonora, this "war scout" placed the pilot inside a sheathed fuselage from which only his head protruded. The engine, too, was shielded behind the fuselage wall. "A terror alike for land or sea," the *Los Angeles Sunday Tribune* bragged about "America's answer to the war plane activities of the old world."

In fact, it was pitifully inferior to European technology, but this situation had not yet been demonstrated to U.S. officials, let alone the American public. Martin turned out a civilian passenger model of the new plane by opening the cockpit to make space for four seats, adding pontoons, and calling it an "aeroyacht." Chicago farm machinery heir Harold F. McCormick, who had already learned how to fly a Curtiss hydro, was among the sportsmen he courted with the $8,500 machine during a tour of Lake Michigan in July. The depression of 1913 made such millionaires the only prospective customers for flying machines outside the military.

About the same time, Martin began to tinker with a knapsack parachute devised by a carnival balloonist who used the stage name Charles Broadwick. Martin rigged a trap seat on his biplane and hired Broadwick's teenaged acrobat companion, Georgia Anne Thompson—who traveled as Miss Tiny Broadwick, Charles's supposed daughter—to fall with one strapped on her back. Billed as "The Doll Girl," she maximized the shock effect by wearing a schoolgirl's pink smock and white sash. It was a new stomach-turner for exhibition crowds, but more to the point an item of potential value for military pilots, who were far more likely to jump from a stricken plane than little girls. Two years later, Martin received patent number 1,165,891 for a packed parachute, ignoring his old friend Broadwick.

Near the end of 1913, Martin gave a ride in the big new machine to John Brashear, a seventy-three-year-old glass craftsman who had been Samuel Langley's chief source of astronomical lenses. "This is very different from Professor Langley's machine that collapsed in the Potomac River," Brashear said afterwards. "But Langley had the idea. All he lacked, I am convinced, was the engine. Aeronautics has taken marvelous strides, but I believe my old companion was one of the greatest airmen, although he died of a broken heart after the Potomac catastrophe because he believed he had failed. I am sure he would have recognized in the fine aeroplane of Mr. Martin his own machine, improved and equipped with a splendid engine but with his original ideas."

Martin, who was starting to consider himself a creator and almost always had something to say to that effect for the morning papers, made no recorded reply.

WITH THE OUTBREAK of world war only months away, Glenn Martin's work was soon dominated by military concerns. In February, word leaked to the press that his Los Angeles factory was building "under conditions of utmost secrecy" two warplanes for testing by the Army and Navy, each based on the 1913 biplane design. "His activity in this field is accepted as evidence that both the Army and the Navy, lately criticized freely for antiquated methods in aviation, are entering upon a new era," wrote the *Los Angeles Examiner*. He flew the Army model to San Diego's North Island military flying center, where the head of the First Aero Corps judged it to be "unquestionably the finest yet made in this country."

Part of the Army's enthusiasm came from the fact that its small fleet of Wright pusher biplanes was turning into a death squadron, with six of twelve officers killed so far in aircraft crashes having died in the latest Wright Model C design. An Army board of investigation concluded that the Model C was "dynamically unsuited for flying"—that is, it tended to fall apart in midair—and condemned it. But though the Martin plane was priced below Wright and Curtiss machines, Martin lacked the personal cachet of those names. Able to turn out only two airplanes a month, he also could not match their capabilities for volume production. The Army was interested enough, however, to order a two-seater version—called the "Martin TT Tractor"—for further tests as an observation or bombing plane.

As described by the *Los Angeles Tribune*, the TT represented an obvious turn away from mere pleasure craft. "The new Army machine has more I-beam steel than Mr. Martin ever has used before, with the result that the body is rigid and almost wholly absorbs the vibration of the engine," the paper reported, not explaining that this would help military pilots to endure longer flights, with better ability to shoot a gun straight and less likelihood of their plane rattling apart over enemy lines. The TT had "a carrying capacity of four passengers, or two men and 500 pounds additional weight"—the latter being its intended configuration of pilot and observer plus munitions. Finally, the new plane had dual steering mechanisms, "so that if the aviator in either control should be disabled"—that is, killed or wounded by enemy fire—"the other could manage the biplane." This was clearly a fighting machine, unlike anything else ever made in the United States.

Hungry for federal contracts he hoped would soon be pouring in from Washington, Martin badly needed a way to dramatize what he could do and raise cash for development work that the government was not yet in the habit of supporting. He reached back to the kind of work he knew best and staged a "Battle of the Clouds" for the opening of the Pomona Speedway in April 1914. It was pure carnival hokum, with two Martin planes staging a mock attack on a fake fort in the infield. But it apparently impressed the military officers he invited from San Diego as much as the grandstand mob.

Working as a factory hand for Martin at the time was Lawrence Bell, who would go on to found another famous aviation weapons company. He described the "Battle" in down-to-earth terms years later:

> In this Battle of the Clouds, we, the manufacturing company, put on an exhibition which we charged to see. You had to sell tickets and all that sort of thing, to get on the field. You must remember—bombing airplanes had never been used in a war, so we had to invent the war and the need for it and all that sort of stuff. We tried to put on a bombing contest.
>
> We had a fort, and out in front of the fort, we dug ten or twelve holes in the ground at various places, and put black powder there which we could ignite electrically. Then we put loose dirt on top, so that when we ignited it it went woof and blew up some dirt several feet high. It didn't make any noise, of course. You didn't see the bomb come in.
>
> So Martin used to fly over the fort. He'd drop the bomb. The bombs were oranges—he'd have some oranges in his pocket, and he'd get about the right place, and he'd drop an orange. Then, of course, we would stand around with the electrical contact and we would ignite the powder charge that was closest to where the orange hit, so it wouldn't look too much like a fake. Of course, as each of these bomb explosions went off, we had a trick scheme where we had the fort scenery [collapse].
>
> We weren't taking too much poetic license in dropping oranges, because nobody knew what a bomb ought to look like anyway.

Orville Wright demonstrates his flying machine for military buyers at Fort Myer, near Washington, D.C., in September 1908. (*UPI/Bettmann*)

Samuel Langley (right) with his assistant, Charles Manly, during the 1903 trials of the ill-fated "aerodrome." Manly, a young graduate of Cornell's engineering college, perfected a lightweight motor for Langley's un-airworthy plane. (*Smithsonian*)

Hardware store in Liberal, Kansas, managed by Glenn Martin's father, about 1890. The Martin family later moved to California to escape the hardscrabble plains. (*Library of Congress*)

Glenn Martin (right) in the early 1890s with his sister, Della, a paranoid schizophrenic who spent most of her adult life in a California state hospital. (*Library of Congress*)

Jack Northrop, age five, in Lincoln, Nebraska. Like the Martins, the Northrops would soon leave the Midwest to seek better economic conditions. (*Northrop Corp.*)

Abandoned Methodist church in Santa Ana, California, where Glenn Martin's first flying machine was constructed, 1909–10. (*Library of Congress*)

Local mechanics at work on the first Martin plane in the church hall. (*Library of Congress*)

Glenn Martin's airplane "factory" in a Santa Ana cannery building, 1911.
(*Library of Congress*)

Glenn Curtiss at the controls of his seminal biplane, which Glenn Martin closely copied.
(*Library of Congress*)

LEFT: Glenn Martin and his mother, Minta, in his airplane at Balboa Beach, California, in 1912. (*Library of Congress*)

BELOW: Glenn Martin posing in his airplane loaded with a bundle of newspapers, about to be delivered by air as a publicity stunt. (*Library of Congress*)

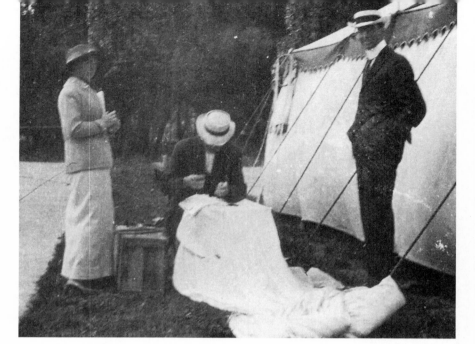

ABOVE: Glenn Martin (standing, right) with Tiny and Charles Broadwick beside a carnival tent, 1913. Charles Broadwick is assembling a backpack parachute, later patented by Martin, to be used in dropping Tiny from Martin's plane during their act. (*Library of Congress*)

LEFT: Glenn Martin (center) with Mary Pickford in a scene from *The Girl of Yesterday*, 1916. (*Library of Congress*)

Glenn Martin in a "fighter" version of his Curtiss-type biplane, about 1913. It was a pathetic attempt to match European technology, which had vastly outpaced American efforts as war approached. (*Library of Congress*)

Glenn Martin prepares to embark from the lake shore near Chicago in his "aeroyacht," 1913. (*Library of Congress*)

Malcolm (left) and Allan Loughead at the controls of the F-1, about 1917. (*Lockheed Corp.*)

BELOW: Allan Loughead (right) and a joyride customer. (*John Lockheed*)

LOUGHEAD SEAPLANE
10 PASSENGER TWIN-MOTORED
MANUFACTURED & OPERATED BY
LOUGHEAD AIRCRAFT MFG. CO.
SANTA BARBARA

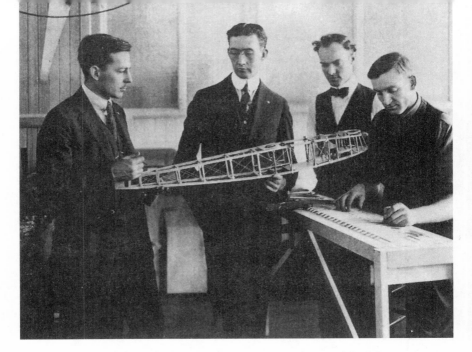

Donald Douglas (left) and Glenn Martin (center) with a model of the MB-1 fuselage. Produced too late for World War I, the MB-1 and its derivatives created high profits for the Martin Company and established the firm's reputation for building bombers. (*Library of Congress*)

The first office of Donald Douglas's aircraft venture in Los Angeles, a barbershop on Pico Boulevard in 1920. (*McDonnell Douglas*)

ABOVE: A Martin MB-2 drops a phosphorus bomb on the German battleship *Ostfriesland* during a test organized by General Billy Mitchell in the summer of 1921. (*Library of Congress*)

RIGHT: The proud, dapper, single-minded Billy Mitchell, ca. 1920. (*Smithsonian*)

BELOW: A Martin MB-1, first built as a mail plane but then reconfigured for the Army, in 1921. (*Library of Congress*)

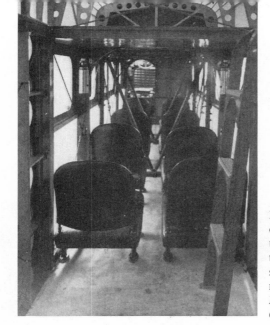

Interior of the main passenger cabin of the Martin twelve-passenger Army transport, called the GMP, supposedly intended for carrying machine-gun squads along the Mexican border. (*Library of Congress*)

Charles and Anne Lindbergh in their Lockheed Sirius. (*Lockheed Corp.*)

Ken Jay, Allan Loughead, Jack Northrop and pilots (right to left) pose with their Dole Race Vega, 1927. The plane was lost during the tragic contest, but the design made a small fortune for the nascent Lockheed company. (*Northrop Corp.*)

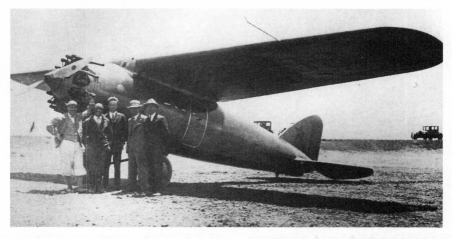

The prototype Martin B-10 bomber, a revolutionary design that brought huge profits to the company when it was exported during the latter half of the 1930s. (*Library of Congress*)

A Northrop Alpha, showing the distinctive "trousers" used to streamline its fixed landing wheels. (*Smithsonian*)

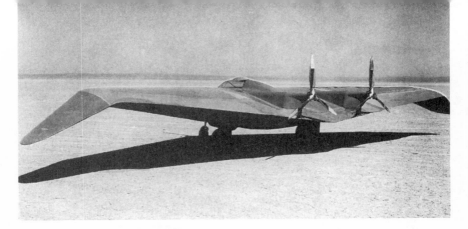

ABOVE: The Northrop N-1M flying wing prototype of 1940, strongly resembling the seagulls Jack Northrop was said to have watched for aerodynamic clues as a youth. (*Smithsonian*)

LEFT: Jack Northrop posing with his obsession, the N-1M. (*Smithsonian*)

United Airlines stewardesses form a chorus line below a Douglas DC-3, 1939. (*United*)

A Lockheed P-38 Lightning warplane, showing the tail that is said to have inspired the Cadillac fins of the 1950s. (*Smithsonian*)

RIGHT: A Northrop Gamma outfitted as a bomber for export to China in the mid-1930s. (*Smithsonian*)

Glenn Martin shakes FDR's hand. Next to the President sits William "Big Bill" Knudsen, president of General Motors, who held authority over aircraft production in 1940. (*Library of Congress*)

Glenn Martin disembarks from a luxurious China Clipper during the 25th-anniversary celebration of his 1912 Catalina flight. The Martin Company lost heavily on the Clippers it built but basked in their publicity as the ultimate in overseas air travel. (*Library of Congress*)

Robert Gross (left) and Glenn Martin confer during a World War II production plant tour. (*Library of Congress*)

Donald Douglas on his yacht, the *Endymion*, in 1943.

The dining room aboard Glenn Martin's motor cruiser, the *Glenmar*, a lavish vessel carried on the company's books as a work boat. (*Library of Congress*)

Glenn Martin and his mother, Minta, chauffered to fly on a new Martin 2-0-2 airliner, ca. 1948. Considered a bellwether project in cutthroat postwar competition for commercial business, the 2-0-2 soon dragged the Martin Company into bankruptcy and forced Glenn Martin to give up control to new managers approved by the military. (*Library of Congress*)

Northrop flying wing bombers, soon to be scrapped, at the Hawthorne factory in 1949. (*Smithsonian*)

Donald Douglas, near the end of his career, with a DC-8 jetliner. The company lost heavily as it tried to compete with Boeing in building the first jet transports, leading finally to a merger that ended the Douglas family's historic dominance of the industry. (*McDonnell Douglas*)

In essence, the Glenn L. Martin Company, a manufacturer of airplanes, was teaching the military and the public how to use the machines as weapons. What was obvious to anyone who had flown an airplane—still a minuscule group of a few surviving exhibition pilots, rich sportsmen, and junior military officers—was evidently hard for anyone else to visualize. Even within the military, it would take a world war and then years of bureaucratic skirmishing before the high command would fully embrace the combat value of warplanes. Martin's business could not wait.

In July 1914, the Signal Corps banned the use of all pusher airplanes, fearing that they stalled too easily and the pilot was too apt to be crushed by the engine behind him in a crash. The Wright Company, which had grown technologically stagnant after Wilbur's death in 1912 and increasingly obsessed with patent litigation, was thus at a dead end. It would not produce a tractor plane until the end of 1915, when Orville was no longer president. Glenn Curtiss was in sound condition with his new hydroplanes, which were prized by the Navy. As for Glenn Martin, the Signal Corps decision turned him from a struggling second-tier builder into a prime source. The Army ordered seventeen Martin Tractors in 1914 alone, at a base price of $8,500 each. His carnival days were over.

In the *Los Angeles Evening Herald* of August 7, 1914—three days after the German invasion of Belgium—under the headline "Aviator Tells of New Terrors for Armies and Navies," the twenty-eight-year-old Martin wrote about the war that was unfolding in Europe. It was a clairvoyant statement, considering how far his vision predated such eminent proponents of air war as Giulio Douhet and Billy Mitchell:

> The aeroplane will practically decide the war in Europe.
> Veritable flying death will smash armies, wreck mammoth battleships and bring the whole world to a vivid realization of the awful possibilities of a few men and a few swift winging aerial demons.
> For the old time war tactics are no more. The generals who realize this quickest and fight first with the flying death, will win.
> Aeroplanes, darting with cyclonic speed, will demoralize armies and navies and leave the earth forces terror stricken. For no matter how brave the soldier may be fighting the enemy he meets on the ground, the aerial peril is a new and terrible factor in war with which armies have yet hardly more than a mere theory.
> Aviators—each with only one life to give for his country—will

amaze the world with their acts of stoical heroism. Aviators, steering into sure death, will grapple with the enemy's aeroplanes thousands of feet in midair. New names will be added to the temple of fame. They will be the names of humble pilots of aircraft.

It is possible for one man, driving an aeroplane laden with high explosives, to dive like a plummet upon the bows of a great warship and destroy it. He gives one life; the enemy gives many.

It is possible, and probable, that pilots of aircraft will thus descend upon citadels of enemy's strength and by losing one life write their names upon the pages of history with their heroic blood.

In this war three distinct types of aeroplanes will be used for three separate phases of war duty.

First, there is the reconnoitering machine, flying 75 miles an hour, carrying two men, one the pilot, the other the observer. The observer operates the wireless apparatus, making notes of the country, of the enemy's position, strength and other data, wirelessing the more important facts to his general back there with the waiting army.

And as this reconnoitering machine dashes on, at least 4000 feet high, beyond the range of the most powerful army rifle, the general, through wireless dispatches from the aeroplane observer, completes his war map and lays his plan.

Second is the flying death machine, carrying two men and 600 to 800 pounds of explosive bombs. This type of machine has a speed of from 45 to 75 miles an hour.

This aeroplane will be the terror of the army, for it will drop not only the known explosive shells, scattering shrapnel and hurling fire bombs, but it will introduce a new kind of projectile into modern warfare.

This projectile is called the needle-shell [fléchette]. It is the most devilish man-killing device yet produced. It kills by wholesale. It comprises a cigar-shaped piece of steel, with a needle point and a fin-like tail to insure accuracy.

These projectiles do not explode, but dropped by the basketful upon an army in any formation whatever, from a height of 400 feet, each weighing about half a pound, drive themselves into men and horses, killing, killing, killing.

Russia has adopted these needles. France and Germany and England, too, I understand, have made demonstrations of them in secret. These demonstrations were made upon herds of goats with terrible results.

The third class of aeroplane is one that is extremely light, with a speed of from 60 to 100 miles an hour, carrying two men and no added weight save fuel and light rapid-fire guns. The duty of these aeroplanes is the most dangerous of all—they must protect the scout planes against the enemy's aeroplanes.

A battle between two airships in midair means death. This machine, with its rapid-fire gun, could rake an enemy's column. Aeroplanes already have figured in this deadly work upon the Austrian army, harassing the soldiers as they marched into Servia. This type of aeroplane could dislodge soldiers in ambush in the mountains or canyons and harass whole armies at night.

Another engine of devilish import was invented by an American named Steinmetz in Philadelphia. It is an explosive fire bomb to be trailed by a long cord from swift-flying aeroplanes. Catching the cord over a dirigible or the wings of an aeroplane, the instrument of death is rapidly drawn up to the doomed craft, where it explodes by contact.

Watch the aeroplanes in the European War. Watch the aeroplanes make new history.

Except for the phrasing—which sounds uncouth because of his carnival background and because the academic terms of twentieth-century military discourse had not yet been coined—Martin could hardly have been more precise. But even the former "Flying Dude," the conqueror of Catalina, and mock consort of Valeska Suratt could not foresee the wealth and power that was on the horizon for his coveted industry.

CHAPTER THREE

SOUTHERN CALIFORNIA

SOUTHERN CALIFORNIA possessed two elements of crucial value to early airplane builders: balmy weather and adventurous millionaires. The first gave them an edenic laboratory for their tender machines, the second represented a source of otherwise skittish capital. Hard-line political conservatism that would in later decades underpin a rich concentration of armament factories in the region did not yet have a firm visage.

The westward migration from hardscrabble farms that threw Glenn Martin into this garden also delivered another young man of strikingly similar background, though of much straighter personality. He was better educated than Martin and driven by technical fascinations rather than mercantilism, but war and technology would push them both to the same high ground.

"My earliest recollection of aviation is going down to Los Angeles to see a Frenchman named Didier Masson," John Knudsen Northrop said near the end of his life. Like so many other country children starved for entertainment, he was drawn to the air carnivals, whose performers could make the same lasting impression that television stars would several generations later.

After Jack Northrop's birth in 1895, his father, Charles, transplanted the family from Newark, New Jersey, to Nebraska to take a job in the book and picture section of a Lincoln department store. It was a dreary migration, but these were depression years for most Americans, not the Gay Nineties. His mother, Helen, had been previously married and widowed, so the move may have served the additional purpose of escaping unpleasant associations. Northrop's clearest memory of life in Nebraska was of his father saving money for a rare family outing to see the Big Blue River, fifteen miles by rented surrey from their home. When they finally made the trip and found a muddy creek "about four feet wide and three feet deep," Northrop knew he was not destined to be a Midwesterner.

By 1904, his parents were ready to uproot again, despite possessing barely enough money for necessities. With a surge of romantic optimism that reflected how open the continent still seemed to even its poorest inhabitants, they picked Santa Barbara, California, because they had heard how beautiful it was. They arrived destitute. Charles found lowly work as a semi-skilled carpenter and managed to move his family out of a poor boardinghouse to a canyon that debouched among eucalyptus trees onto the beach, just below a private estate. It was not the site of a dream home, however. He got permission to pitch two tents there, building wooden floors under them and calling one a kitchen and the other a bedroom. It was a Robinson Crusoe abode—magical perhaps for a nine-year-old boy, but extremely humble.

Like many lower-class American boys, Jack Northrop depended on the YMCA for organized recreation. A gym instructor there formed a Boy Scout troop in 1910, which the young Northrop joined out of love for the nearby mountains, despite the fact that camping was for him a nightly necessity. "The difference between a knot that will slip and one that won't is a very vital thing in anybody's life, I'm sure," Northrop opined guilelessly about his Scout lessons many years after becoming one of the nation's premier weapons industrialists. This brand of unabashed corny wisdom was as typical of Glenn Martin, Northrop, and their cohort as an affinity for radical technology, suggesting that the two traits somehow went hand in hand.

Again like many poor children, Northrop was shy beyond his circle and did not enjoy the few other plain social events, such as school dances, that accessible to him. He was thin and sandy-haired, with smooth Scandinavian features that projected a youthful earnestness well into middle age. But he must have seemed rather exotic to his high school sweetheart and future wife, Inez, daughter of the artist Alexander Harmer and a descendant of Spanish colonists. It was the kind of marriage that could open doors otherwise tightly shut against an impoverished young man whose family dwelled in a tent.

At Santa Barbara's public schools, Northrop was lucky to find practical courses in algebra, geometry, physics, chemistry, and mechanical drawing. They were the limits of his formal instruction. At the age of sixteen, he saw Masson fly and was infected with the idea of building airplanes. As for literature or history, his interest was negligible, resulting in near illiteracy. He had to suffer the indignity of watching his two half brothers from his mother's first marriage go off to Cornell and Berkeley, while he took a job as a garage mechanic after graduation in 1913. They had an inheritance to spend on college, but Charles Northrop could not afford such luxuries for his own son.

Perhaps because his older half brother soon began to enjoy a reputation as a prominent architect, Jack quit the garage after two years and worked as an architectural draftsman until 1916. He was never a designer, merely an assistant who rendered sketches into technical drawings. In effect, he was acquiring the loosely defined set of mechanical skills that could still help turn a naturally gifted technician into a sophisticated engineer. This would be the last American generation to profit from such wide-open access to the ranks of advanced technology.

Like Glenn Martin the auto dealer, it seems possible that Northrop could have continued on from here in a familiar, modestly prosperous way. His father eventually became established as a building contractor, an occupation Jack could have fit into quite easily. But in 1916, with war roiling in Europe, two young aviators arrived in Santa Barbara with a plan to build hydro-aeroplanes for the U.S. Navy. They rented the back half of the garage where Jack had worked, started to put together a mammoth twin-engined flying boat, and called themselves the Loughead Aircraft Manufacturing Company. For three days, Northrop walked back and forth past the garage in theatrical nonchalance before asking for a job. They hired him as their "chief engineer."

MALCOLM AND ALLAN LOUGHEAD, like Glenn Martin, were carnival performers who had become enmeshed in the birth of the aviation industry. Their older half brother, Victor, was a disciple of John Montgomery and had written books portraying Montgomery as the foremost aviation pioneer of the era. The family was thus an exemplary mix of the technological audacity and abstruse claptrap that fueled the industry's early growth.

Born in Niles, California, southeast of San Francisco, in 1887 and 1889, respectively, Malcolm and Allan were raised by their mother, Flora, a former teacher who supplemented her income by writing features for the *San Francisco Chronicle*. While they were still very young, she had separated from their father, who worked in a San Francisco hardware store; and she moved to a thirty-five-acre fruit ranch in the Santa Clara Valley. The boys' formal education was unnotable—Allan did not even finish grammar school. John Montgomery, the glider builder whose inventions were launched from hot-air balloons in Santa Clara during the spring of 1905, was the celebrity who attracted first Victor and then the two brothers to aeronautics.

Victor had changed his name to Lougheed (closer to a phonetic spelling of the Scots-Irish name) and was working in Chicago as an automotive engineer when he read about the Santa Clara exhibition. He began a correspondence with Montgomery and in 1907 met with him in California to talk about whether powered propellers could be mounted on the gliders. Lougheed was also an author of practical handbooks about automobiles—he was one of the founders of the Society of Automotive Engineers—and considered himself a scholar, so his predilections blended naturally with those of "Professor" Montgomery, who taught geometry at a Santa Clara Jesuit school and wrote pseudoscientific articles about flying. When Victor published *Vehicles of the Air* in 1909—a popular how-to book about the development of airplanes that denigrated the Wright Brothers and lauded Montgomery—he created the kind of reference work that could advance their business potential together.

In 1910, Lougheed persuaded a wealthy Chicago sportsman and distributor of White automobiles, James Plew, to finance the construction of flying machines based on Montgomery's gliders. Plew also bought a Curtiss biplane for good measure and hired a Curtiss-trained pilot. When he began looking for a competent engine mechanic willing to tinker with airplanes, Victor recommended his half brother Allan, who happened to work for the White agency in San Francisco. For six dollars a week, Allan moved to Chicago with the assignment of helping to fasten a motor on a new Montgomery glider. Before he left, he told his skeptical friends that airplanes were going to be "the safest means of transportation at forty to sixty miles per hour, and the cheapest"—the sort of statement that could make a young man sound like either a genius or a crackpot.

The latter reputation would dominate for many years. His immediate task quickly proved hopeless because of the Professor's totally unsuitable design. Montgomery was later killed at the age of fifty-three while attempting to soar in a sister craft. Plew went on to lose two Curtiss planes in crashes, a spate of expensive if not uncommon bad luck that convinced him to abandon aviation in favor of autos. In the meantime, Allan Loughead learned to fly and became familiar with the construction of Curtiss machines. During the summer of 1911, he married and signed on as an exhibition pilot with the International Aeroplane Manufacturing Company of Chicago, an overblown enterprise organized by another auto dealer who also sold Curtiss-type biplanes. He traveled the county fair circuit until September, when his rain-soaked machine tangled in telephone wires near the village of Hoopeston, Illinois. Uninjured, he nonetheless agreed with his bride to give up circus flying, though not the rest of the aviation business.

In early 1912, Allan sold his motorcycle for a hundred dollars and bought two tickets to return to San Francisco. He carried with him a pastiche idea for a hydro-aeroplane, even ordering a set of wing ribs from the shop foreman at International before he left Chicago. He picked up work again as an auto mechanic and soon persuaded his brother Malcolm, who was a test supervisor at the White factory in San Francisco, to help build the flying machine. They rented a vacant garage at the corner of Pacific and Polk on the waterfront, where they worked in their spare time until outside financing was needed to finish the plane. Well-known among the region's automobile owners, they quickly found a backer in Max Mamlock, head of the Alco Cab Company, who added $1,200 to their own $1,800 investment. Other friends put up another $1,000 to reach the $4,000 needed for completion. In June 1913, a hydro-aeroplane they called their "Model G" slid into San Francisco Bay near the foot of Laguna Street.

The Lougheads' Model G (there were no models A through F, only discarded drawings) was strikingly similar to the tractor "aeroyacht" produced by Glenn Martin's factory in Los Angeles during the same period. The Martin plane was a few feet smaller in wingspan and length, about 150 pounds lighter, and thus a bit faster, but otherwise the two aircraft were close cousins. Pilot and passengers sat in tandem open cockpits in a real fuselage. Ailerons were mounted between the biplane wings. Propulsion came from an eight-cylinder eighty-horsepower Curtiss engine in the fuselage nose. The whole structure rested on a long pontoon that resembled a toboggan sled divided into watertight compartments, with floats beneath each wing tip for balance. The Model G was not a toy—it weighed more than a ton when fully loaded and could cruise at more than fifty miles per hour. As a first venture for novice builders, it was highly ambitious, suggesting that the brothers were proceeding with confidence in already proven technology.

There is no record of personal contact between Glenn Martin and Allan Loughead, though both were in Chicago during the summer of 1911 and had close associations with the Aero Club of Illinois. Given the social nature of aviation in those days, it seems certain that some cross-pollination of design ideas occurred. The coincidence of the Martin and Loughead designs would suggest, at a minimum, that neither young man was justified in making the sweeping claims for invention that each would relish in later life.

With the Model G—also called the "Alco Number One" in deference to Mamlock—at the center of their moneymaking schemes, the Lougheads tried to establish a charter service around San Francisco Bay. Aviation's reputation was still too attached to the bloody air shows, however, for many ordinary people to risk ten dollars for a ride. The brothers thought they sensed a change in the social atmosphere when Ferdinand Theriot, a San Francisco sportsman in the manner of Frank Garbutt, staged a society fair in San Mateo and arranged for them to take his friends on sightseeing flights. But on the first trip, Allan struck a levee with Theriot on board, smashing a propeller and pontoon. No one was injured, but Mamlock could not afford to be sued by the likes of Theriot and so took the opportunity to close out his investment by seizing the aircraft. After locking it up in a San Francisco warehouse, he informed Allan and Malcolm that if they needed it again, the key was for sale.

There were many sensible entrepreneurs like Plew and Mamlock who took their knocks in the early aviation business and got out. Objective evidence at the time, especially during a recession year like 1913, suggested strongly that they were following the wisest course. But there were a few others who stayed in regardless of setbacks or calamities, whether because they were true believers or because they had nothing better to do. One is tempted to conclude that some, like the Lougheads, would have slept in the streets rather than give up their outré trade, which took on the nature of an obsession. The Loughead brothers did not resort to crime to get back their Model G, but for the next year and a half they floundered in desperate attempts to amass enough cash to continue.

In 1914, they prospected for gold, one of the few ventures even riskier than flying. There were not many nuggets left in California's Mother Lode country for two amateur miners, so Allan soon returned to work as an auto mechanic, while Malcolm took on the tenuous assignment of setting up a Curtiss sales agency in Hong Kong. He was on a freighter in the Pacific when a British destroyer intercepted the ship and seized his sample biplane as contraband. He surfaced next in Sonora, Mexico, where he served briefly as chief engineer for the air force of Plutarco Elías Calles, a politico-military associate of Alvaro Obregón. The "air force" consisted of one Martin scout, possibly a much-mended reincarnation of Didier Masson's *Sonora*. But employment in Mexican aviation was even less reliable than gold mining. He returned to San Francisco when he learned that Allan had found a new investor.

Fresh funding materialized because of the imminent Panama-Pacific Exposition of 1915, which celebrated the opening of the Panama Canal and its benefits for San Francisco as a shortcut between the East and West coasts. When the city began taking bids for a passenger-flying concession at the fair, Allan put advertisements in the local papers for a partner to refinance the Model G. The ads drew the attention of a baker named Paul Meyer, who had made a small fortune running a restaurant during the Alaskan gold rush. Meyer bought Max Mamlock's warehouse key, freeing the Model G for rejuvenation in the Loughead-Meyer Flying Service.

Though the Lougheads were underbid by another local company, they were granted the concession in September 1915 after their competitor lost his aircraft in a crash. With Allan as pilot and Malcolm doubling as mechanic and carnival barker, they made more than 600 ten-minute flights at ten dollars per passenger. One notable visitor they were not able to coax on board was Henry Ford, who told them that he liked to keep his feet on the ground. Nonetheless, the exposition turned them into minor celebrities, and they cordially bought out Meyer's stake, with $4,000 left at the end to carry on their business. With this nest egg and new plans for a large flying boat that might interest the Navy, the two enterprising, irrepressible brothers moved down to Santa Barbara in the auspicious spring of 1916.

*   *   *

IT HAS BEEN SAID that warfare exaggerates prewar conditions. The condition of American aviation before the United States entered World War I was barely more than that of a bright child—a rather eccentric one if the tortuous Loughead saga was at all representative. The commercial enterprises of the Wright brothers and Glenn Curtiss had gained a degree of international respectability, but even they were far removed from the seats of power occupied by the leaders of steel, railroad, oil, and automobile companies. As for Glenn Martin and the Lougheads, they may have been known within aeronautical circles, but to almost anyone else they would have seemed exceedingly marginal, probably laughable.

By 1916, a few of the young aviation companies had at least cut their teeth on solid designs and gained a feel for their shaky market. They sensed that airplanes were not like cars, that the government would have to play a strong part in their economic success. The year 1916 would in fact see the start of a relationship between government and industry compelled by war that would overwhelm all other issues facing these new companies. They would never again be exactly private in the classical sense of free enterprise.

In mid-1915, Secretary of the Navy Josephus Daniels formed the Navy Consulting Board, a group of unofficial advisers from engineering and industrial societies, to help guide military expansion when American entry into the European war looked inevitable— the British liner *Lusitania* had been sunk by a German U-boat on May 7, killing 1,198 passengers, of whom 128 were Americans. (In February, the *Lusitania* had carried three Curtiss hydros from New York to Liverpool for delivery to the Royal Navy. The German government had lodged a formal complaint with the State Department in late January about such weapons transfers.) As a Progressive, Daniels was wary of big business and already at odds with the steel industry over Navy prices. A subpanel called the Industrial Preparedness Committee surveyed thousands of companies in 1916 with an eye toward wartime production. The project was dominated by Howard E. Coffin, the forty-two-year-old vice president of the Hudson Motor Car Company and former president of the Society of Automotive Engineers.

Coffin was something of an archetype in American business culture, an ambitious reactionary who would keep reappearing at high levels for the next twenty years, despite a predilection for dwelling in the gray areas between legal and illegal activity. With no sense of camp, he described himself as "a descendant in direct line from Tristram Coffin, of Nantucket, who emigrated in 1630 from England, seeking religious, social and economic freedom in America." He tolerated government intervention in commerce only as long as it helped fill the pockets of deserving men. Like advisers to the throne throughout history, Coffin and his colleagues were driven by a self-serving combination of patriotism and a desire to advance their own positions.

They were not above bullying minor interests such as the nascent airplane industry, though as president of the SAE Coffin had championed standardization of parts as a way to protect small manufacturers. War was a duty and an opportunity, with emphasis on the latter. "Twentieth century warfare demands that the blood of the soldier must be mingled with from three to five parts of the sweat of the man in the factories," Coffin said, calling World War I "the greatest business proposition since time began." He foresaw an elite class of technocratic managers running a mobilized economy. Though utterly commonplace today, these notions were a significant departure from the past, when armies had handled their own procurement of weapons and supplies, relying heavily on federal arsenals instead of private contractors.

Equating private interest with public need was a constant theme among preparedness campaigners, who were actually breaking down the traditional distinctions between private and public institutions. They even argued that industrial preparedness would lessen any pressure for federal ownership of weapons plants. By spreading armament orders around the country, furthermore, "the boogaboo of the Munitions lobby" might be diffused, as Coffin crassly put it.

In August 1916, as part of the annual budget legislation for the Army, Congress created with little debate a Council of National Defense that formalized the procedures and membership of the Navy Consulting Board. The bill's language had been written under the guidance of Secretary of War Newton D. Baker—a Progressive who favored government-industry cooperation in regulating the economy—and then protected by the chairmen of the various military affairs committees. Without giving up their positions and salaries in private business, industrial leaders thereby became public officials advising the President on mobilization. The underlying principle was that private industry would be the primary source of arms when the nation declared war.

As might have been expected for such a radical shift in military supply, the uniformed establishment opposed the plan, along with civilian groups that feared the re-emergence of profiteers who had plundered the U.S. military since the Revolution. There were also many factions—inside and outside the military, of all political parties—that were not enthusiastic about the centralizing tide of modern technology and warfare. Congress therefore set up a joint military-civilian board to study the issue, but it wound up rubber-stamping the conclusions of the Navy Consulting Board.

The Council of National Defense "opens up a new and direct channel of communication and cooperation between business and scientific men and all departments of the Government," President Woodrow Wilson proclaimed in October 1916, trying not to jeopardize his popular image as a neutral peacemaker even as he prepared the nation for war. Though American bankers and manufacturers were by now fully engaged with Allied forces, public opinion was still sharply divided on the question of intervention in Europe, setting political limits on the degree of institutional cooperation needed for Wilson's prewar planning. Much the same situation would reoccur on the eve of World War II.

Other players did not have to be so circumspect. "It is our hope that we may lay the foundation for that closely knit structure, industrial, civil and military, which every thinking American has come to realize is vital to the future life of this country, in peace and in commerce, no less than in possible war," wrote council member Coffin to the DuPont family in December 1916, privately heralding the start of a partnership that would vex the nation for the rest of the century.

Led by businessmen who were not eager to change the status quo where earnings were concerned, the council's operational unit—the National Defense Advisory Commission, or NDAC—divided up into panels that reflected the private sector's power structure. Big companies and trade associations naturally dominated the system. In practice, NDAC members awarded contracts to themselves and their friends, violating military regulations and antitrust laws under federal sanction.

The NDAC was thus able to force some headway in organizing supply, but demand was beyond its reach and authority. With more than a half-dozen bureaus independently procuring for the Army alone, chaos was inevitable (between April 1916 and June 1919, the Army poured some $14.5 billion into the economy, out of total war costs of about $32 billion). As for the fledgling aviation industry, the problem was not chaos but relative insignificance. Circus performers and auto mechanics were hardly in the same orbit as the masters of late-nineteenth-century business organizations who ruled NDAC. Wealthy balloon enthusiasts who founded the Aero Club of America in 1907 had regarded men interested in flying machines as not "clubbable"—a form of class discrimination that still pertained.

Moreover, as late as 1916, the military services were still largely ignorant about what flying machines meant as weapons. Not until the summer of 1914 had Congress created the Aviation Section of the Signal Corps, mostly in response to declarations of war in Europe and a growing fear that the United States lagged far behind in aircraft and engine construction. (In 1913, the Smithsonian had sent two university professors with an interest in aviation on a fact-finding tour of Europe. Their report helped spawn the National Advisory Committee for Aeronautics, or NACA, in a rider to the Naval Appropriations Act signed by President Wilson in March 1915. This technical panel, modeled after a similar group in Great Britain, consisted of four airmen from the Army and Navy; one member each from the Smithsonian, the Weather Bureau, and the Bureau of Standards; plus five persons from outside the government. The chairman was the secretary of the Smithsonian, Charles Walcott, who had tried to sell Samuel Langley's doomed projects to Theodore Roosevelt in 1898.) The new Aviation Section consisted of sixty unmarried lieutenants and 260 enlisted men eligible for special aviation ratings and flying pay—thus presenting manufacturers with a full-fledged customer, at least on paper. But Congress could not cause a revolution in military thought. In December 1914, the Chief Signal Officer told the House Military Affairs Committee that "as a fighting machine the airplane has not justified its existence."

Nonetheless, the North Island military aviation center in San Diego attempted to study new aircraft and come up with specifications for testing commercial models. The Martin TT was part of this process, one of twelve different bids in 1914 for a reconnaissance airplane. By 1916, the Signal Corps had established a working relationship with a number of manufacturers, though the quantity of aircraft procured remained low because of small appropriations from Washington. In the whole country, only about 400 aircraft were built in 1916.

A turning point in money matters came with the same legislative cycle that created the Council of National Defense. In June 1916, Congress authorized some $500,000 in deficiency funds over and above the Aviation Section's annual $300,000. The Chief Signal Officer also began to change his mind in the face of stark news from Europe, declaring that the need for various types of military flying machines was now "obvious." He proposed forming squadrons of a dozen planes—eight for observation, two for chase or transport, and two for bomb-carrying. War would prove this mix that favored reconnaissance over combat to be totally wrong, but it was better than nothing.

A unit of eight Curtiss "Jenny" biplanes, called the First Aero Squadron, was thus part of General John Pershing's muddled invasion of Mexico against Pancho Villa in 1916. It took just six weeks to smash or otherwise render useless every one of these "Suicide Machines," as the soldiers called them, without a single combat engagement. The experience was worthless from a tactical standpoint and spawned the Army's first aviation scandal, but at least it taught some quick lessons about maintenance in the field that only a few obscure mercenaries like Didier Masson had yet learned in the Western Hemisphere.

The Signal Corps tried to learn more about air combat from what was happening in Europe but found itself blacked out by the warring powers' secrecy measures. Foreign embassies repeatedly turned down requests by the War Department to send aviation experts to the front as observers. The strained ambiguity of Wilson's political neutrality thus worked against his own military. What little useful information there was came from American manufacturers working on export contracts for the belligerents. When a congressional committee asked for a statement on the relative position of United States aviation compared to Europe, the Signal Corps was forced to admit that it could not give a meaningful answer.

By early 1917, the Army's Aviation Section had a grand total of seven "combat" squadrons—four in the U.S., two in Panama, and one in the Philippines. Compared to Europe's active air units, they were strictly recreational. On the eve of U.S. entry into the war, American airplane builders were thus in a tender position vis-à-vis the government and the future of their product. They certainly were in no position to dictate the terms on which they would participate as weapons builders. Neither they nor the government had any coherent picture of what was about to unfold.

The war and its exigencies would proceed to engulf them, as it did almost every human enterprise, creating new organizations and favoring personalities that would not have otherwise found expression. Neither Southern California nor any other good place for building flying machines would ever be the same. Nineteen sixteen, not 1946, saw the beginnings of a military-industrial complex.

CHAPTER FOUR

THE GREAT
PROPOSITION

IN AUGUST 1915, Glenn Martin navigated his green Stoddard-Dayton with the nickel trim and mohair top through the Los Angeles heat to pick up an engineer he had just hired from the East. Martin's star was higher than ever following his role as the roguish pilot in a Mary Pickford film called *The Girl of Yesterday*. Playing an aspiring aviatrix, America's thousand-dollar-a-week sweetheart is compromised by Martin after they crash in the hinterlands. But mere kissing had once again posed real-life problems, forcing the twenty-nine-year-old daredevil to plead offscreen, "I don't think my mother would like it."

Today his difficulties seemed less intractable. As the design and production of flying machines had become more rigorous with the military issuing its own specifications, Martin reluctantly admitted that he could no longer slip by on cut-and-try. When his best designer quit the company to work for Glenn Curtiss, he wrote to one of the leading American aviation technologists, Jerome Hunsaker—also a twenty-nine-year-old Iowa native—who had developed an aeronautical engineering curriculum for the Navy at the Massachusetts Institute of Technology and had been one of the two Smithsonian experts who reported from Europe in 1913.

For a variety of good reasons, Hunsaker was not optimistic about the career outlook in airplane building. He nonetheless recom-

mended the raw young man now waiting in the Baltimore Hotel at
Fifth and South Los Angeles Street, a dewy twenty-three-year-old
graduate named Donald Wills Douglas.

"I knew Glenn from seeing his pictures," Douglas recalled many
years later. "I stood around the lobby until he walked in, then went
up to him and said, 'Mr. Martin?' He looked surprised and said,
'Yes?' I said, 'My name is Douglas.' He looked at me and sort of
blanched. I think he had the idea that anybody who came out of
school, from MIT, would have a long beard. I was said to be very
youthful-looking then."

The meeting of Glenn Martin and Donald Douglas brought together a single-minded mercantile ambition with one of the nation's
first trained aeronautical talents. Martin, especially, was lucky. The
professional union would not last long, but it would produce one of
the few significant American warplane designs of World War I and
put both young men firmly on the course of weapons building.
Their personalities were vastly different, but their talents melded
fortuitously at a pivotal moment.

Douglas was decidedly not from the threadbare backyard garage
school that had spawned Martin, Northrop, and the Lougheads. He
had been born on April 6, 1892, in Brooklyn, New York, the second
son of William Douglas, a Scotsman from Albany who worked on
Wall Street as a bank cashier, and Dorothy Hagenlocker, a domineering German immigrant remembered frankly by her grand-
daughter as "a bitch." Despite the catastrophic depression of the
1890s—583 banks failed nationwide in 1893 alone—the family remained modestly prosperous and owned their own home near Prospect Park, which in those long lost days was a stable middle-class
neighborhood with decent public schools and unthreatening streets.
When Donald was thirteen, they moved into a Manhattan apartment along Riverside Drive.

As a boy, Douglas showed a literary bent, writing poems and
starting a children's newspaper, *The Riverside Jester*, with his school
friends. There was a whimsical side to his character that could rise
to the surface at important moments. Using a home printing press,
he published *The Handy Magazine*, apologizing in the first issue for
the lack of a lower-case "t" with the phrase "noT enTirely The faulT
of The prinTers." At fourteen, he turned out a book of his own
poems, titled *Ye Scroll*, asking readers not to make "judgment terse"
on his well-intentioned "verse." Summers were spent on Long Island Sound, sailing his father's small boat at a Connecticut yacht

club. During one of these vacations he was deeply impressed by a newspaper account of the Wrights' first flights.

"We had really a very fine life, although I don't suppose my father's income was ever very great," he said warmly of his childhood. Yet there were jagged edges behind the smooth bourgeois facade—his mother lavished all of her affection on his older brother, Harold, and treated his father with icy disregard.

Because of its sprawling military shipyard on the East River and the hundreds of local people who worked there, Brooklyn was a sharply flavored Navy town during Douglas's youth. "My mother used to dress my brother and me up in sailor clothes—the Navy tailors would make them," he said. "I think one of the first newspaper headlines I can remember was when the *Maine* was sunk [1898]." In September 1908, during a trip through Washington with his mother to petition several congressmen for an appointment to the Naval Academy, he watched the Signal Corps trials of the Wright *Flyer* at Fort Myer, Virginia. In June 1909, he saw Glenn Curtiss fly his seminal *Gold Bug* biplane—the twin of the *June Bug* ordered by the Aeronautic Society of New York that sparked the Wrights' patent lawsuit—around a racetrack at Morris Park in the Bronx, headquarters of the society. For the American public, these were two of the most provocative demonstrations of the first decade of flight, and witnessing both as a teenager was a formative experience for Douglas. In the fall of 1909, after attending Trinity Chapel, a strict Episcopal high school in lower Manhattan, and then preparatory classes in Annapolis, he followed his brother into the ranks of midshipmen.

"I came from probably the average kind of family that sent boys to the Academy, money-wise and in social standing," he said. "I was somewhat influenced by the fact that my brother went there two years before I did, and I looked up to him. Actually, I had no desire to go into the Navy. In fact, I had some idea that I was going to write. I think the real thing that quenched me was my brother dying."

When Harold Douglas graduated from Annapolis, he transferred to the Coast Artillery because of poor eyesight. While on a minesweeper in the Gulf of Mexico south of New Orleans, he was stricken by appendicitis. He was not taken to a hospital quickly enough, and he died.

"I never had trouble of any kind, scholastically or with discipline. I had nothing to regret in any way. But I just got the feeling that I didn't want to live under those circumstances all my life. I just didn't like the idea of the discipline over me. I thought there wasn't enough freedom. I commenced to see some of the facets of naval life after you became an officer, and I didn't like them."

Repulsed by the notion that his brother might have died because of Navy rigidity, Douglas resigned from the Academy in 1912. His mother was furious, but by now he was self-confident enough to ignore her. He applied for jobs with Glenn Curtiss and Grover Cleveland Loening, another aviation pioneer-manufacturer, but was turned down by both. In the fall of that year, he enrolled at MIT, where he had wanted to study all along.

In terms of mental discipline and cultural austerity, the jump from Annapolis to Cambridge did not require much of a stretch. Then as now, engineering studies left little time for anything else, deepening the social isolation of students who came with adolescent vocational fixations. Douglas seldom even read a newspaper. "Unless somebody hit me with a rock, I wouldn't know what was going on. I was very insular. I didn't talk over the future of aeronautical work then. I didn't talk over the future of anything with anybody."

There were no courses about airplanes yet, so he studied mechanical engineering. Formed chiefly around the established industries of New England, MIT had a pragmatic faculty whose major interests and income derived from links with regional businesses. Because there was little automotive work in New England, Douglas could not find a class on internal combustion engines at the illustrious technical college. Other than Hunsaker, whose courses specifically intended for postgraduate military instructors did not open until the autumn semester of 1914, there was little encouragement for a career in building airplanes.

"Nobody persuaded me to go into aeronautical engineering," Douglas claimed. "The only person I really talked to about it was Jerry Hunsaker, and he tried to dissuade me from it." His interest was sustained by a fascination with the subject he termed "mystical," a not uncommon reaction in those days to the prospect of flying.

Following his intuitions, Douglas graduated from MIT in the spring of 1914—a recession period in the United States—then stayed another year as Hunsaker's research assistant at an annual salary of $600. According to Hunsaker, Douglas was broke and took the job only because it was a ready source of cash. They operated the institute's first wind tunnel—copied from the British National Physical Laboratory—and did consulting work for the Connecticut Aircraft

Company, a small firm in New Haven that was making dirigibles. Some of Hunsaker's wariness about the aviation industry's future may have come from dealing with the company's management—an old-time carnival balloonist and tightrope walker called "Captain" Tom Baldwin and an Austrian eccentric who had sunk his family fortune in zeppelins. (It was Baldwin who had bought a second-hand motorcycle engine from Glenn Curtiss in 1904 to power the *California Arrow* dirigible.) After helping them win a $45,636 contract for the U.S. Navy's first dirigible in mid-1915, Douglas left New England for the West.

What he found there was in every way a new world. At the Martin company, the neophyte engineer confronted what must have seemed to him a bizarre enterprise after Hunsaker's erudition—a technical manufacturing operation based entirely on empiricism and almost completely naive about the theoretical mechanics of motion. "At this time, there was practically no engineering, it was all done by good judgment," Douglas recalled. "When I got there, they made no detailed drawings of anything. So when I told the people I was going to give Martin detailed drawings, they all said 'The boy is crazy—it can't be done.' They just drew a picture—no real stress analysis or anything like that." In other words, it was a cottage industry where no wisdom held sway beyond the direct experience of its craftsmen. Hunsaker's cautionary advice must have been ringing in the young man's ears, so far away from home.

"The first day when Martin picked me up, he immediately took me down to Nigger Slough, a body of fresh water where he flew seaplanes. He had a biplane up on horses [wooden supports]. He got into it and jounced around, and everybody looked very interested and looked over the wires and fittings. After he came down, I said very timidly, 'What are you doing, Mr. Martin?' He looked at me, shocked—wondering why I didn't know any better than to ask such a question—and said, 'I was testing it, to see if it was strong enough.' But this didn't represent any load that the plane bore in flight. Nobody there had the foggiest idea about that. That's how crude it was."

Douglas was also perplexed by the personal oddities of his new boss, who lived at home with mother, Minta. Behind the public persona of daredevil pilot and cinema rogue, he found Glenn Martin to be "a strange chap, very meticulous," with "no female interests, as we understood such things."

But a few student years in worldly Cambridge must have prepared him for the existence of such strange chaps, because Douglas quickly acknowledged Martin's talents. "He was not an engineer, but he had good instinctive ideas. Glenn's greatest strength was that he was amazingly well coordinated physically. Most of the fellows who flew in those early days had to be. The airplanes were not very stable or controllable.

"But you could see it in Glenn in other ways—he was a marvelous bowler, billiard player and duck shot. He was a very strange chap, no question about that. Lots of people found him difficult to get along with. I haven't the foggiest idea how he got interested in aviation. He was a young boy down at Santa Ana, and I guess he just had a yen."

Hunsaker summed up Martin with more impatience: "He was pretty close to being described as inventor, experimenter and experimental flyer. His engineering knowledge was zero, and in the early days he had no idea that he needed any or that anybody could help him."

Withdrawing into the task at hand was the only antidote to the peculiarities of the situation. Douglas's co-workers started to call him the "pencil eater." His first design project was a two-seater hydro, the Model S, developed from the earlier T and TT biplane plans. Proper engineering analyses enabled the rapid construction of a far larger aircraft than Martin's crew had ever produced before. Its wingspan was thirteen feet longer than the TT military tractor, it was more than twice as heavy, its flying radius was rated at 500 miles versus 350, and its air speed fully loaded was only a bit slower. At about $12,000 a copy, it was also 50 percent more expensive than the TT, and the four sold to the Signal Corps in 1915 represented a "pretty good business" for the small firm, according to Douglas's modest assessment. Most significantly, it was a quantum leap beyond the personal-sized craft that Martin had turned out until then, representing a final departure from the ethos of "Come Josephine in My Flying Machine." The Model S was not sports equipment.

"My salary was fifty dollars a week in gold, and I got it in gold," Douglas recalled. This was a rich wage in those days, making him feel confident enough to buy his first car. It also imparted enough optimism for him to marry in June 1916 a nursing student named Charlotte Ogg, an Iowa girl who lived in Riverside. Four years his junior, an orphan since the age of nine, she was said to be every bit as domineering as his mother.

The S was soon followed by a Model R, a two-place trainer and reconnaissance plane that could carry a motorcycle. Some two dozen were ordered by the Dutch government, supposedly for use against guerrilla armies in the jungles of Java and Sumatra, though the impossibility of maintaining an air force in such remote locations made China or Japan more plausible destinations. By the middle of 1916, the factory was capable of turning out one aircraft every four days, but Martin still needed other kinds of work to make ends meet. When a local cinema crew began to film a picture about "the future world war," the Martin Company sold them large kites for use as props. In Europe, it was the year of bloodbaths at Verdun and the Somme.

Through his flying school, Martin attracted the financial interest of a rich young auto racer from New York, Caleb Smith Bragg, whom Douglas derided as "the pride of the chorus girls." Outwardly timid and shy like Martin, but strikingly handsome with a sensuous mouth and large eyes, the confirmed bachelor Bragg was attractive to Martin on many levels. A 1908 Yale graduate, winner in 1912 of a race against track legend Barney Oldfield in California, attaché at the American embassy in Paris in 1914, he embodied the kind of upper class dash that Martin would strive after for a lifetime.

To advance his flying lessons after returning to the United States from France, Bragg had purchased a deluxe Model S for his personal use—a lavish expression of generosity toward the Martin Company and its founder. In October 1915, Glenn and Mother Minta attended an aeronautical exhibition in New York, where they posed together for photographs in formal evening attire standing in front of a TT biplane. Bragg joined them on the long train trip back to Los Angeles, where he soon became a vice president and director of the company. He poured his private fortune into the development of a new aircraft engine, for which the Signal Corps was holding a design competition, but the $175,000 project did not bear fruit. Nonetheless, the winsome Bragg now formed a crucial bridge to Eastern capitalists who were piecing together an engine conglomerate. Their imaginations were fixated, too, on "the future world war," especially the business it would spawn.

JUST ONE YEAR after meeting Glenn Martin at the Baltimore Hotel, Douglas saw his promising start in Los Angeles derailed by powers stronger than he or Martin could yet comprehend. They were no different in this regard from thousands of other young men whose lives were swept up by war, except for their special positions in a new industry of unplumbed military significance. For a while they became pawns in a struggle far over their heads, an experience that would leave each determined to establish his independence above all else.

On August 29, 1916, Congress passed an Army appropriations bill that contained $13,281,666 for the "purchase, manufacture, maintenance, operation and repair of airships and other aerial machines" in fiscal 1917—more than forty times greater than in any previous year. Naval aviation received $3.5 million, also a momentous increase. That same month, Glenn Martin's California enterprises were acquired by the Wright Company of Dayton, Ohio, to form the Wright-Martin Aircraft Corporation. Capitalization of $5 million (about $50 million today) was announced by the organizers, consisting of 7 percent cumulative preferred stock and 500,000 shares of no-par-value common stock. Martin was named a vice president of the new company and told he was in charge of all airplane production. It was a dramatic show of financial muscle in a heretofore anemic industry, clearly indicating what the war was going to mean for big investors.

The stage for such consolidation had been set ten months earlier, in October 1915, when Orville Wright had sold his sagging business, its illustrious name, and especially its seminal patent for a little more than a million dollars in cash to a syndicate of Wall Street investors led by mining tycoon William Boyce Thompson. Allied purchases of American-made munitions and myriad other war supplies were accelerating, helping to fuel the greatest national prosperity in years. Syndicate members included Albert Wiggin, whose stock dealings while chairman of the board of Chase Bank were later investigated by the Senate Committee on Banking and Currency; Charles Sabin, who would become chairman of the board of Guaranty Trust Company; T. Frank Manville of Johns-Manville; Henry Sutphen and Henry Carse of the Electric Boat submarine company; and Ambrose Monell of Monel Metal. The board of directors was further salted with major New York bankers from Manufacturers Trust and Girard Trust. They retained Orville Wright as a consultant for $25,000 a year, but he held no deeper role in the company than as a figurehead to attract stock buyers.

Wright-Martin "will bring about, according to those involved in the merger, the production of the best flying machine America has known," reported the *New York World,* which headlined the syndicate's estimate of $2.3 million in annual profits. Wildly optimistic advertising of this sort had already led to one of the first public protests against heavy losses in aviation stocks. Late in 1915, the new Wright Company floated a common stock issue, followed by an offer of convertible preferred stock that came out at 75, rose to 80, and quickly fell to 30. There were angry allegations in the market about misrepresentation and excessive insider trading profits, two complaints that would become typical of future manipulations in the airplane industry. But the stock market was still a lawless free-for-all where syndicates like Wright-Martin could operate with total abandon.

Like many other Wall Street mergers on the eve of U.S. entry into the war, the syndicate's obvious ambition was to reap a share of federal appropriations, not necessarily to cultivate a long-term position in aviation. After the battle of the Somme, in July 1916, when the British first sent masses of aircraft to attack enemy trenches (leading the Germans to coin the term "strafe," from the verb meaning "to punish") interest quickened in using the new weapon as something more than a lone scout. Armies would require lots of them, and investors awakened to the possibilities.

The Thompson group first used the original Wright patent, Number 821393 of 1906, to attract capital to their new company, then to wring royalties from other aviation firms. Their only competition was Glenn Curtiss, who held several controlling patents on seaplanes. Curtiss had recently absorbed the Burgess Company of Marblehead, Massachusetts, another respected builder of hydros, but unlike the Wright brothers he had never used his patents to demand royalties from fellow manufacturers.

Thompson may have approached Curtiss about a merger, but a long history of bitter litigation between the Wright and Curtiss interests made a friendly wartime union of their businesses out of the question. Glenn Martin, on the other hand, had been on cordial terms with Orville Wright for years, even supplying him in 1915 with clandestinely obtained technical information about Curtiss airplanes stationed at North Island, San Diego. The Martin Company, furthermore, was relatively weak and thus easily acquired. Caleb Bragg's financial associations with syndicate members helped to cement the deal. For the "Flying Dude" and his greasy-handed crew, it must have seemed a miraculous opportunity.

Soon after the Wright-Martin merger, NACA examined the air-

craft patent situation as a prelude to removing any legal impedi-
ments to wartime production surges. The auto industry, which had
already been spotlighted in Howard Coffin's NDAC surveys as the
potential recipient of massive warplane orders, was especially concerned about avoiding litigation. "In reviewing the records of the
Army and Navy Departments as to planes purchased during the
eight years prior to our recent heavy appropriations for aerial defense," NACA investigators wrote, "it was brought out that four
different factories supplied all of those planes and apparently contributed most in the development and reduction to practice of the
aviation art. Those named in order of their appearance on the
records are Wright, Curtiss, Burgess and Martin. By a strange coincidence, Curtiss and Burgess joined hands and later the Wright and
Martin interests came together. While there are other aircraft patents, it was found that these two combinations owned and controlled what might be considered the two dominating groups of
patents." The Wright patent was considered to be "possibly basic"
and therefore of central legal significance, as well as greatest commercial value.

To launch its offensive, Wright-Martin sent letters to every notable aircraft builder in the United States on December 18, 1916,
enclosing a license agreement for the use of patent 821393, which
the syndicate claimed was being ubiquitously infringed. They demanded a 5 percent royalty on the price of every complete airplane
and a minimum payment of $10,000 per year from each manufac-
turer. In some cases the royalty would have amounted to over
$1,000 added cost per plane. The letters precipitated a crisis through-
out the fledgling industry, which had been stifled by patent battles
since 1909. In Washington, where Howard Coffin was determined
to blast any obstacle in the path of mass production by auto companies, it brought panic.

Normally ice-bound cogs of government spun hotly against
Wright-Martin. On January 30, 1917, after both the Secretary of the
Navy and the Secretary of War complained of "demoralization" in
the industry and an unbearable rise in the cost of warplanes to the
government because of royalty charges, Coffin called for federal
seizure of the Wright patent. At a summit meeting with NACA
officials on Saturday, February 3, Wright-Martin executives capitu-
lated and "expressed a willingness to cooperate with the government in any way that would promote the development of the
industry." This was the same day that President Wilson severed

diplomatic relations with Germany upon learning that Kaiser Wilhelm had commenced unrestricted submarine warfare. On Sunday the fourth, a letter to Wilson was drafted to be sent to the White House on Monday, recommending legislation to obtain the patent. Congress responded on March 2—one day after headlines revealed the duplicitous Zimmermann telegram, in which Germany had secretly proposed a military alliance with Mexico against the United States—with a $1 million appropriation "to enable the Secretary of War and the Secretary of the Navy to secure by purchase, condemnation, donation or otherwise such basic patent or patents as they may consider necessary to the manufacture and development of aircraft in the United States for governmental and civil purposes." The $1 million figure was derived from what Thompson had originally paid Orville Wright for his business, though patent 821393 was later found to have accounted for only half of Wright's sale price.

But the government never followed through. After war was officially declared in April, Wright-Martin figured that its prized patent was worth more than a million dollars. Instead of outright purchase, the government settled on a cross-license pool with the deceptively collegial name of Manufacturers' Aircraft Association Inc. to neutralize the duopoly of the Wright and Curtiss patent claims. In an arrangement similar to one that had existed for years in the auto industry, but with lower royalty rates, each company that joined agreed to pay the pool $200 for every airplane it built regardless of cost, which would be disbursed to Wright-Martin ($135), Curtiss ($40) and the association operating fund ($25) until Wright-Martin and then Curtiss had each been paid $2 million. In return, they could carry on their business without fear of lawsuits. This, at least, was the official rationale for the pool.

Curiously, the government was not a signed party to the agreement—the association was a wholly private body incorporated in New York. But it was endorsed by all the relevant federal bureaucracies, despite the fact that member companies routinely charged their $200 royalty fees to the Treasury as part of production costs. Even when the military built its own planes, as at the Naval Aircraft Factory in Philadelphia, the government paid the association the same royalty as any commercial firm.

The pool was overseen by three trustees, none of whom could be said to represent the public interest: W. Benton Crisp, a lawyer who had represented Ford and Hudson in milestone auto patent cases;

Albert H. Flint (brother of Charles R. Flint, international arms dealer
and early business agent for the Wright brothers, also known to
muckrakers in the press as the "Father of Trusts"), a railroad executive who had moved into the airplane business; and Joseph Ames,
a physics professor who later became chairman of NACA.

One knowledgeable industry executive of the day felt that Curtiss
"was to some extent possibly forced into the association" by the
government, under threat of having to recognize the Wright patent
once and for all as fundamental. The order of pool disbursements
accomplished this anyway, but at no terribly punitive cost to Curtiss.
Indeed, the government itself later obtained an opinion from three
leading patent attorneys that the payments to Wright and Curtiss
were "in all probability less than the basis of recovery a court would
allow" either company if they had continued to fight each other
over whose patents were primary.

Far more important than finessing the old Wright vs. Curtiss
patent battle, the procedure swept away the legal cloud over mass
production of warplanes that had worried the powerful auto companies. Like the horseless carriage before it, the airplane was prov-
ing too valuable for government and private capital to tolerate
monopolization by a handful of eccentric inventors. The Wright
brothers, especially, were thus consigned to the harmless stature of
founding fathers, rather than becoming obstructive billionaires.
Though he had spent an enormous amount of personal energy on
patent litigation over the years, Orville had no stomach left for big
business. He was a millionaire, apparently content in his fundamental Methodist spirit with what God had provided him.

To no one's surprise, least of all the nine companies that initially
joined it, the patent pool established a powerful new clique in American business. Widely perceived as yet another of the era's economic
Hydras, it soon became known to the public as the "Aircraft
Trust"—a phrase that only a few years earlier would have sounded
ludicrous. This charge was taken up by the Aeronautical Society of
America, the same group of engineers and inventors who had sponsored Glenn Curtiss's 1909 New York exhibition and then grown to
represent the desires of small aeronautical businesses on the Navy
Consulting Board. The term "Aircraft Trust" resonated deeply with
rumors of graft, stock manipulation, and excessive profits, leading
Navy Secretary Josephus Daniels and Secretary of War Newton

Baker to safeguard themselves by seeking approval of the patent pool from the Attorney General.

This was duly delivered in October 1917. But Attorney General Thomas W. Gregory conceded in a letter to Baker that "under the arrangement the interests of the Wright-Martin and Curtiss Corporations, as owners of the principal patents and entitled to the bulk of the royalties provided for in the agreement, are somewhat antagonistic to the interests of the smaller manufacturers who have to pay these royalties"—to say nothing of the interests of taxpayers who would ultimately foot the bill.

Daniels remained uncomfortable with the pool. "I should much prefer to handle the patent situation by purchase within the limit of the appropriation made," he wrote to the Manufacturers' Aircraft Association's general manager in March 1918, "but have not found the owners of the Wright patents willing to agree to a figure which would enable this to be done." He threatened that the government would stop reimbursing airplane builders for their royalty payments unless the association agreed to cut total disbursements to Wright-Martin and Curtiss in half, from $4 million to $2 million.

"Owing to the great and growing requirements of the Government for airplanes, under the royalty of $200 per plane, the limit of $4,000,000 would be paid by the Government alone during the next few months," he argued. "I consider this excessive and inadmissible." The association calmly went along with his demand, since "great and growing" warplane orders had already made a great deal of money for major investors. Total royalties paid during the war actually amounted to only $1,831,200.

By August 1918, even the Senate Military Affairs Committee had picked up the antitrust cry, reporting that the Aircraft Association was "vicious" and "designed to reap large profits by taking advantage of the necessities of the Government." The committee was not taking quite so radical a stand as this language implied, however, since by then the production of planes for the European front was such a fiasco that contractors were no longer being required to join the pool.

THE WRIGHT-MARTIN SYNDICATE'S brazen attempt at the end of 1916 to bleed heavy royalties from the entire aircraft industry thus backfired within six months as Howard Coffin successfully orchestrated the creation of a patent consortium. During the course of negotiations with the government, Wright-Martin officials revealed that they were spending a half-million dollars per month to develop a superior airplane engine as their primary product. Toward this end the syndicate had acquired the Simplex Automobile Company of New Brunswick, New Jersey, where William Boyce Thompson happened to be a director.

Simplex operated a large foundry and would be the production site for Hispano-Suiza engines (called "Spanish-Swiss" because they had been designed by a Swiss engineer in Barcelona immediately after the outbreak of war in 1914), for which a French patent-holding firm had granted Wright-Martin an exclusive license in the United States. Ostensibly because of the Hispano-Suiza patent's foreign registry, the syndicate was able to obtain a concession from the U.S. government whereby the engine license would not be thrown into the Manufacturers' Aircraft Association pool, thus preserving its lucrative exclusivity. As it turned out, Howard Coffin made sure that the Hispano-Suiza never got in the way.

Using the Wright and Martin names, the Dayton and Los Angeles airplane plants, the Hispano-Suiza license, and the Simplex factory, the syndicate should have been in a reasonably promising position regardless of the fate of its patent gambit. "With a Wright-Martin body, equipped with a Hispano-Suiza engine, and the whole thing constructed, instead of merely assembled, in the Simplex plant, the incorporators of the new concern expect to turn out for both foreign and domestic sale a flying machine which will far surpass any previous American product of the sort," the *New York World* had trumpeted when the merger was announced. But the Hispano-Suiza motors—superlative machines that had been custom-made by hand in Europe—proved difficult to adapt to American production methods based on standardized interchangeable parts. Formal tolerances were in the heads of foreign craftsmen, not to be found on technical drawings. It was rather like asking Yankee carpenters to construct copies of a European cathedral.

A federal investigation would later find that the syndicate focused on turning out Simplex automobiles at the New Brunswick plant during the crucial start-up months of its aircraft engine work, both as a hedge against whether the United States would enter the war and an attempt to generate badly needed cash. Although the syndicate had accepted an order from the French government in mid-1916 for 450 engines and imported a team of European machinists to nurse the project forward, it was unable to deliver any units until August 1917, when they were already becoming obsolete.

The resultant loss of millions of dollars dried up development funds for aircraft production, leaving Glenn Martin with nothing to do. The entire maze of patent strategies and subsidiary deals was way beyond his provincial concept of the airplane business. But it was he that would change, not the maze.

EVEN WITHOUT ITS engine-development fiasco and its bruising confrontation over patents with the government, Wright-Martin faced a hazardous period for starting an aviation company. The Army, struggling to sort out requirements for American combat aircraft from secondhand impressions about the sporadic air war abroad, had not yet formulated any doctrine, policy, or technical guidelines for building an air force. For that matter, neither had the belligerents in Europe, who were learning how to kill with airplanes as the war stumbled forward. At first, flying machines had been thought of only as reconnaissance aids, but pilots quickly began to carry bombs along on such missions. Fighters soon rose from enemy territory to intercept the observation planes. And the fighter began to transform into a more complex machine that could also harass troops on the ground. In the course of this evolution, the British tended to draw upon single-seat fighters, while the Germans gleaned more from their experience with two-seat "battlefield cooperation" planes that had multiple roles. Every air corps had its own technical prejudices, of course. It was all quite confusing, especially from a transatlantic vantage point.

Confounding the matter further, the U.S. military's procurement apparatus was becoming more and more subservient to NDAC. In particular, an Aircraft Production Board, proposed by NACA chairman Charles Wolcott, was hastily formed in April and May 1917 to consider mass production problems. It was led by the peripatetic Howard Coffin—who was a personal friend of the Signal Corps' top officer, but who knew nothing of airplanes or their construction— and quickly moved to consolidate warplane work in proven designs at factories capable of high-volume production. The military fell in line, if only because it had no better idea of its own.

Such factories, of course, did not exist except in the auto business. Neither of Wright-Martin's airplane plants, in Ohio and California, fit this category. Nor could the company afford to build new ones. The Dayton plant was shut down in March 1917 and the Los Angeles factory soon followed. (The loss of his old factory had no effect on Glenn Martin's immediate fortunes. In May, with lots of time on his hands, he bought his first Cadillac.) As the syndicate began finally to crank out Hispano-Suiza engines in the latter part of 1917, the company recovered somewhat from earlier losses, but retrenched to maintain only its New Brunswick assembly line.

"At that time, we were [capable of] delivering one plane per day to the government," Glenn Martin recalled years later about the demise of his Los Angeles plant. "But we were ordered by the Aircraft Production Board to close, because we could not promise a rate of more than three planes per day within six months. It was stated by the Board that they had more sources of supply than they had need for aircraft. Therefore, they would close the small plants and concentrate production in a few plants in the East."

The board had no legal authority to "order" the Martin factory to shut down, though this was an accurate perception of its advisory power. In all matters of aircraft procurement, it had the last word. By making clear that high-volume contracts would not be forthcoming from the government, the same end was accomplished. Coffin's claim that there were "more sources of supply than they had need for aircraft," if actually stated, would have been an ugly lie in any major war, when there can never be enough weapons. Allied aircraft factories in Europe were in fact strained to the limit. Its only logical purpose could have been to thicken the bureaucratic smoke covering the board's intention to funnel contracts to the auto industry.

DONALD DOUGLAS DID NOT LINGER in Los Angeles to watch these baroque forces play themselves out. In December 1916, soon after syndicate president Edward Hagar—a Portland cement magnate who would last less than a year at the job—ordered Martin to replace him as chief engineer with another young talent dispatched from the East, Douglas left for Washington. He landed on his feet as head of the Signal Corps' new in-house engineering department, a civilian job offered by a former classmate at MIT, Virginius E. Clark. Widely considered the Army's brightest aeronautical engineer, Lieutenant Colonel Clark was frantically compiling information about the state of U.S. airplane manufacturing and needed anyone with up-to-date experience.

Douglas's first assignment was to catalogue East Coast producers—a task that only confirmed the Signal Corps' appalling lack of readiness. On April 6, 1917—his twenty-fifth birthday—the United States declared war on Germany. As aviation historian I. B. Holley wrote, American flying officers "had to choose their tools before learning the nature of the job to be done."

It took a month or so for these officers and their political leaders to accept that the herculean American construction program then envisioned—some 22,000 aircraft and 45,000 engines funded by an unprecedented $640 million appropriation bill (introduced in Congress on July 6 and signed by the President on July 24)—was not even viable propaganda without an emergency infusion of technical knowledge from Europe. The occasional advice of various Allied tacticians and engineers who had been trickling into Washington for conferences with the Signal Corps was usually worthless by the time they arrived. Thus, in June 1917, a panel of twelve military and industrial "experts"—called the "Bolling Commission" after its leader, Raynal Cawthorne Bolling, a reserve officer who in civilian life had been general counsel for U.S. Steel—steamed to Europe and, after several weeks of inquiries, began shipping back actual samples of warplanes they thought should be produced in the United States. Bolling had already helped the Army find justifications for the $640 million appropriation, and Secretary Baker thought his business background might help him deal with aircraft producers abroad. To lend a semblance of practicality, the committee was accompanied by ninety-three factory men from the auto industry who were to learn European shop techniques and later act as instructors in the U.S.

There was no time left now for American ingenuity, which had fallen to the level of child's play compared to the battle-driven sophistication of European designers. Manfred von Richthofen, the German ace called the "Red Baron," was in his deadly prime during the spring and early summer of 1917. In the United States, the musty old Board of Ordnance and Fortification, which had stiffed the Wrights' first inquiries to the War Department in 1905, was soon abolished as a "sheer waste of time."

The planes selected by Bolling were the obvious choices anyone with so little time would have made—rather like shopping for Europe's finest wines in a single weekend. (After the war there would be allegations by industry apologists, supported by some documentation, that the British, French, and Italian governments had tried to protect the future competitiveness of their aircraft industries by asking for relatively menial assistance from the United States, but casualties along the battle fronts in 1917 would make this seem an awfully cynical interpretation. Business considerations were not taken lightly, however—the French, in particular, seemed more anxious at first about royalty payments from American manufacturers than about getting deliveries under way. Raynal Bolling never had to answer for his work—he was shot by a German soldier when he and his chauffeur drove too close to enemy lines near Amiens, France, on March 26, 1918.) The weaponry included the British De Havilland DH-4 biplane for long-range reconnaissance and day bombing; the British Bristol and French SPAD (Société Pour Aviation et ses Dérivées) biplanes for fighting, both powered by Hispano-Suiza engines; and the Italian Caproni triplane for night bombing. The SPADs were especially beautiful machines in the era of ungainly-looking contrivances—compact, strong, finely handcrafted, they had the look and feel of racers. Lieutenant Colonel Clark had a steady hand in the selection process, thus drawing Donald Douglas into intimate contact with the technology when it reached American shores.

Luckily for America's homegrown weapons industry, Douglas had by now lost his naiveté about who was most likely to profit from this privilege. "We had a building [part of the Smithsonian complex] where we kept a sort of museum of these airplanes," he said, "and we collected all the information on their performance characteristics. This was the period when the automobile industry entered the picture, in that the Aircraft Production Board was headed largely by automobile people who had no regard or respect for the American aircraft industry.

"One day, in our little museum and engineering office, one of the top members of the Board came in, called on Colonel Clark and said within my hearing: 'Don't let any of these American aircraft people in here to see these airplanes.' So I took it on myself to see that the American aircraft industry knew everything they should about them. I took documents home at night, made copies, and mailed them to my friends in different companies. Without that, they wouldn't have known anything.

"Many engineers and technicians from the automobile factories went abroad to see all the technology. None of us had that opportunity. My feelings were bitter. We had no direct connection at all, in my work, with the Aircraft Production Board, other than Clark getting orders to do certain things."

Douglas thus became a self-appointed agent of industrial espionage, a role that would have earned him a prison term had modern military secrecy laws been in existence. One need only consider the security breach today of an engineer trafficking classified weapons data among private manufacturers—such as occurred at Boeing Company and RCA Corporation in cases prosecuted by the Justice Department's "Operation Uncover" during the late 1980s—to appreciate the audacity of Douglas's freelance work, as well as the depth of his bitterness. For the rest of his life, he would never wholly trust government incursion into his business, no matter how deeply he depended on it.

The Production Board's decision to throw much of American warplane production to the auto industry, while involving a bald conflict of interest, had one plausible justification. If there was to be any credence at all in the dramatic American pledge of sending tens of thousands of aircraft to Europe, then established mass-production facilities were the only hope—or so it seemed to planners who had never faced such a crisis before. Yet some of the largest contracts went to new companies set up by automobile interests that had never made anything before. And besides the problem of teaching auto workers how to build a totally different product, the very nature of auto assembly lines—which depend on frozen designs and a minimum of rework—subverted adaptation to the most important dynamic in warplane production: overnight obsolescence.

None of this was understood by Washington in 1917, as experience with each Bolling aircraft selection soon made clear. At the end of July, the commission designated the single-seat SPAD-13 as one of the two best fighters in Europe. During the second battle of Verdun in August, air-combat innovations demonstrated by Germany's new two-place Albatross biplane led the Allies to believe the SPAD was obsolete. Yet in September, a sample SPAD arrived from Europe and was sent directly to Curtiss, which tooled up for full-scale production. Later the project was abruptly cancelled, then just as suddenly renewed, but was never fully under way, wasting time and labor right up until Armistice Day.

Similarly, the DH-4 was known to be obsolete and superseded by an improved model, the DH-9, almost as soon as Bolling's team picked it (Clark later claimed to have recommended only the DH-9). One of the principal improvements was relocation of the fuel tank, which on the DH-4 was in a position to torch both pilot and observer in a crash or if pierced by incendiary bullets—a feature that led flyers to give the DH-4 its infamous sobriquet, the "Flying Coffin." But a sample DH-4 that arrived from England in mid-August 1917 served as the production prototype for "Liberty Planes" in the United States.

"These programs, with their variations and schedules of deliveries, appear to be grotesque in the light of actual facts," wrote Charles E. Hughes, former Chief Justice of the Supreme Court, in an October 1918 investigative report of the troubled aircraft program ordered by President Wilson. They might have been grotesque, but they were very profitable for a few manufacturers.

A few military officers, including Virginius Clark, were aware of the debilitating problems but could not act to solve them in the face of the Army's own disorganization. The Aircraft Production Board's relentless demand for quantity, quantity, quantity, obliterated all other considerations.

JUST BEFORE THE Wright-Martin plant in Dayton was abandoned in March 1917, four local businessmen with ties to the auto industry persuaded Orville Wright to squander his name again on a new company. They signed up several Wright-Martin department heads and incorporated the Dayton-Wright Airplane Company with an announced (though unpaid) capital stock of $500,000 three days after President Wilson declared war. Replying to a startled query from Glenn Martin about his new associates, the either gullible or specious Orville wrote in May that "they are going to try to carry out some of my ideas in creating a sport in aeronautics."

Sport was not what the investors had in mind. A brand-new factory capable of high-volume war production was built near Dayton on property partly owned by one of the incorporators, Edward A. Deeds. Deeds was president of National Cash Register and cofounder of Delco, a supplier of self-starter ignitions to the auto industry. On the strength of these influential business credentials and the Wright friendship, he was appointed by Howard Coffin to be the Aircraft Production Board's procurement chief. In August, he was also commissioned a colonel in the Signal Corps as head of its

Equipment Division, thus taking virtual control of Army warplane production. He proceeded to steer two lucrative cost-plus contracts to his Ohio partners—one for 4,000 DH-4s and another for 400 J-1 trainers designed by Charles Day of the Standard Aircraft Corporation (who had left Martin's employ in 1912).

The U.S. Attorney General would later find Deeds "guilty of censurable conduct" in such transactions, but weak statutes prevented criminal prosecution. The J-1s, scheduled for completion by December 1917, were not delivered until March 1918. The selection of the Standard design, which was equivalent to the well-known Curtiss Jennies but had never been produced in quantity or extensively flown, can perhaps be explained by the fact that the treasurer of the Manufacturers' Aircraft Association was also Standard's president. When tests revealed dangerous vibrations and fire hazard due to an engine mismatch, 1,600 of the planes were condemned in June 1918 after an expenditure of $17.5 million.

The first Dayton-Wright DH-4 did not reach France until May 1918. In July, the month of the second battle of the Marne, the Air Service reported that all DH-4s that had been received abroad had serious defects that required rework. It took until early August, after the failure of the last major German offensive of the war, to launch the first DH-4 sortie over enemy territory.

A second order for 4,000 DH-4s went to Fisher Body Corporation, an auto firm which also was unable to get into quantity production until August. Dayton-Wright's profit on the $30 million DH-4/J-1 deal was at least $6.35 million. (Though profits were nominally fixed at 12.5 percent of costs, they were calculated from an estimated or "bogey" cost figure. The DH-4 bogey was $7,000 per plane, but the actual cost was under $4,400. The company even won an extra 25 percent of this "saving.") This was the kind of work that the Wright-Martin syndicate had banked on, but their timing and political connections had been fatally flawed.

"I have viewed with much regret the troubles of the Wright-Martin Company," Orville Wright wrote to his letterhead partner, Glenn Martin, who had more or less dropped out of sight rather than play the role of a puppet. "Although I never had any figures concerning their business, it would seem to me that they have lost at least several millions in the past year. I think I could have saved them some of this loss, but all new people taking hold of the business are inclined to think that they know a good deal more about it than those who have had experience. I hope things will come around better for the company, so that the stockholders will not lose the money they have put into it."

THE HARDER-BOILED Martin must have winced at this pathetic expression of goodwill. By now completely disillusioned, he fell back on the shoulder of his old California friend and longtime benefactor, Frank Garbutt. As a founder of the Los Angeles Athletic Club, Garbutt maintained lively contacts in the world of sportsmen-entrepreneurs. He introduced Martin to Alva Bradley, owner of the Cleveland Indians baseball team, who quickly convened seven other Cleveland industrialists to finance a new Martin Company in that city. Though Martin's tenure with the Wright-Martin syndicate had been a personal embarrassment, it had at least served to toss his name into the rarefied circles of major capital.

Cleveland also happened to be where Secretary of War Newton Baker had once been mayor. Moreover, Baker's assistant secretary in charge of procurement, Benedict Crowell, was a prominent Cleveland businessman. (Such ties had already led Baker's brother to organize an aircraft company there.) Many of Martin's former California employees flocked to the new venture, which was incorporated on September 10, 1917, with an authorized capital of $2.5 million, largely unpaid. A factory was hastily built eight miles from the downtown business district. There was still time to cash in on the war.

Wright-Martin continued to operate under its now totally meaningless name. Management devolved to a retired general, George W. Goethals—an engineer who had completed the Panama Canal and been in charge of the Army's chaotic procurement system—and a director of the Manufacturers' Aircraft Association, George H. Houston. From the start, the company had attracted all the right insiders, but somehow never enough business.

Glenn Martin did not join the association until March 1919, apparently because he too saw it now as an "Aircraft Trust," a front for Wall Street and Detroit-Dayton interests. In any case, the patent debate had been diffused and would never again attain prewar significance (by law, the Wright patent expired in 1923). With hundreds of millions of dollars in federal contracts up for grabs, men like Bradley were eager to play games much more fast-moving than baseball.

The Cleveland capitalists had at least one thing in common with the Main Street businessmen of Santa Ana who had backed Glenn Martin in 1912. They insisted he give up piloting. This time he agreed and never steered a flying machine through the air again. As for Mother Minta, who had remained in Los Angeles while her son learned what big-time Eastern finance was all about, there was now the happy prospect of rejoining him as the primary social companion of Cleveland's newest industrialist.

WITH CONTINUALLY IMPROVISED schemes for churning out thousands of airplanes, the Aircraft Production Board flailed about for expedient measures that inevitably involved technical shortcuts. Foremost was the decision to build only copies of European designs. Of equal consequence was the reverse-logic move to adapt all of these designs, regardless of type, to a single engine—called the "Liberty" in keeping with the era's most popular catchword—created and mass-produced in America.

The reasoning behind the Liberty engine was simple enough to blind its proponents within the government to its serious flaws. The Wright-Martin syndicate held an exclusive license to build Hispano-Suiza motors—the better of two existing power plants, both of foreign origin, worth copying—outside the Manufacturers' Aircraft Association pool. Even if these engines proved to be produceable in mass quantities, which was far from certain during most of 1917, they would therefore cost more because of royalty payments. Since all automakers knew how to build internal combustion engines and were also used to putting one type of engine in many different cars, an entirely new American power plant seemed the best alternative.

The Liberty was descended from an engine developed by the Packard Motor Car Company in 1915 and 1916. It had been tested only in racing cars and was too heavy per horsepower for aeronautical use. Edward Deeds recruited E. J. Hall of the Hall-Scott Motor Car Company to work with Packard's chief engineer and vice president, Jesse G. Vincent, telling them to come up with an improved design that "must embody no device that has not already been tested and proved in existing engines—you must avoid all experimentation." He then confined them in a suite at the Willard Hotel in Washington on May 29, 1917, until they emerged with drawings for a new eight-cylinder engine two days later.

It just so happened that the Delco ignition system, never before adapted for aeronautical work, would be used on the Liberty. Deeds at the time owned 17,500 shares in Delco's parent company, United Motors Corporation, of which he was also a vice president and director.

Development of the Liberty was thus a fait accompli before the Bolling team left for Europe to select airplanes that would use it. After rubber-stamp approval of the engine design by the Aircraft Production Board and the Joint Army and Navy Technical Board, Packard, Ford, and several other auto firms began mass production. Eleven thousand Liberties were pledged for February 1918, but by early spring only several hundred had been delivered, partly because of problems with the Delco ignition and a faulty oiling system. The original eight-cylinder engine was soon abandoned in favor of a twelve-cylinder design, with attendant delays.

"The production of Liberty motors to date is, of course, gravely disappointing," the Senate Committee on Military Affairs reported on April 10, 1918. "Government officials having the manufacture of the Liberty motor in charge have made the mistake of leading the public and the allied nations to the belief that many thousands of these motors would be completed by the spring of 1918. Information of this sort, not borne out by the facts, has been injurious, and its constant dissemination the committee regards as misleading and detrimental to our cause." Regarding the aircraft program as a whole, the committee noted that "a certain aloofness in dealing with persons possessing information based upon experience" and "an apparent intention of confining the actual production to a restricted number of concerns" had "contributed to the failure."

By the end of the war, 150 twelve-cylinder Liberties were reaching the end of assembly lines every day, hailed by officials with nothing else to show as America's great technological contribution to victory. A few months after the Armistice, Brigadier General Billy Mitchell, the Air Service's most experienced combat officer, called the Liberty "not a suitable motor for air work."

While the rationale for launching the Liberty program may have sounded attractive to the strapped auto executives—and marvelously profitable to Deeds and Vincent, among others—aeronautical engineers like Donald Douglas knew all along that it was deadly.

"The Board ordered Colonel Clark to put the Liberty engine in the Bristol fighter," Douglas said, "which was a two-seater with an [original] engine around 200 horsepower. We found that we'd have to make very extensive changes in the airplane—there was a big jump in power and weight with the 400 horsepower Liberty. Clark said his orders were that we couldn't change a thing in back of the firewall.

"I called his attention to the fact that there were practically no safety factors remaining if we didn't. He said, 'I'm sorry, that's my orders.' I said, 'I'm sorry, I'm leaving. Goodbye.' "

Twenty-seven Liberty-powered Bristols were delivered by Curtiss, which began a $19 million cost-plus order for 2,000 units in November 1917. Some 739 engineering changes were incorporated in the airplane as Curtiss engineers struggled to accomplish what Douglas had refused to do. All that were flight-tested crashed, triggering cancellation of the program by July 1918 after an expenditure of more than $3 million. As for Clark, he was well rewarded for obeying orders, becoming chief engineer at Dayton-Wright after the war.

For the rest of his career, Douglas never lost his antipathy for the auto industry. "I think I met Howard Coffin in Washington," he remembered with disdain. "I knew Deeds more than anyone else. It was not a good picture."

Indeed, growing public awareness of a gross scandal in aircraft production led to Deeds's ouster as Equipment Division chief in January 1918. A month earlier, along with a front-page account of a speech by Secretary of War Baker in which he blamed "pessimism" about America's war production effort on "seditionists and spies," the *New York Times* had published the text of a letter written by a French general in July 1917. At first suppressed by American military censorship and then delayed by "intervention" of the American ambassador in Paris, according to the *Times*, the letter pleaded for "great haste" in U.S. warplane production. A prominent article about the letter in the French press, the *Times* noted, had emphasized the "harm and danger" in America's unmet promises. Airplane production problems may have seemed rather theoretical in Washington and New York, but along the Western Front they were life-and-death matters—a fact that did not dawn fully on the American people until almost a year after entering the war.

Deeds was succeeded by William C. Potter, a businessman associated with the Guggenheim mining interests (and also a member of the Wright-Martin syndicate), who was in turn replaced by John D. Ryan, president of Anaconda Copper. The arrival of these men with no aeronautical experience whatsoever continued to mystify Congress and a few journalists. By March, Coffin was still blaming delays in the aircraft program on "winter weather." To their credit, however, Potter and Ryan opened the doors to native aeronautical talent that had been largely excluded until then.

The new regime was under heavy pressure to correct past blunders. Flying in the face of Wilson's "Fourteen Points" peace proposal, Germany's long-anticipated offensive to end the war was under way in early 1918, bringing large formations of warplanes over the battlefields. Between the last week of March and the first week of June 1918, the Germans made spectacular gains along the Western Front, a tide that would not begin to turn until mid-July. Britain would lose 1,032 of 1,232 planes on hand in two months of fighting. Despite such numbers, combat in the air was still not a predominant factor in the war, where generals favored monumental infantry and artillery clashes. If the slaughter had continued into 1919, airplanes would have taken a larger part, as the next generation of generals would see.

In May 1918, U.S. Army aviation was removed from the Signal Corps and given its own offices. Coffin resigned as chairman of the Aircraft Production Board. Deeds was ultimately recommended for court-martial, but Secretary of War Baker blocked the action. As in all wars, battlefield reality had a purgative effect on the bureaucracy.

Douglas left his government job rather than wait for the system to fix itself. He quit Clark's office in December 1917 and went back to Glenn Martin as chief engineer at the new Cleveland plant. His salary there was a very generous $10,000 a year. Martin had been among the "American aircraft people" who benefited most from Douglas's espionage, enabling him to sell a new biplane to the Signal Corps that was one of the few combat planes of the war period conceived at home—a twin-engined fighter-bomber designated MB-1, sometimes called a "Corps d'Armée" plane after the French term for observation craft that fought in close consort with ground forces.

A fixed-price deal for six aircraft was worth $300,000 upon signing in January 1918 (after Deeds's departure), with the contract clearly stating that the government intended to buy large quantities of the MB-1 if it proved satisfactory. By April, construction of a 61,000-square-foot factory was completed on a seventy-one-acre flying field. The prototype flew on August 17, and in late October the Army told Martin to begin building fifty MB-1s on a cost-plus basis. Like every other American-designed combat plane, however, the MB-1 came too late to reach the front before Armistice Day, November 11. On November 13, the order was reduced to ten. A day later, it dropped to four. Nonetheless, subsequent work brought total earnings to more than a million dollars over the next eighteen months, saving the company from the sudden oblivion that greeted many other aircraft manufacturers when peace broke out.

THE WAR WAS a bumpy ride for Martin and Douglas, but it did not kill them. The butchery fields were far away, as distant as the connection between a soldier's death in Picardy and the failure to build an airplane in New Jersey. Rather than putting them in a grave at St. Nazaire, the war delivered both into pastures much greener than they had known before. Martin found himself well known among nationally prominent financiers, and Douglas gained priceless savvy from his stint in the Signal Corps engineering office. Given where they started from, they could not have asked for better turns of fate.

The same cannot be said for Jack Northrop and the Loughead brothers, whose war experience was more typical of the provincial businessmen and self-taught technologists who composed the American aviation industry in 1916. They stumbled through the next few years on the same thin shoestring, relying on their youth and intuition while others tapped the riches of Washington and Wall Street.

With their $4,000 stake from the Panama-Pacific Exposition, plus the backing of a well-connected local machine shop owner and venture capitalist named Burton Rodman, Allan and Malcolm Loughead had begun work in Santa Barbara during the early months of 1916 on a ten-passenger flying boat. Their design was not especially original, but it was big, with an upper wingspan of seventy-four feet, a triple tail, and two 160-horsepower engines. Conceived by Allan, who was probably driven by hopes of multiplying the per-flight earnings of the old Model G tourist plane, the machine would have been an ambitious project for the most proficient companies in the world, let alone the poorly capitalized and understaffed Lougheads. By summer, realizing the tenuousness of their dream, they hired Jack Northrop, a local draftsman who claimed to know some pertinent mathematics. They had nothing to lose.

Northrop performed the same magical feats of stress analysis for the Lougheads that Douglas had accomplished for Glenn Martin, allowing the assembly of a structure far larger than seemed intu-itively possible. His memory of the experience was almost identical to Douglas's:

"When I got the job I found that they really had no engineering, as such. They were building a float for a seaplane, but there were no layout drawings of the airplane—it was just being built 'by guess and by golly.' I was given the job of designing the wing truss. I hardly feel I was qualified, but at least the wings didn't come off. Fortunately I knew enough about math, stress, structures and strength of materials.

"I had a draughting board over in the corner and a T-square and triangle, and did what engineering was done. I calculated the stresses, the wing loadings, and what size the brace wires should be, because neither of the Loughead brothers was capable of doing it.

"Anyhow, it worked."

It worked, but it was not finished and given its maiden flight until the end of March 1918. By then, the Navy had chosen the Liberty-powered Curtiss HS-2L flying boat as a standard, ordering more than 600 planes. This should not have surprised anyone in the business, given Glenn Curtiss's primogenitive association with naval aviation, but the Lougheads dwelled far outside the circles of power. Allan nonetheless traveled to Washington with blueprints and pho-tographs of his creation, hopefully called the F-1 using a Navy des-ignation for experimental seaplanes. His objective was to gain an audience with Jerome Hunsaker, who had become chief of Navy aircraft design. Mr. Rodman had shouldered the full development cost of a very expensive flying machine and was not about to let the Loughead brothers run another sightseeing circus with it.

Wartime Washington was a treacherous city for the most pow-erful capitalists in the country, let alone a provincial mechanic try-ing to sell the U.S. Navy an airplane from snapshots. After weeks without penetrating Hunsaker's ring of subordinates, Allan man-aged an interview with one of California's senators. A meeting was thus arranged with Rear Admiral David W. Taylor, member of the Aircraft Production Board and the Navy's chief of construction and repair, who greeted the obscure airplane builder by complaining that he was "being pestered by every Tom, Dick and Harry who ever built a coffin, a toilet seat or a chicken coop."

The imperturbable Allan spread before the admiral his photo-graphs of the F-1, along with Northrop's three-view technical draw-ings. Taylor, an illustrious naval architect and strong partisan of military aviation, saw that the investment behind the F-1 was substantial enough to take seriously. He also must have realized immediately that the F-1 design was germane to a special Navy project about which Allan Loughead was ignorant. "Have you shown these to Lieutenant Commander Hunsaker?" he asked, with no further explanation.

Hunsaker was not enthused about having an unknown mechanic with a few political connections pressed on him from above. He, too, recognized the credibility of the F-1 documentation, then disparaged the plane's uncouth features. Yet he asked detailed questions about the aircraft, especially regarding its performance and capacity. Allan was forced to admit that only a few test flights had been made, handing the impatient Hunsaker the chance to declare that the Navy was not about to buy a hangar queen. Then, out of nowhere, Hunsaker offered the Loughead company a $90,000 contract for two HS-2Ls and spares on a wartime cost-plus-12.5-percent basis. In what must have seemed to Allan a curious sign of further interest, he also arranged for an evaluation of the F-1 at San Diego, if the Lougheads would assume the cost and risk of flying it down from Santa Barbara.

Thus, in May 1918, Allan and Malcolm, along with a mechanic and a newspaper reporter, flew the F-1 nonstop to San Diego, an impressive trip of 211 miles in 181 minutes. The brothers contacted the base's commanding officer, Earl Winfield Spencer Jr., who assigned two of his pilots to take the plane up. Spencer, well-to-do son of a member of the Chicago Stock Exchange and first husband of Bessie Wallis Warfield—the future Duchess of Windsor—was just the sort of affluent aviation enthusiast that the Lougheads perennially tried to recruit for their business schemes. For the next week, Allan coached the pilots, made several flights with Spencer, and pitched the notion of being partners in an airplane company after the war. The Navy men sent a favorable report on the F-1 back to Washington, but did not behold it as a career option. Hunsaker ordered the North Island station to retain the aircraft for further tests, yet still offered no contract.

In June, the Loughead company turned its attention to building the two HS-2Ls, expanding its work force over the next three months from fifteen to eighty-five. A few men were hired from the defunct Martin plant in Los Angeles. Jack Northrop, who had been drafted into the Infantry that spring and then transferred to the Signal Corps, was furloughed back to Santa Barbara to work on the seaplanes. The company received verbal permission from the Navy to spend $4,500 for additional floor space at the old car garage, but despite the liberal manpower and plant expansion, the aircraft were not finished in time to see any service. The company was never reimbursed for its $4,500 shop investment, because the Lougheads had not bothered to ask the government for a written agreement.

By August, the Navy notified the company that evaluation of the F-1 was over and that it could be retrieved from San Diego. Though there were obvious reasons for turning down the big seaplane—it was heavier and slower than the HS-2L, did not use the standard Liberty engine, and was not worth the investment for quantity production with the war waning—the rejection was a bitter pill for the company. A crew was sent to crate the aircraft and ship it north by railroad flatcar. Burt Rodman pitched a tent for the boxes on a vacant lot next to his house, keeping a close eye on the wilted flower of his investment.

With both the end of the war and the end of the HS-2L cash flow in sight, the Loughead brothers had few choices other than trying to make the old F-1 pay off in some way. "Somebody—I don't know who—had the brilliant idea that they could get a great deal of publicity and therefore a lot of money to build a big airplane company through a transcontinental flight," recalled Jack Northrop. "So the [F-1] was rebuilt, using the wings and the engines with a [new] fuselage and a landing gear."

This desperate notion most likely came from Allan Loughead, who remained convinced of the superiority of his design. He also believed the Army might be interested in the plane's potential as a strategic bomber, a concept just coming into its own with German and British aviators in 1918. As further evidence either of Allan's salesmanship or Rodman's willingness to gamble with HS-2L profits, some $10,000 was invested in the conversion from F-1 to F-1A as Allan began to pitch the scheme of taking off from Santa Barbara, refueling in Chicago, and landing on the White House lawn. He would not be the heroic pilot, however. Like Glenn Martin, he was grounded by investors who valued his name more than his nerve. Three freelancers were hired for the crew.

On paper, the F-1A was capable of an average speed of ninety miles per hour over a range of 1,200 miles. The Lougheads figured the flying time across the continent at thirty-five hours, or an elapsed time of three days for the entire trip. Operating in a dreamscape of optimism, they arranged for supplies to be on hand at Deming, New

Mexico, and Cairo, Illinois. Santa Fe Railroad workers promised to burn flares along the tracks to help with navigation. When the Armistice made it obvious that the military was not going to be shopping for new bombers, the Santa Barbara postmaster was called upon to supply a letter for delivery to the Postmaster General in Washington, D.C., signifying the dawn of transcontinental airmail. Movie actress Mary Miles Minter, a Santa Barbaran whose mother had sunk $5,000 in the Loughead enterprise, penned a letter for President Wilson congratulating him on his Fourteen Points peace proposal. And another prominent resident who thought himself to be a personal friend of Navy Secretary Josephus Daniels wrote the following jaunty message:

"This will introduce the gentlemen who are making the first aerial flight from Santa Barbara, California, to Washington, D.C. Any courtesies you may show them will be appreciated. We trust this will be the forerunner of a permanent mail route in the near future between the two cities. Allow me to congratulate you for the efficient service your department has rendered during the world war."

The voyage was scheduled to begin from a plowed field—there were, of course, no airports—on November 20, 1918, but three days of rain mired the 6,000-pound aircraft in mud. After two farm tractors were unable to pull it out, an eight-horse team hitched up to its undercarriage and dragged it to a nearby meadow. On November 23, under the pressure of the Lougheads' publicity blitz, the F-1A finally took off into a gale that was blowing down houses and trees in Palm Springs. Six hours and 400 miles later, a broken rocker arm in the right engine forced the plane onto the desert floor near Tacna, Arizona.

Repairs were made, and on November 26 the crew managed to hop to Gila Bend for further work and refueling. That afternoon, as the plane was taking off again in front of an excited crowd, a car swerved into its path and forced the pilot to abort, blowing out one of the landing gear tires. On November 27, embarking once more, an engine died when the plane was twenty feet off the ground and the F-1A upended violently on the soft desert sand. Two of the crew were injured, the third crushed by a shifting gas tank. This was the end of the transcontinental line, 500 miles from Santa Barbara—an entirely normal turn of events in aeronautics circa 1918.

Driving a Hupmobile, Malcolm Loughead took three days to reach the crash site. He disassembled the wrecked airplane, loaded the salvageable parts onto horse wagons, and hauled them to the railway platform at Gila Bend. Back in Santa Barbara, the F-1A was rebuilt into a hydro resembling the F-1, available for five-dollar joyrides with Allan and Malcolm at the controls. Burt Rodman withdrew permanently from the brothers' enterprises.

The Great War and its great business propositions had led the Loughead Aircraft Manufacturing Company back to where it started, which was precisely nowhere.

CHAPTER FIVE

THE RAT RACE

''No American-designed combat plane flew in France or Italy during the entire war. The foreign planes built in this country failed to arrive in Europe either on schedule or in the promised numbers, until what had started out as a triumphant exhibition of American know-how turned into a humiliating series of Congressional and other investigations.''

General Hap Arnold's indictment of the nation's aircraft production record in World War I stands as the clearest measure of how much the industry would change during the next twenty years. By the end of 1918, at least $100 million had been spent in quest of a winged cavalry of American warplanes that never materialized. According to the congressional testimony of Air Service officers, the only machines in use at the front by U.S. forces at the end of the war were 527 planes ''bought or borrowed'' from foreign sources and 196 American-built DH-4s. Some of those remaining in France were stacked and burned in a ''Billion Dollar Bonfire'' that further inflamed public outrage.

''None of us in France could understand what prevented our great country from furnishing machines equal to the best in the world,'' wrote Eddie Rickenbacker, the American ace. ''Many a gallant life was lost to American aviation during those early months of 1918, the responsibility for which must lie heavily upon some guilty conscience.''

Many a gallant aviator was lost by every nation at war, of course,

113

regardless of the quality of their equipment. What the Americans could not furnish themselves, they tried to get from others. Rickenbacker himself was best known for his exploits in a French-built SPAD. Particularly galling to the proud young American pilots, especially after the wave of propaganda in 1917 about aerial armadas chasing the Hun back to Berlin, was that American-built airplanes were considered low-class material—obsolete even when well made. It was like being issued longbows when everyone else carried crossbows.

The official figure for warplane production between April 1917 and November 1918 was 13,894. Records of the Manufacturers' Aircraft Association claimed that 14,255 planes had been built in the United States between July 1917 and December 1918, most in the last few months of the war. The report of a 1920 congressional investigation put U.S. production between January 1 and September 30, 1918, at 7,749 planes, noting that this figure included 1,413 useless DH-4s, 1,660 worthless J-1s, and scores of discarded Bristols. Over the years, apologists would argue that a production rate of 21,000 planes a year was reached near the end, a remarkable achievement for a country that had produced only several hundred aircraft in 1916, none of which were remotely state-of-the-art. But even allowing for the usual controversy of such counts, the difference in orders of magnitude between production and deliveries to the battlefield indicated scandalous inefficiencies at best, and criminal profiteering at worst.

To be sure, the aircraft program was not alone in such mismanagement and corruption. Some $200 million was appropriated to develop tanks for the Army, but of 4,400 put under contract, a grand total of fifteen reached France—after the Armistice. Airplanes, however, had been highly dramatized and thus embodied the general failures of supply.

"War means waste and extravagance and, unless expenditures are of a questionable character, criticism does not ordinarily follow," a bipartisan House investigative panel wrote, charging that "incompetence, inexperience, blundering, or personal interests were permitted to delay or thwart aviation production and thereby jeopardize the winning of the war with attendant unnecessary loss of lives." This was strong language in the aftermath of war, "the mightiest struggle of all history," when official bodies usually glorify the nation's sacrifices whether noble or not.

There was, of course, burgeoning sentiment that the Great War had been a great waste. The Allies mobilized a total of 42 million men, with a casualty (killed, wounded, captured, missing) rate of 52 percent. The Central Powers lost 67 percent of some 23 million men. For certain countries, the ratios were even more ghastly. France suffered losses of 73 percent of 8.4 million under arms, Russia 76 percent of 12 million, Austria-Hungary lost 90 percent of an army of 7.8 million. For these unfortunate countries, an entire generation was gutted. By comparison, the United States fared well, mobilizing 4.355 million with a casualty rate of only 8 percent.

Whatever the losses, the new horrors of combat and the massive economic sacrifices of supporting it seemed out of proportion with the reasons for fighting, which had been abstruse from the start. People were easily disposed toward believing that sinister economic interests could foment war in order to take profitable advantage of it. The evidence that aircraft producers in particular had managed to divert a great many tax dollars helped create a political imperative to shut off the flow after Armistice Day.

The money spent had not evaporated. This was a fact to which Glenn Martin and even the Lougheads could attest. For them, the government's eagerness to write lucrative cost-plus contracts for weapons that had no impact on the battlefield was a memorable business lesson. In a 1940 study of the financial background of American aircraft producers, historian Elsbeth Freudenthal offered a gloss on World War I that was too cynical to win acceptance in the mainstream, but accurately summarized the experience:

Army aviation was really a "school run by the government for a whole group of businessmen," she found. "They matriculated in the Army Signal Corps in 1917, having graduated from automobiles, cash registers, and other fields. They were given key positions through which they learned how to develop holding companies and patent pools, what criminal statutes to avoid, and the intricacies of cost-plus and fixed-price contracts. The government entrusted them with over one billion dollars for Army aircraft to fight the war. As their part of the bargain, they spent the money, and the Army forces on the battle front received 196 defective observation planes. They graduated in 1918 or were eased out earlier, hurdled investigations which scanned their activities, and took a leading part in the aviation industry for many years after the war."

Freudenthal was specifically concerned about the roles of Howard Coffin, Edward Deeds, and various investment barons, not Glenn Martin or Donald Douglas. But no one at any level in aviation could have missed the wartime lessons cited, which sound remarkably relevant seventy years later after revelations of contracting fraud in modern "defense buildups." The resultant mood of men like Martin and Douglas who had seen the way the world really worked was perhaps best expressed by their peer Grover Cleveland Loening, whose name was evidence of an upbringing that had not sought to impart skepticism about big business.

"The more cost, the more plus—that was the theme song of the aviation contractors, particularly the automobile crowd who had such great plans for such large orders of airplanes," Loening wrote. "Anyone who has any doubts upon how much the World War meant in its boosting of aviation need only be reminded of the stagnation that existed just prior to the start of hostilities." After the Armistice, "I realized that this great opportunity was gone forever. I should not say I was bitterly disappoitned, because we all wanted the war to end, but the truth of the matter is that it was a terrible blow."

Within months of the Armistice, the American aircraft industry reverted to its prewar stagnation, crushed by across-the-board federal contract cancellations worth $100 million. No company was spared the financial shock of watching its orders evaporate, often in a matter of days. Even by the optimistic tallies of the Manufacturers' Aircraft Association, production nose-dived from 4,345 planes in the last quarter of 1918 to twenty-six for the same period a year later. Ninety percent of the production facilities built up during the war were liquidated in 1919. Major investors quickly lost interest in the business (one notable exception being Wright-Martin, which changed its name in 1919 to Wright Aeronautical Company and became a principal engine builder), leaving the founders to fend again for themselves. On top of the drying up of federal appropriations came the stock market tumble of 1920, part of a worldwide downturn as economies struggled to revive from wartime chaos. Total investment in the entire U.S. aircraft industry was reckoned at the time to be no more than $5 million.

Through most of 1919, Glenn Martin's Cleveland backers stuck with him, because the MB-1 was still bringing them a relatively smart return. The Army had first sought a new three-seat "Corps d'Armée" or gunner airplane for supporting troops on the ground. Donald Douglas brought his insider knowledge of foreign technology to bear on the problem, borrowing heavily from British De Havilland designs. The result seemed promising enough in terms of

power and weight that the Army changed its mind and decided that
the MB-1 should be a bomber, though its payload capacity was only
marginally suitable for this mission. Bombers had demonstrated
their promise during the very last months of the war and were now
in vogue, but the U.S. Army did not yet possess one.

By any name, the MB-1 was a killing machine. Manned by a crew
of three, it had a biplane wingspan of seventy-one feet, was forty-
five feet long and weighed more than three tons when empty. Powered by two 400-horsepower twelve-cylinder Liberty engines, it
could reach a maximum speed of 105 miles per hour and climb to
over 10,000 feet. It carried five machine guns and a thousand
pounds of bombs. Here for the first time from American drawing
boards was a formidable offensive air weapon with a single
purpose—mass destruction.

"After the Armistice, we really got more business than we'd had
before," Douglas recalled, overlooking the sudden cancellation of
an order for fifty bombers as soon as the war ended. Because of all
the structural changes necessary in order to accomplish the plane's
new purpose, Martin's postwar contract turned into an enormous
accounting headache for both the company and the government,
spawning litigation about billings that dragged on for the next six
years. Of the ten Army MB-1s, the first four were used only for
observation and training because they were too far along in construction when the decision was made to turn the gunner plane into
a bomber. The next three, after extensive redesign, were meant to be
true bombers, but only one was equipped for this purpose, because
of further zigzags in Army thinking. Building a weapon as deadly as
the MB-1 had distinct political implications that were not congruent
with postwar disarmament.

Each of the last three was also redesigned: one, called the "GMT"
for Glenn Martin Transcontinental, configured as a long-distance
version with a new fuel system and extra gasoline capacity; another,
the GMC, fitted with a cannon on its nose; and the final model, the
GMP, altered to seat ten passengers in a cabin with glass windows.
This tenth MB-1 was intended to transport machine-gun squads
along the Mexican border, at least in the minds of Army officials
who needed some justification for buying it. Due to slack management and myriad engineering changes, the cost of an MB-1 ba-
looned nearly 70 percent, from about $50,000 to $84,000.

In addition to these ten MB-1s for the Army, the company also
built two for the Navy to carry torpedoes, eight similar machines for

the Marines, and six mail planes for the postal service. "It was a rather good plane for its day—very good," Douglas declared. "This was the only thing Martin was building at the time." Because the basic price of MB-1s was derived from wartime values inflated by urgency, the company probably gained at least twice the nominal 12.5 percent profit on each plane through 1919.

High cost and technical problems gradually cooled the military's enthusiasm for the MB-1, which in turn cooled Alva Bradley's enthusiasm for investing in the Glenn L. Martin Company. In mid-August 1919, Brigadier General Billy Mitchell, who was then director of Army aeronautics, received an unusually blunt warning from the chief of his engineering division, Thurman Bane. "We must close the present contract with Glenn Martin Co. or go to jail," Bane wrote. "The thing has been running so long now and the costs have mounted up so high that it will be very inadvisable to do any more work at the Martin plant on the present contract."

In public statements and testimony before Congress, however, Mitchell continued to tout the MB-1 as "up to any bombardment airplane of this class anywhere," even though tests were showing it to be unsatisfactory as a bomber. Mitchell was trying to sell a program, not just an airplane. To further his radical vision of a powerful Air Force independent of the Army and Navy, he had already set up an aerial Forest Fire Patrol on the West Coast and a Border Patrol along the Rio Grande to help stop illegal crossings. These were far more useful for public relations—that is, for helping to counter resentment of military aviation that festered after the wartime production scandals—than for their stated purpose. Still, being obliged to use obsolete and defective surplus planes from the war was an affront to Mitchell's sense of polish and technical perfection. Like Glenn Martin, he had a penchant for stunts when conventional forms of persuasion did not bear fruit.

In July 1919, for example, he sent two Army lieutenants and a crew of mechanics on a 9,000-mile circuit of the continental United States in one of the unequipped MB-1 bombers. That it took three and a half months to complete was somehow irrelevant—the *New York Times* gave it front-page coverage. Emboldened by the way this trip caught the public's attention, Mitchell staged a "Transcontinental Reliability and Endurance Test" in October. Of sixty-four aircraft entered, most of which were war-surplus, five (all DH-4s) were involved in fatal accidents that killed seven aviators. Two MB-1s— the GMT and another unequipped bomber—failed to finish. Instead of positive publicity for his beloved Air Corps, the test raised hackles in Washington and dramatized the perils of flying. At the end of October, Congress rejected a $15 million request from the War Department for the purchase of new American-built planes.

In early 1920, with the end of MB-1 work in sight and a deepening trough in the general economy, Martin's Cleveland investors withdrew from further support of his company. As he had done so many times in the past, he went directly to the public via friendly newspapers to ward off any damage to his reputation. "Only a failure of the United States government to place orders with our successful airplane designers and builders will cause our aircraft industrial strength to slip back into the position it occupied three years ago," he wrote on page one of the *Cleveland Plain Dealer*. "A vital point is being overlooked by the American people," he insisted. "It is immediately evident that the industrial strength of the United States must be at the war strength all the time, as production of aircraft cannot be developed under one year from any given date." The government must therefore "stimulate and aid in the application of aircraft industrially, and also aid in foreign trade, furnishing sufficient outlet for industrial aviation and guaranteeing a continuity of production at the required rate."

Once again, Glenn Martin was several decades ahead of conventional thinking. The notion of maintaining a robust industrial base in peacetime for weapons production was alien to most Americans and would not be accepted at high levels of government until after World War II. Martin knew from direct experience, however, that warplanes could never be turned out overnight on an emergency basis like boots and bullets.

By this time Martin was on personal terms with Billy Mitchell, whom he soon visited in Washington to plead for more bombers. "I have just been going over the proposition of bombardment airplanes with Glenn Martin," Mitchell wrote to Thurman Bane in February, "and took him over to the Bomber [MB-1] which we have here and went over the disadvantages of it." The disadvantages were so extensive—"get the weight lower, all the oil, gas, etc. out of the fuselage, the navigating personnel back out of the nose, the bombs inside, an opening in the landing gear to permit placing water torpedo equipment on when necessary, dihedral [stability], increased lift, balanced ailerons, etc."—that a whole new aircraft was clearly what Mitchell wanted.

Although not solid enough to bring back the Cleveland investors, the prospect of building a new plane for Mitchell was sufficient to keep Martin's embers glowing. He went hat-in-hand to bankers and friends, much as during the old days in California, asking them to help him buy out the Bradley group. Since the company's total capitalization at this point was less than $300,000, he did not have to rely upon the kindness of Wall Street heavies who might use him again for their own hidden purposes. He did turn, however, to one of Cleveland's shadier financial institutions, Guardian Trust, which placed an officer on his board of directors. Guardian would collapse in the early 1930s—"a direct result of unsound banking practices and mismanagement . . . over a period of years," Senate investigators would find, though the bank's loans to Martin would not be implicated.

Bane wrote back to Mitchell at the end of February about "the Glenn Martin situation." The engineering division had been trying to interest the company in building, "a better Bomber," but Martin was so distracted by "reorganization difficulties" that he could not make a commitment. "I understand from Mr. Bell [Lawrence Bell, Martin's highly capable factory manager], who is here today, that Glenn Martin has taken over the whole outfit and owns the entire Company now."

Martin had once again been saved from oblivion, thanks to a die-hard faith in aviation that he was somehow able to transmit to risk-loving lenders even during the gloomiest times. This was doubtlessly his greatest personal asset, a skill respected even by Jerome Hunsaker, who observed that Martin was "a much tougher sort" than other industry founders who perhaps had stronger technical aptitudes. "He dealt with financial and business people," Hunsaker said, "and he dealt with this competitive rat-race that he got into."

But in the process, Martin lost his second most valuable asset, Donald Douglas. The restless young engineer had seen enough to know that he might as well try the business on his own, since Martin seemed to possess no special formula for steady good fortune. Douglas's defection in March 1920 was a serious blow, because the new Army bomber was to be an evolutionary step from his MB-1. Bringing in a new chief engineer would consume precious time. Faced with a gaping hole at the top of the engineering staff, Martin was forced to equivocate with the Army at the worst possible moment.

"Glenn Martin admitted in the presence of General Menoher and Col. Gillmore [chief of the Air Service and head of supply, respectively] that it would take him nine months at least to complete the contract for three new Bombers," Bane wrote to Mitchell early in March. "He further admitted that he did not have enough engineering data yet to accept a contract and that it would take him forty-five days to complete his engineering data on this job. I should like to state that, in my opinion, his estimate of nine months is optimistic."

Mitchell's response was a blistering attack on Bane's cautious management of the Engineering Division, not a dismissal of the hobbled Martin company. "We want, if there is any erring, to have the mistake on the side of over-development rather than over-conservatism," he scolded Bane. Thus, a spectacular contract for twenty MB-2 bombers—worth $1,003,737, plus $188,870 for spare parts—was awarded to Martin on June 19, 1920, "in the rush of the close of the fiscal year [June 30]," as Bane later described it. In Mitchell's haste to obtain a proper bomber, he approved a deal that the U.S. Comptroller General would later find "remarkable" for what it gave away to a private company. Glenn Martin was back in the big time, thanks to Billy Mitchell's cost-be-damned command.

GIVEN THEIR PERSONALITIES ALONE, Mitchell and Martin were unlikely allies. Mitchell was a gregarious, hard-charging gallant who had been the ranking Air Service officer on the European Front—the recipient of a Distinguished Service Medal and a Croix de Guerre. During the Meuse-Argonne offensive in September 1918, when he commanded a force of 1,481 British, French, Italian, and American warplanes against the Germans' St. Mihiel salient, he had glimpsed the future of air superiority. Propelled by this indelible experience, he adopted what one historian called the "joy stick approach" to military aircraft procurement, buying whatever he personally admired. A hidebound older generation of military leaders had squandered millions of lives in the hideous trenches, radicalizing junior officers like Mitchell into embracing new combat doctrines and weapons that might obviate the slaughter on the ground. With slick dark hair and rugged visage, he was remembered by Hap Arnold at this stage of his career as having "an undercurrent of angry impatience."

Martin, by contrast, was a prissy businessman who rarely went anywhere without his mother. Tall, very slender, narrow-shouldered, with bulging forehead and heavy black hair, he showed a preference for Continental-style hats and artistic round eyeglass frames. He liked to carry two twenty-dollar gold pieces in his pocket for security. Once asked what he read, he said the *Saturday Evening Post* and the *Literary Digest*. At social occasions he was more likely to cluck over the table setting's inadequacies than converse with his hosts. "He was not the kind of man that I would spend a holiday with willingly," Jerome Hunsaker once said. "He was a little hard to understand. He was a bachelor—and he had to live up to mother's idea that he was the great man of this generation."

But in their notions of the inevitable supremacy of air power in war, the promise of commercial transport, and the need for a government-industry partnership to ensure both, Mitchell and Martin were identical twins. In a world where these were still minority views, they naturally gravitated toward each other. "Adequate aerial defense can only be assured by the development of an aeronautical industry in peace time devoted largely to commercial manufacture which can be diverted immediately when hostilities are threatened to the quantity production of military planes," Mitchell wrote in a prophetic memorandum of September 1919. His rescue of Martin can be scored as the first manifestation of how the government would eventually equate its security with the prosperity of certain favored contractors, regardless of cost or who was in charge.

THERE COULD HARDLY have been a more dismal season to launch an aviation company than the spring of 1920. Federal orders were sparse, the commercial market was flooded with war surplus, and Wall Street was panic-prone. Donald Douglas always insisted that he was happy with his job in Cleveland, that he had no problems with Glenn Martin, that ambition simply compelled him to go into business for himself. But a Navy officer who became a prominent aviation industry executive in the 1930s, Eugene Wilson, opined in later years that Douglas and other key Martin employees who would soon leave Cleveland "couldn't get along with Glenn—and you couldn't blame them—so they hauled off and went out to California."

"I didn't discuss my plan with Martin—not until I presented it to him as my decision," Douglas said. "Then, he thought I was foolish. He tried to dissuade me. He wanted me to stay, and promised lots of things if I would. I said, 'Well, Glenn, I just think it's time for me to make a break.' "

Divorces are not so polite. Douglas was loath to admit that the

MB-1 had been a rocky program, one that had fallen short of the Army's expectations—hardly his fault alone, but still a mark against his immediate career. The company stayed solvent because of a lavish wartime contract and the ability to spin off several derivatives to other federal customers. Martin would naturally have been uneager to install a new chief engineer for the transition to MB-2 production, though the Army evidently found this acceptable (he hired another former Army engineer from MIT, Clarence Hanscom). Douglas, after all, had snubbed the War Department when he quit Virginius Clark's ill-fated Bristol redesign effort.

In any case, it is safe to assume that Douglas was quite uncomfortable at the company, given its precarious financial condition, its peculiar boss, and the pressures of his own job. Being forced to cooperate with his old Army engineering office while it lurched from one MB-1 variant to another must have been infuriating as well. What is more, Charlotte Douglas had gone back to California with their two young sons in January, creating pressures of a more intimate sort. By contrast, Lawrence Bell, an equally talented man who had considered leaving Martin with Douglas, decided to stay. He was promptly rewarded with a vice presidency.

"Why did I think I could come out to California, raise money, and succeed in the business?" Douglas mused. "Well, I'd been there, of course, and I liked the country, and I considered the climatic advantages, and the fact that I felt that the financial people out there were a little more likely to be the ones who would put money into [an aviation company] than people in the East. Of course, at that time aviation was pretty low. There wasn't much work. Just a few government orders. So you couldn't draw too good a picture.

"But I came out with a good deal of information that I'd prepared on civil transports, and that was what I attempted to raise our money on to begin with—not for military work but for transportation." Glenn Martin laughed at this plan, saying there was "no better way of going broke than building commercial planes in Southern California."

If Douglas was being honest with himself in this recollection, then his return to California was almost pure whim. The information on "civil transports" was gleaned from his MB-1 work, specifically the so-called GMP (Glenn Martin Passenger) version built for the Army. The GMP was merely an MB-1 stripped of all military gear, with the top of the fuselage lifted to make more headroom, plus crude windows and seats. Flying any distance in such a machine would have been hellish—one of many reasons why civilian air travel was still just a gleam in a few businessmen's eyes. Carrying mail was a better possibility, but the Post Office's fledgling routes were dominated by surplus DH-4s supplied gratis by the War Department.

Douglas's rather dreamy notions about commercial flying were collected in an article published by the Society of Automotive Engineers just before he left the Martin Company. Because of his later accomplishments, the essay has acquired historical significance beyond its actual content, though its divergence from the thinking of Glenn Martin and Billy Mitchell is noteworthy. "Many people argue that the Government cannot afford to allow this great and new industry to languish because of lack of business, but must foster and further it by subsidy or continued military orders," he wrote, seeming to address Martin in particular. "That such a course would be welcome at this critical period is undeniable, but total dependence of existence on this problematic possibility would be fatal." Martin might have agreed in principle, but he never expressed such sharp skepticism about federal contracting.

For a hypothetical airline running three GMPs between Cleveland and Detroit, Douglas predicted a 30 percent profit on investment, perhaps in an attempt to make civilian transport seem as attractive to Wall Street as wartime weapons building had been. In contrast to the $84,000 paid by the government for a GMP, he priced the machine at $40,000—no doubt closer to its true cost. The rest of the article was a meandering survey of potential uses for non-military aircraft—fast freight service, aerial prospecting and cartography, law enforcement—most of which had been promoted by Glenn Martin and others for many years.

And so the father of two small children left a $10,000-a-year job (about $100,000 today) in the middle of an economic depression to start a commercial air transportation company in Los Angeles. It would be either a brilliant or an idiotic move, depending on the kindness of fate. Boyish and disarmingly casual, Douglas himself may not have fully appreciated the gamble he was taking. The gods, of course, were going to be very kind to him, but he would find it necessary to make a much more intimate deal with Mars than he ever imagined in the spring of 1920.

AFTER THE IGNOBLE failure of their attempt in late 1918 to fly across the continent, the Loughead brothers found themselves back in the five-dollars-a-head sightseeing business, on a par with hackney drivers. The 1919 summer tourist season brought a steady if slim income, with occasional grotesque touches such as an aerial wedding during which the best man vomited wildly from airsickness. There was some carriage trade—the State Department chartered the F-1 to take King Albert and Queen Elizabeth of Belgium on a ride from Santa Barbara to picnic at Santa Cruz Island's Pelican Bay, a ten-dollar fare that won the brothers something called the "Belgian Order of the Golden Crown." And working for the movie industry was, even in those days, often more profitable than building weapons for the government. Glenn Martin made $700 a day on the Mary Pickford film set. But the Lougheads were at the bottom of their profession and they knew it.

One prominent event must have been especially depressing to them during that low year of 1919. In May, three Navy flying boats attempted the first crossing of the Atlantic Ocean. Lord Northcliffe, publisher of the London *Daily Mail* and organizer of a munitions pipeline from the United States, had offered $50,000 in 1914 for a successful nonstop transatlantic flight. Dreams of winning this prize had been part of the inspiration behind Allan Loughead's F-1 design. Soon after the United States entered the war, Rear Admiral David Taylor had prodded Jerome Hunsaker to conceive a seaplane able to intercept German U-boats in the Atlantic. The *Daily Mail* contest and the Taylor project appealed to Navy Undersecretary Franklin D. Roosevelt, who supported an elaborate scheme to send three coastal patrol planes—completed between October 1918 and April 1919 by Curtiss under a contract worth $700,000—from Newfoundland to Portugal. Sixty-eight destroyers and five battleships were deployed across the ocean as markers and weather stations.

As finally configured, the Navy seaplanes did not have sufficient range for a nonstop crossing. Only one of the three starters made it to Lisbon, after a nine-day pause in the Azores. Of the remaining two, one capsized and sank under tow about a hundred miles short of the islands. The other was damaged while trying to sail stern first in rough water into an Azores port. The successful plane came back to America by freighter and toured forty-three cities to drum up support for the fiscal 1920 Navy aviation budget, but Congress slashed it anyway from a request of $85.7 million to $25 million.

Of interest to the Lougheads was the unmistakable resemblance of the Curtiss planes to the F-1, especially their short hulls, triple tail surfaces, and outrigger booms. The NCs were much larger and heavier, using four engines instead of two, yet no one would have missed the look-alike connection. There was no clear-cut piracy—the F-1 itself borrowed from existing designs—but at last Allan thought he understood Hunsaker's three-month scrutiny of the F-1 at San Diego during the war and the token award of two Curtiss HS-2Ls.

By the end of 1919, competition among war veteran barnstormers had dragged the price of joyrides down to a dollar or less. Once again, Malcolm Loughead quit the airplane business, this time to market an invention he had been tinkering with since 1916. The move, which made his fortune and brought far more cachet to the family name than aviation would for many more years, was to the auto industry in Detroit, where he formed the Lockheed Hydraulic Brake Company—using the phonetic spelling of Loughead for the first time. Within four years his patented four-wheel braking system was adopted by Chrysler, eventually becoming a universal standard. He never built flying machines again.

Sorely missing his brother's mechanical skills, Allan Loughead sold the F-1 early in 1920 to an amusement company that hoped to fly day-trippers to Catalina Island. The plane and its engines could only have been near exhaustion, underlining Allan's reputation as a gifted salesman. The venture, or the F-1, failed to get off the ground, and the pride of the Loughead Aircraft Manufacturing Company was gradually picked apart by vandals on the Santa Barbara beach.

With the cash from this sale, profits left from the HS-2L work, and the support of two remaining outside investors who had originally given money to the Lougheads in order to secure draft-exempt jobs for their sons, Allan put together yet another aeronautical enterprise. For the past year, he and Jack Northrop had been developing plans for a small sport sesquiplane (a biplane whose fixed upper wing is wider than the movable lower wing, which is used for turning) that they hoped might capture the elusive market for personal flying—an inexpensive machine that could be towed behind a car and parked in a home garage, just like a boat. The effect on this imagined market of war-surplus Jennies and DH-4s, dumped by the government for a few hundred dollars apiece, was as yet unnoticed.

Northrop's idiosyncratic design talents were given free play for the first time. For the fuselage they chose a still-novel monocoque body, in which two molded plywood half-shells were glued together around concentric wooden hoops. In monocoque construction, most twisting and bending stresses are carried by the external skin rather than by internal braces, thus saving weight and space. This technique was of European origin, dating back to the Deperdussin monocoque racer of 1912, but they had filed a patent application in August 1919 (granted three years later) for a process that greatly simplified the production of a rounded shape. Instead of gluing many strips of plywood over a form, three spruce sheets were soaked with glue and placed in a concrete mold that looked like a bathtub. Under a tightly clamped lid, a rubber balloon was then inflated in the cavity to press the plywood against the mold. Twenty-four hours later, the smooth half-shell was ready to be joined to another to create a cigar-shaped fuselage. The monocoque body appealed to Northrop's nascent obsession for streamlining and would become a permanent part of his shop-based engineering repertoire.

A second novel feature of the S-1 was its lack of ailerons. For lateral control, both halves of the entire lower wing rotated on a pivot bolt where they joined the fuselage. They could also be deflected up to ninety degrees to act as air brakes, or landing flaps. Company folklore would maintain that Northrop discovered this unconventional system after watching seagulls swoop down on fish that he scattered along the waterfront, but it seems more likely that Allan Loughead was trying to keep the aircraft as cheap to build as possible. They may also have been trying to avoid paying royalties to the Manufacturers' Aircraft Association if the S-1 ever went into mass production. As with the legend of Araminta Martin holding a lantern while her son toiled at night on his first flying machine, the fish story suggests more about Jack Northrop than about creating the S-1. He worked without benefit of any formal engineering education.

Allan Loughead's scheme was to introduce the S-1 at the San Francisco International Aircraft Exhibition (sponsored by the Manufacturers' Aircraft Association) on April 20, 1920, much as he had relaunched the old Model G at the Panama-Pacific Exposition in 1915. In early March they learned that the Greene Motor Company of Birmingham, England, from which they had mail-ordered a twenty-five-horsepower engine, had gone bankrupt. Within several weeks, using parts scavenged from auto garages, their shop foreman built a crude engine from scratch, grandly named the XL-1. The avid heart of Allan Loughead can be heard beating behind the sales pitch for this jerrybuilt engine, found in a 1920 brochure:

> The Loughead Aircraft Motor represents the highest development of
> the two-cylinder horizontally opposed, water cooled, valve-in-head

power plant. Designed especially for the Sport Biplane and produced in our own factory, this motor meets the long felt need for a lightweight, dependable motor of conservative power, suitable for use in the lighter types of aircraft. Every precaution has been taken to eliminate the possibility of engine trouble, the bugbear of aviation which so often interrupts a flight and necessitates forced landings in uncertain fields. Dual inlet and exhaust valves are used. Two independent Bosch magnetos furnish the ignition, and two high pressure gear-type oil pumps insure proper lubrication to all parts. The highest grade of workmanship and materials are used throughout. Simplicity of design, interchangeability and accessibility of parts make this the logical motor for the layman aviator.

A few days before the exhibition opened, the S-1 was flight-tested near Redwood City. Handling characteristics were good, but a homemade propeller had to be hastily mounted in place of a store-bought one to produce more lift. The plane weighed just 375 pounds, with a maximum velocity of seventy-five miles per hour and a landing speed of twenty-five miles per hour—exciting statistics for Everyman's airplane, though flying in anything other than flawless California weather would have required Rickenbacker nerves. The retail price was $2,500—not cheap, but not just for the Caleb Bragg types. During the six-day show "the baby," as it quickly became known, attracted many admirers, including repeated inquiries from a young Lieutenant Hap Arnold, then stationed in San Francisco as the head of Army aviation in the West. But the plane remained the sole property of Allan Loughead.

For the next few months, he towed the S-1 around Northern California's summer carnival circuit, touting it as "small, reliable and economical" and "within the reach of every automobile owner." But airplanes never had much in common with automobiles, and even most country boys intuited that they could never be pilots. Without a single sale, Allan soon ran out of cash. The $29,800 investment in the S-1 was lost, along with the Loughead Aircraft Manufacturing Company.

Allan soon took his penchant for salesmanship and promotion south to Hollywood, which was already a mecca for talented pitchmen. More than anything, he knew how to meet people and make deals with their money. Having promoted himself out of existence in Santa Barbara, he put his best side to work in a profession that was always generous to ambitious Californians with a gift for fast talk—real estate. He found a job developing tracts of barley fields bounded by Fairfax, La Brea, Melrose, and Beverly Boulevard—a richer gold mine than the barren Mother Lode he and Malcolm had picked through after going broke in 1914.

He also became an agent for his brother's burgeoning hydraulic brake company. And in the summer of 1922—with mechanical help from Northrop, plus cash from Cecil B. De Mille and Charlie Chaplin's brother, Sydney—he ran a bizarre concession at Catalina called "The Thrill of Avalon." The thrill this time was not a flying machine, just a car body bolted to two Navy surplus pontoons, pushed through the water by a 175-horsepower aircraft engine and propeller. Five minutes cost one dollar for adults, fifty cents for children. Even for summer amusement crowds, this nightmarish vehicle must have gone too far over the threshold between thrilling and deadly. With a wife and two children presenting down-to-earth demands, Allan soon settled back into real estate and middle-class anonymity.

Northrop, too, retreated into a more conventional existence, working for his father's contracting firm in Santa Barbara. He handled purchasing and whatever designing or drawing was necessary for building houses and small stores. At the age of twenty-six, the truth was that his ambitions were in tatters.

The great opportunity of the Great War must indeed have seemed gone forever. Allan Loughead had been personally involved with aviation since 1910, organizing the private construction of three custom-made aircraft and two Navy seaplanes. Now, at thirty-three, he had nothing to show except a string of irate investors. It had been six years since Jack Northrop tried to break away from his poverty-stricken childhood by entering the airplane business, a choice that could only have enraged his strapped father—as Glenn Martin's early ventures had appalled Clarence Martin. Now he was forced to take his paycheck from the man who had not been able to send him to college. It is at such junctures in the lives of notable people that one must suspend analysis and let them drift foward in time. This is what Allan Loughead and Jack Northrop did for the next six years— they drifted.

IN THE SPRING of 1920, while Loughead was scouring the ranchlands to find a customer for the unsellable S-1, Donald Douglas was trying to enthuse Los Angeles enterpreneurs about an aerial transportation company whose only asset was Donald Douglas. A writer for the *Los Angeles Times*, Bill Henry, who had been Glenn

Martin's advertising manager in Cleveland, introduced him to local venture capitalists. "I have perhaps vain hopes of interesting
capital in Southern California in an aircraft venture," Douglas
wrote on April 1 to his old mentor Jerome Hunsaker. "California
has long been a place where I wanted to live not only because of
personal reasons, but because I have felt that if there is to be any
civilian aeronautics it will be there that it will first attain real success." Always the realist, Hunsaker wrote back with regrets that
the Navy would not be doing business with him through any of
the existing contractors.

"I think I'd saved a thousand dollars from my wartime salary,"
Douglas remembered about his return to Los Angeles. He also took
$2,000 that his wife had recently inherited from her stepfather.
"The depression was starting in 1920. This is what I didn't know. I
really didn't know that the stock market crashed in '20, and when
Bill Henry and I started to go around to the big men to put money
into the company, they mentioned that there was a depression. We
got a negative reaction entirely. Some of these people would say,
'Well, you think that you're going to make 20 percent profit'—this
would be in our prospectus—'but what's the good of that? We can
buy Liberty Bonds yielding about five percent and no tax. So we're
not interested in anything unless it makes a terrific profit.' " In the
back of their minds, perhaps, were reports of wartime profits in the
30 to 50 percent range.

Douglas rented a house on a half-acre lot in Ocean Park for forty
dollars a month, a rather steep outlay. Nearly crazed with fear of
going broke, he dug up the yard and planted potatoes, which rotted
in the ground after late spring rains. "I wasn't much of a farmer," he
said. "You can't even make alcohol out of dead potatoes. I don't
know what I did wrong." This was the second time panic had
gripped him in California, even more deeply now than when he had
watched Glenn Martin test a new airplane in 1915 by prancing
across its wings.

"At about the time we got near the end of that rope and hadn't
raised any money, there was a chap by the name of Davis who got
in touch with me. He wanted to build an airplane to fly nonstop
across the country." David R. Davis, yet another wealthy California
sportsman who had become infatuated with aviation, was no
dilettante—he would later patent a high-lift wing design that was
mass-produced on World War II bombers. He was savvy enough to
recognize Douglas's civil transport ideas as premature, yet eager to

put up $40,000 for the creation of a transcontinental plane that would carry him to fame as co-pilot.

"Of course, I told him that I could do it," Douglas said. Allan Loughead would have approved.

In June, the Davis-Douglas Company rented the back room of a barbershop on Pico Boulevard. A sign in the sidewalk window alerted passersby, or at least anyone who stopped outside for a shoeshine, to the location of the firm's "Engineering Department." The low-rent nature of this start-up is a vivid clue to how outré aviation still was. Davis gave Douglas a quarter share of the business—enough to persuade five former colleagues to leave Glenn Martin in Cleveland, underlining Martin's perennial difficulties with shop morale. They were later joined by Martin's chief test pilot, who had taken the MB-1 on its maiden flight. When the time came to start building components, they rented loft space in an East Los Angeles woodworking mill. Sections were lowered down an elevator shaft and then trucked to a blimp hangar on the outskirts of town for assembly.

In early February 1921, the Liberty-powered wood-and-fabric biplane was almost lost on its maiden flight when it failed to get airborne off the short field adjacent to the hangar. Named the *Cloudster* in black humor, it was successfully flown later that month by Davis and the ex-Martin pilot. By mid-March it was up to design performance levels, setting a West Coast altitude record of 19,160 feet. It was also the first airplane ever to lift a load greater than its own empty weight—an accomplishment of crucial relevance to the fuel requirements of a nonstop cross-continent voyage. Hap Arnold heard of the project and asked to fly the new plane, a request the company anxiously granted because "it meant we had the Air Corps with us," Douglas said. The horizon began to look slightly wider than Pico Boulevard.

While David Davis prepared for his flight from Riverside, California, to Long Island, Douglas worried about more work. He was a businessman now, not just an engineer. The Navy had already begun to outline proposals for new aircraft in the fleet, hinted at by Hunsaker in his reply to Douglas's letter a year earlier. "I had heard that [the Navy] had some requirement for a torpedo plane coming up," Douglas said, "and I figured that the development from [the *Cloudster*] might make a pretty good torpedo plane. So Davis authorized me to go to Washington and see if I could negotiate a contract."

To Davis's amazement, Douglas returned from the capital in April with a provisional $119,550 contract from Hunsaker to build three experimental torpedo planes. To Douglas's horror, Davis greeted the news by ending their partnership. "So I came back, and Davis said he wasn't interested anymore—I don't know why," Douglas said, though he must have known that sudden commitment to such a major project might shock his senior partner. "I have a feeling he didn't have enough money to risk anymore. He said, 'You can have the works—I'll take your note for what assets I'm turning over to you.' It was two thousand some-odd dollars. That was all that was left. The [Cloudster] was his, and all that was left were a few hand drills and wrenches and vises, and a little bit of material, for which I gave him a personal note. And I took over the Davis-Douglas Company, lock, stock and barrel."

Douglas finally had what he wanted, his own company. He also acquired the immediate problem that had scared off David Davis—satisfying the Navy's provision that he raise capital to support construction of the planes. "At this time, I went downtown to finance this contract," Douglas remembered. "We didn't have any money. We had nothing." During the *Cloudster* project, Douglas had earned only "enough to live on, maybe $100 a week." Commercial banks were not inclined to back the venture, whether he had a Navy contract in his pocket or not. So in return for a quarter share of the new company—which amounted to one quarter of nothing—Douglas convinced Bill Henry to arrange a meeting with the local godfather of risky business and publisher of the *Los Angeles Times,* Harry Chandler.

"Chandler in those days was really the chap who made the town progress," Douglas said of the notorious land speculator, water czar, and right-wing propagandist. "He was in many, many things. And of course, by this time we had an actual contract. This was a little different from the original scheme of just starting the shop with nothing but an idea."

The stout, rugged Chandler had also been an aviation enthusiast since the earliest days, sponsoring the 1910 Los Angeles air show and hiring pilots like Didier Masson to carry copies of his newspaper to other California towns. Like everything else he touched, airplanes were a means to increase the wealth of Southern California, and thus the power of Harry Chandler. In turning to this infamous strongman for financial support, Douglas put himself at the mercy of someone who would not hesitate to crush him if he failed. It was clear evidence of desperation, and perhaps of a certain moral ambivalence that cannot be explained away as youthful naiveté.

"Chandler had this funny little office down there in the old Times building—a great big old room in it, and a bunch of old black leather furniture, with a little ante-room outside, possibly about 15 square feet, with a male secretary. Harry would come in there about 3:00 or 3:30 in the afternoon, and he'd stay until the paper was put to bed late at night. His office opened when he came there. People who wanted to see him would come in; no appointments were given. You'd sit and you took your turn—the first fellow in got the first chance to talk to Uncle Harry. So the place would be filled before Harry got there, everybody coming in early to be sure they'd get to see him that day.

"When you'd get in, you'd sit down and discuss your problem with him. He was a very kindly chap. Of course, I'm sure that in certain sections he had the reputation of being a money-grabber, but that was not my experience with him. People always wondered afterwards how come Harry didn't have an interest in the company. He never asked for it. That was Harry." Of course, a man in Chandler's position did not stoop to ask for an interest in a business as minuscule as Douglas's. The young airplane builder was now in his web, interest or no interest.

The proposition that Douglas put to Chandler was to co-guarantee a $15,000 bank loan. "I don't know whether you know anything about building airplanes," Chandler replied, "but Los Angeles needs more business." On a scrap of paper he wrote the names of nine of the city's biggest financial kingpins. "If each of these men will guarantee one-tenth of the loan, I'll guarantee a tenth." Sizing up Douglas as too timid, he told Bill Henry to stay close to him.

Within four days, carrying Chandler's almighty imprimatur, they obtained the additional guarantors. The last name on the list was Joseph Sartori, head of the powerful Security Trust & Savings Bank. Henry asked if he thought it was a good loan. "I guess it's all right," the banker said. "You have at least a hundred million dollars' worth of guaranty behind the $15,000."

"So we financed this $120,000 contract on $15,000, plus $5,000 I borrowed from my father," Douglas said, "plus extending our credit with supply companies. Chandler was the key man in helping us get started. Once we got his name, the rest were easy." That Chandler insisted on nine more names showed just how risky the deal was perceived to be.

In July 1921, a few weeks after David Davis was forced to abandon his transcontinental dream when the *Cloudster* developed engine trouble over Texas, the lowly Douglas Company was incorporated in California—a momentous development for a region that would one day dominate the nation's aerospace industry. The *Cloudster* was never reoutfitted for its intended purpose (two Army lieutenants made the first successful cross-country flight, from New York to San Diego, in May 1923, flying an airplane purchased in Holland). It was sold several times—once being used to airlift beer into Tijuana, Mexico—and finally abandoned in 1926 on a Baja California beach, rather like the junked Loughead F-1.

The *Cloudster* design, however, was the progenitor of Douglas's weapons-building fortune. His Navy torpedo planes, designated "DT," were simply ruggedized versions of the *Cloudster* mounted on floats, without its long-range features. "I adapted the Cloudster for this plane [the DT]," Douglas said. "You design an airplane for one purpose, it proves successful, then you try to find the things you can do with it." DT-1 was completed in October 1921, first flown in early November, and accepted by the Navy at San Diego in mid-December. During the construction of the next two, the company's bare-bones financing did not fill the gaps between payments from the Navy, forcing Douglas to rely occasionally on his quarter-share partner's *Los Angeles Times* paycheck for grocery money handed out to the crew. Harry Chandler's helping hand was no doubt heavy on his shoulder.

By the end, however, they had earned a lot more than grocery money. "That contract proved profitable," Douglas acknowledged in vast understatement. "We made a net profit on the $120,000 contract of about $40,000. Then the Navy wanted to purchase more."

In competition with four other manufacturers' prototypes (including Curtiss, the Navy's mainstay), the DT was chosen for mass production in 1922. The Navy asked Douglas to build eighteen DT-2s—"a tremendous order for those days"—which were slightly modified versions of the original, at $16,900 each. The Navy also awarded eleven to the Dayton-Wright Company, which had been acquired by General Motors after the war, but this did not prevent the firm from soon going bankrupt—surely much to Douglas's delight. A repeat order for twenty more planes went to Douglas in 1923. In 1924, the Norwegian government bought a manufacturing license. Donald Douglas never had to plant potatoes or borrow from someone else's paycheck again.

"I went down to the Security Bank, discharged my note to Chandler and the rest of his people, and told Security what I'd need for the new contract," Douglas said of the shift to what passed in those days for mass production. "By this time, of course, we were getting into real money—three or four hundred thousand dollars—which, to my surprise, the bank was willing to lend me.

"After all, I'd just made a 33 percent profit."

The Navy's procurement of DT-2s from a brand-new manufacturer naturally aroused the curiosity of the Army Air Service. It was, after all, the plane of the moment. Whether out of a traditional sense of competition with their Navy counterparts or an ongoing search for ways to justify and dramatize their own existence, several Air Service officers began to lobby during the spring of 1923 for a flight around the world. In June, the War Department authorized them "to take necessary actions to procure all available data on the Fokker F-5 transport and the Davis-Douglas Cloudster . . . and to procure one of these planes for test."

The wording of this directive was a clever deception, since the foreign-built Fokker was totally unsuitable and the *Cloudster*, modified as the DT-1, had been out of production for two years. What the Army really wanted was its own DT-2, which is exactly what Douglas submitted in July—calling it the DWC, for "Douglas World Cruiser." That the DWC could be culled directly from the DT-2 assembly line was evident when Douglas promised that he would deliver it for $23,721 within forty-five days of signing a contract. The only major modification was to the fuel system, which was increased from a capacity of 115 gallons to 644 gallons.

"The world cruise was started largely by a group of young Air Service officers," Douglas said. "They had learned about this Navy torpedo plane and the forerunner of it, the Cloudster." In August, Douglas received a contract to build a prototype, followed in November by a $192,684 order for four DWCs plus spares. "This wasn't in competition," Douglas recalled. "This was a negotiated contract"—meaning that the Army paid Douglas whatever he thought he needed to do the job. Thus, even with spares that included fourteen extra pairs of pontoons and almost two complete sets of airframe replacement parts, Douglas maintained his profit level of at least 33 percent.

Far more than keeping up a deep cash flow, however, the DWC project and the round-the-world flight that took place in the spring and summer of 1924 sealed the Douglas company's reputation as the industry's *enfant doré*. The spectacular 28,945-mile, 175-day journey of four aircraft and eight men from Seattle westward through Southeast Asia, India, the Middle East, Europe, and across the North Atlantic (completed by two of the crews) was a mass sensation closely tracked by the press and cinema newsreels. Melo-drama was established early on, when the flight commander's plane crashed against a foggy mountainside in Alaska. Presumed dead by the outside world, he and his co-pilot walked for ten days through blizzard conditions to reach the nearest village. Another plane was lost when it capsized under tow off the coast of England. Though eclipsed by hero worship of Charles Lindbergh three years later, the adventure crystallized for the first time the ability of air travel to neutralize geographic and cultural distances. To Lowell Thomas, it was a new voyage of Magellan.

"The development of the world cruisers was probably the best thing that occurred for us," Douglas said. "It made us known. That opened up an opportunity for doing Air Corps business that hadn't existed before. Up until that time, Douglas was fairly localized. We were well-known in aviation circles, because of the torpedo plane, but we were not well-known in the Air Corps. Many of them knew me, of course, because of my time in Washington, but after the world flight airplane we got into more Air Corps business."

Indeed, in the latter part of 1924, Douglas completed a $60,000 development project for an observation aircraft slated to replace the Army's war-surplus DH-4s. Production of this reliable and easy-to-fly biplane would eventually total 879—including 108 sold to for-eign governments—over the next dozen years. It turned the Douglas Company from a bright young shop into a giant of the industry, a multimillion-dollar venture that neither economic depression nor political change in Washington would ever much perturb.

In 1920–1921, his first full year of business, Douglas built six planes for $130,890. By 1926, the company was producing 120 planes worth $1,662,723. Between 1922 and 1928, Douglas deliv-ered 375 aircraft—314 for the U.S. Army and Navy, most of the rest for foreign air forces—valued at $7,161,000, earning $1,250,000 after taxes. The incredible transformation in just five years from barbershop storefront to international aviation manufacturer was fueled not by a commercial market, which was essentially nonexistent, but by a new system of enormously profitable contracts with the federal government. It was with Donald Douglas that World War I aircraft procurement was refined into conventional business practice.

FOR GLENN MARTIN, Douglas's meteoric success could only have deepened his own depression. By the spring of 1921, the production of twenty MB-2 bombers for the Army was almost finished. Martin had run into technical trouble with the plane, though Billy Mitchell was still cheerleading the service's only bomber. As part of his publicity blitz for an independent Air Force, Mitchell had spearheaded a debate about whether battleships, then the quintessence of naval power, could be sunk by airplanes. The MB-2 was to be the vehicle for his challenge to the Navy, a series of bombing trials conducted off the mouth of the Chesapeake Bay that summer.

"Now, the very important thing just at the present time is to get those Martin airplanes, have them work, and be delivered," he lectured Thurman Bane on March 1, 1921. "We are having a lot of trouble with the one at Langley [a test field near Washington, D.C., named for the former Smithsonian secretary]—can't make the motors turn up, and apparently the ship won't lift the load it is designed to lift. We are going to get those battleships, and must not leave any stone unturned to put the matter over. We want to have the twenty Martins delivered at Langley just as soon as it is possible."

Martin had no quarrel with Mitchell, but he wanted much more from the Army than the original order for twenty MB-2s. This expectation had been formalized in an extraordinary paragraph of the June 1920 MB-2 contract: "It is understood by both parties hereto that the [$1,003,737 payment] is not of itself sufficient to induce the contractor to undertake the work herein contracted for, without the possibility of additional remuneration from the Government. Is is therefore agreed that one of the considerations of this contract is said element of possibility of additional remuneration, which is at the same time calculated to afford every encouragement to the contractor to expend its best efforts to make the [MB-2] so superior to the contract requirements that a material contribution will thus be made to the science and art of aviation and as a result of which the Government will consider it advisable to reproduce such articles in quantity."

Martin clearly thought he had an ironclad promise for mass production. At the end of April 1921, however, the Army dealt him a stunning blow by opening the production contract to competitive bidding—a tradition in federal procurement that had been suspended during the war. The next thirty-five MB-2s went to the L.W.F. (Lawson, Willard, and Fowler) Company of College Point, Long Island. Though this flew in the face of Martin's MB-2 contract language, it was legal, since the competitive bidding system empowered the government to acquire the design rights to airplanes it developed at private companies. The practice generated much bad blood, but was viewed by the military as a way to preserve an industrial base by spreading appropriations among hungry manufacturers.

A series of letters that Glenn Martin wrote to the editor of *Aviation* magazine, Lester Gardner, shed light on the early volatility of the industry, as well as Martin's grim outlook at the time. They also refer to certain problems in the government-industry relationship that have survived intact to the present day.

My dear Lester, [he wrote on April 25, 1921] I have been all over the bomber proposition with Col. Bane and with Col. Gillmore, and it is now a settled fact that they probably will not want any more than the 35 bombers, contract for which is being prepared for the L.W.F. Company. . . . Flint's [Albert H. Flint, president of L.W.F. and one of the original directors of the Manufacturers' Aircraft Association] past record for unreliableness does not seem to make much difference in their present scheme of calculation, and it is evident that they will have to have the experience of trying to get something for nothing and quality service where it does not exist.

We have let 71 men off our pay-roll and will let about 30 more off this coming Saturday night, as we have finished the manufacture of all parts on the entire [MB-2] order, have assembled and delivered 15 of the bombers and will finish the balance at the rate of a machine every eight days.

I have not yet been able to turn up any business, and it is evident that I will be left almost flat by the month of June.

You know, Lester, it hardly seems possible, in a civilized country, that a group of supposedly intelligent army officers could willfully overlook the value and economy of producing aircraft in our factory, especially when they agree, after the inspection of Col. Bane, that there are no operations in our scheme that are not required to produce the kind of a bomber they want, and if they had a mind to they could prove to their own satisfaction that we bought our materials as cheap as they can be bought in the market, and it is already an admitted fact that we are paying our labor less money per hour than is paid for labor in Flint's plant or in anybody else's plant in the United States that is delivering any kind of aircraft material.

We have felt, here in Cleveland, that because of our untiring effort and organizing ability, that we would be able to produce an article so superior to the usual run of airplane material that we would at once gain such an advantage over our competitors, considering quality, that we would be very hard to compete with, and I am certain that if the present policies continue, without some very immediate reparation, such aircraft industry as may exist in America will be broken down and all the government will have to show for their policy will be a hopeless mess, with no progress whatever in the art.

I have talked with some of my confidential friends in the [War] department and I find that my advertising and efforts made in trying to help the airplane industry in general have been held against me with decided effect, various members of the Manufacturers Association pointing out that the Martin Company necessarily must get very high prices to pay for page ads in aeronautical magazines, Saturday Evening Post copy, etc., etc., and many other overhead expenses conducted by this organization which they do not conduct, and which the other manufacturers estimate cost us approximately $100,000 per year. It has even been pointed out that our stationery is of a very finished and high-class order, which further gives the impression that we have business ideas equal to the U.S. Steel Corporation, etc., etc.

However, I do not think they will need to worry much about my expenditures from now on, for, of necessity, if I have no business I must advertise accordingly. On the other hand, I shall not allow unsuccessful so-called aircraft manufacturers to bias my successful factory management plans, and feel that there is a day of judgement coming when the political inside worker, who does not build satisfactory aircraft, cannot possibly continue doing business. In the meantime, I shall keep at it and see if it is not possible to dig up some odd jobs to tide over the waiting spell, as I am sure that some of these days the government may again want bombers, and when they do we shall be in the running stronger than ever.

On June 23, Martin again wrote to Gardner that

we are still having considerable difficulty in finding work for the shops, and must say that we have been up against a particularly disastrous combination in the attempt to break the price of the bombers.

In [the Army's] effort to procure bombers, they have had seven Government accountants in the plant for about fifteen days, and have applied the old definition of cost used fifteen years ago, wherein they estimate that the cost of manufacturing an article is material, labor and direct expenses, such as supervision and paper, pad and pencil, and that nearly all other natural overhead expenses, except light, heat and power, are not elements of cost and must be paid for out of "profits." Naturally, all advertising expenses of every nature, all photographic work, all hospital service or welfare of any nature, and even the local advertising in the papers for help, are stricken from the elements of cost, also all traveling expenses, etc.; in fact, we must run our business out of profits. On this basis, the Government feel they are very charitable when they allow us a new contract at cost and 20 percent profit; the 20 percent profit going a long way toward meeting the 23 percent of overhead necessary in the operation of our factory, but not considered a part of cost in the definition of cost.

I have one very nice consolation, however, that while I will not make any money on my next order, I did make some money on the last one [MB-2], and Flint will go broke on his order for 35, if they are ever accepted.

I have had a mighty fine letter from Billy Mitchell, and cannot express my appreciation and elation over it. . . I think Billy is a wonder, and sympathize with him in the battle of Washington and Langley Field, but hope that he will see the day when his opposition will be less harsh and his vision will be more thoroughly appreciated.

Martin could not have been in too dire straits regarding his personal finances, because he went on to tell Gardner about his recent purchase of a nineteen-room house and 2.7 acres on Lake Shore Boulevard in Bratenahl, one of Cleveland's poshest neighborhoods. He would later "sell" this house to the company, which would carry it on the books as a business expense.

On July 1, three weeks before Mitchell's bombing test against the Navy, Martin again wrote to Gardner with more bad news about the MB-2:

We haven't a penny's worth of work in the house; we have discharged over 300 employees and are planning on development work. We have, however, retained an organization of about 90 people—our best development men and the most expert of the workmen, and feel that we are capable of doing some development work for both military and commercial purposes. . . .

Meanwhile, Curtiss will be building 50 of our bombers, at $18,900 apiece, with $222,000 worth of spare parts business, and the L.W.F. Company will be building 35 bombers at $23,000 each. Both the Government and the above two manufacturers are going to be very sick of their tasks before any of these bombers are delivered.

It is quite impossible for me to take the business at their figure, although in my part of the contest I had offered to build the last 50 airplanes for the unbelievably low figure of $22,485, expected to give back to the Government a part of the money that we have made on the last contract. Now that the contract has been let, it is sure a relief to know that our plant is not laboring under the delusion that they could build the ships for this price.

I am exceedingly anxious to get into commercial work; either in the building of airplanes or the building of anything else on which a living can be made. I shall be busy from early until late planning on what we can manufacture that somebody would like to have.

Martin ended this letter with a PS to Gardner to find out "if it would be proper" to bring his mother along to watch the Mitchell bombing trials. The answer was definitely not.

On July 21, 1921, a formation of seven MB-2s took off from Langley Field's grass runway carrying special armor-piercing bombs. Concern about the bombers' lifting capacity had been focused on this specially designed ordnance, which was needed to attack the prime target of the day, the captured German battleship *Ostfriesland*. With its four-layer steel hull and watertight bulkheads that had withstood naval gunfire in the 1916 Battle of Jutland (the war's only major naval engagement between Britain and Germany), the dreadnought was widely regarded as unsinkable even when undefended and dead in the water. Former Navy Secretary Daniels boasted that he would stand bareheaded on the bridge of any battleship and expect to be safe.

The Navy insisted on an orderly sequence of bombings followed by pauses to inspect the damage, but Mitchell's planes ignored the rules and dropped their load on the supposedly impregnable vessel, sending it to the bottom in just twenty-two minutes. It was hardly a realistic test, but it ended one of the great officers' clubroom arguments of the day—flying machines could indeed sink modern capital ships. "Day of the Battleship Ended," the press trumpeted, precisely the reaction that Mitchell had hoped for. Though his campaign for an independent air force similar to England's and France's still went nowhere, a Navy board soon recommended that the service obtain more planes to base upon warships, foreseeing a future when the fleet would steam under cover of its own air protection—in other words, carrier groups. This was of immediate consequence to the fortunes of Glenn Martin and Donald Douglas. The full import of the test, however, would not be understood until Pearl Harbor.

Mitchell made one other notable use of his bombers, one that was never celebrated like the *Ostfriesland* sinking. In the fall of 1921, when federal troops marched into West Virginia to end a coal miners' strike, two Army air squadrons were ordered to Charleston. The force consisted of DH-4s and the same MB-2s that had participated in the battleship trials. The bombers only flew supply runs for the operation, but Mitchell's willingness to threaten American citizens with weapons that were already associated in the public mind with the horrors of modern warfare revealed the cyclopic intensity of his ambition.

While the bombing trials provided priceless publicity for the Glenn Martin Company, turning the MB-2 into a legend, the damage of losing additional work to his competitors took its toll on Martin's spirit. "Business is such a mess," he wrote to Gardner in November, "and when the air is so full of deceitful, misleading crooks and various other pests there doesn't seem to be very much satisfaction in competing." Though it meant nothing to his own pocketbook, Martin must have gained some satisfaction from watching L.W.F. go bankrupt on the MB-2 and Curtiss lose $249,000.

For the next thirteen years, the Martin Company would do no business with the U.S. Army, depending instead on the Navy to meet its payroll. This long estrangement was due in large part to Martin's disillusionment over watching the Army hand the MB-2 to his competitors, though he would engage in the same cutthroat bidding tactics to wrest production work away from them. There were other more objective factors, however, such as the defection of key talent to Douglas and bureaucratic antagonism against Billy Mitchell. The zealous airman was demoted to colonel and then, after publicly criticizing the military for "incompetency, criminal negligence, and almost treasonable administration" in the fatal loss of a Navy dirigible, court-martialed for insubordination in 1926.

Perhaps of overarching significance was the Washington Conference of 1921–22, during which restrictions on the use of bombing planes were debated. The American Advisory Committee to the conference issued a resolution condemning bombardment of unfortified towns, but delegates could not decide on a practical measurement—such as tonnage of naval warships—for limiting airplanes. (The United States' Air Force was comparatively small at this time, with 537 planes. Britain's possessed 1,048, and France's, the largest in the world, had 1,722. The British feared that the French force was pointed across the Channel and therefore favored some kind of peacetime limitation.) Billy Mitchell suggested with tongue in cheek that the "only practicable limitation as to the numbers of aircraft for military purposes would be to abolish the use of aircraft for any purpose." Everyone recognized that while commercial planes could easily convert to military models, restricting the growth of civil aviation was not feasible. Though the nations attending the conference were thus unwilling to sign a treaty on limiting bombers, the talks threw cold water on Martin's hottest product.

"I want to warn you—I don't think I am very crazy," a clearly shaken, defensive Glenn Martin told the Cleveland Chamber of Commerce just before Christmas 1921. "It's just possible I might be some type of a nut, but I'll leave that to you to decide for yourselves." They gradually did.

BETWEEN 1924 AND 1926, as the postwar economic depression eased and pacifism faded, the government conducted several prominent investigations of the military and commercial aircraft business. Federal expenditures on all aspects of aviation, including the Army, Navy, Post Office, and NACA, totaled $433 million from 1920 to 1924, underlining the need for new policy. These inquiries and the pro-industry legislation they spawned finally turned the page on the Great War, clearing the way for frenzied investment in airplane companies that was part of the wild bull market of the late 1920s.

Festering postwar suspicions about excessive profits, mismanagement, and favoritism in the aviation business—between 1919 and 1924, the Army spent $22 million for "motors, planes, and accessories," plus $20 million for "research and engineering," yet had only 754 aircraft of all types in commission by mid-1924, many of which were still war surplus—gained momentum from the graft scandals of President Warren Harding's Republican administration. By October 1924, critics of the industry persuaded the House of Representatives to convene a bipartisan Select Committee of Inquiry into Operations of the United States Air Services, which became known as the Lampert Committee after its chairman, Florian Lampert, a Wisconsin Democrat who was a member of the House Patent Committee. During the next year, the panel questioned scores of witnesses, including Glenn Martin.

Martin came to make his pet case for keeping production of an aircraft solely at the company that developed it—that is, for eliminating competitive bidding. This was a position favored by NACA and the largest manufacturers, who believed that contracts negotiated directly between customer and builder—such as the Douglas World Cruisers—resulted in the best products. Naturally, it was not favored by smaller companies that lacked the capital and personnel to maintain strong engineering development offices that could bargain with the government. They argued that negotiated contracts fostered favoritism that kept them from growing larger through federal orders.

When asked his opinion on whether the government should parcel out work to preserve the industry as a whole, Martin's answer was blunt, with no trace of sympathy for beginners. The MB-2 was still a sharp bone in his throat. "There is a misnomer as to industry," he said. "There are some so-called manufacturers who have objected to standing on their own feet and could not produce a satisfactory article. The Government has attempted to maintain as many manufacturers as possible, and they have gone into some of these manufacturers' places and with their assistance have tried to help them along, to keep them going. I cannot quite agree with that."

The committee examined a number of situations in which firms that had expended considerable resources under negotiated development contracts eventually lost competitive production orders to lower bidders who had spent nothing on development. The MB-2 imbroglio was perhaps the best known of these cases. Martin admitted that he had snatched production of seventy-five "SC" seaplanes developed in 1923 by Curtiss for the Navy by bidding $20,000 each compared to Curtiss's $32,000. Martin had still made a handsome profit, but he insisted that in order to build the SCs his engineers had to redesign the planes to suit Martin Company shop practices. Somehow gauging that the distress such episodes were causing the industry was not worth the savings to taxpayers, the committee concluded that the current system of competitive bidding was destructive. The government should therefore stop claiming ownership of design rights and modify its long-standing insistence on competitive procurement.

In the course of Lampert's investigative work, the old MB-2 contract was turned inside out, much to Martin's chagrin. Into the public record went the fact that Martin had tried to charge the government for "overhead" items such as entertaining visiting Army officers, employees' income tax, bribes to railroad crews, and gold service buttons. These and other charges disallowed by the govern-

ment resulted in Martin having to refund $22,461. Overhead pad-
ding would become a perennial feature of the business, though so
deeply buried in the mountains of paperwork surrounding military
contracts that it would take subpoena power for congressional committees to uncover such items as dog kennel fees, wives' hotel expenses, and private limousine fees.

The committee also grilled Martin about his first deal with the
Navy after losing the MB-2. In April 1922, the company had won a
contract for six experimental observation monoplanes, called MOs,
designed by Naval Aircraft Factory engineers. After only token testing, the Navy ordered thirty of the planes at $26,072 each. As pro-
duction and further testing proceeded concurrently, it became
evident that the MOs were defective. Martin was not blamed, but
the episode called attention to the company's shaky technical footing at the time.

The Lampert inquiry began as a forum for industry critics. When
it issued its final report in December 1925, it wound up certifying
many positions sought by industry apologists, such as relaxed competition. In the same month, another prominent government
panel—appointed earlier that year by Republican President Calvin
Coolidge as an antidote to Billy Mitchell's criticisms and the Lampert Committee's supposed sympathy for them—released its findings. Called the President's Aircraft Board, or Morrow Board after its
chairman, J. P. Morgan partner Dwight D. Morrow (an Amherst
classmate of Coolidge and friend of Harry Guggenheim), this group
chiseled in establishment stone all of the aircraft industry's fondest
wishes. Howard Coffin was among its nine members, who included
a general, a rear admiral, and two powerful members of the congressional military committees. Coffin had taken up the issue of
industrial preparedness almost as an ideological crusade after the
war—whether out of guilt, greed, or patriotism can only be guessed.

Glenn Martin also appeared before the Morrow Board, which
listened to statements from witnesses without questioning them in
depth. Once again he urged that mass production of an airplane
remain with its designer. The government should stop building aircraft itself, such as at the Naval Aircraft Factory. It should stop
cultivating its own design expertise ("we want cooperation from the
experts of the Government instead of competition"). It should protect a manufacturer's proprietary rights. It should provide steady
funding for new planes. It should adopt tax rules that recognized the
risk in a volatile new manufacturing art. It should make progress

payments during the course of work instead of lump sums upon delivery of finished goods. In other words, it should leave its money at the front desk and then get out of the way.

Howard Coffin was sympathetic to these suggestions. "Mr. Chairman, I would like to ask Mr. Martin if he does not think that the Government would be able to buy its automobiles on a better basis if the Studebaker Company, for instance, was asked to build Dodge cars during one year, to build Packard cars during the next year, and a third kind of car during the third period. Do you think the motor cars, under those conditions, could be reduced in price or bettered in quality? That is about what you are being asked to do in aircraft, is it not?"

"Yes," Martin answered on cue, "that is a very good comparison. . . . In order to shift the make of cars, they would have to have a tremendous overhead of engineering and other development, and it would be impossible, almost, to state how much those cars would each cost. I would say it would be probably five times the present selling price of the cars." Coffin did not ask Martin how he had been able to bid $12,000 below Curtiss on the price of a Navy SC and remain solvent.

Not surprisingly, the Morrow Board's halfhearted investigators went even further than the Lampert Committee in rubber-stamping industry proposals. Some conservative historians, such as John B. Rae, have suggested that the only alternative was socialization of aviation, but this was an impossible choice during the Coolidge era. Privatization was the guiding spirit of the day. Congress had already handed over aerial postal service to private operators, paving the way for heavy subsidies. The business of government was to help business.

Three major pieces of legislation followed the Lampert and Morrow reports in 1926. First came the Air Commerce Act in May, regulating interstate flying, providing for inspection of private aircraft, and generally bringing commercial aviation under the law. The Naval Aircraft Act and Air Corps Act followed in June and July, respectively, setting up five-year expansion programs whereby the Navy would own 1,000 first-line warplanes and the Army 1,600 by 1931. At the time, the Army possessed 1,394 planes of all types, with only 605 considered up to date.

"After the Morrow Board, aviation started to become respectable," Donald Douglas, who did not testify at either the Lampert or the Morrow hearings, said. "They started to get some fairly respectable appropriations. The Morrow Board was very important, because as a result came the Air Corps Act, which permitted negotiation." Douglas, of course, appreciated how lucrative negotiation, as opposed to competition, could be.

As part of the Air Corps Act, a procurement bill was passed that reflected a political compromise between those who favored the Lampert-Morrow recommendation to relax competitive bidding and those who were suspicious of what might happen if military officers bargained directly with sole-source contractors. Partisans from each side interpreted the bill to their own advantage, creating a fog of ambiguity. The bill contained a requirement for competition at the design stage and gave the service secretaries wide latitude to negotiate with contractors for production. In practice, design competition became relatively trivial, since sole-source negotiation for mass production is what both the military and the industry liked best.

"It is a fortunate situation," President Coolidge observed about the new law, "when the needs of the Government can be met by affording an orderly stimulation of the industry upon which we depend to supply our needs."

THE COMBINATION OF a sharp upturn in the general economy and sweetheart legislation in Washington brought a flood of investors back into the aviation industry for the first time since 1916. The surge went from top to bottom, even rescuing Allan Loughead and Jack Northrop from obscurity.

In 1923, Northrop had been forced to look for new work when his father went broke constructing a Santa Barbara office building. "When my father was wiped out and had to go out of business, we both needed to find something else," he recalled. "Douglas had become well known in the Los Angeles area and it appeared that I should at least make an effort to get back into the type of work I loved."

The glad-handing Allan Loughead, who had already forged an acquaintance with Donald Douglas in Los Angeles, introduced Northrop to the prospering businessman-engineer. Douglas offered to subcontract the production of DT pontoons to Loughead and Northrop, but they were in no position to establish a new shop. Instead, Northrop signed on as a Douglas Company employee. "My first job at Douglas was assembling wooden wing ribs for the World Cruisers," he said—a rather lowly task. "I was so thankful to have anything to do that would bring in some income that the period in the shop went quite quickly and was a pleasant experience."

Douglas knew about Northrop's diamond-in-the-rough technical acumen, however, and soon moved him up. Either no one queried him about his skills or he oversold himself, because his first assignment was above his skill level. "I think one of the worst times I ever had in my life was the morning that I was permitted to come to work in the Engineering Department," Northrop said. "I was told to design the fuselage fairing on the World Cruiser. Now, the fairing was simply a form and support for fabric, which surrounded a welded steel tube fuselage. It made the fuselage a reasonably good streamlined shape. It so happened that my work with Loughead had never included designing fairing for any kind of a steel tube fuselage—I had no idea how to go about it.

"I fussed and fiddled and stalled around all morning, getting more frightened all the time. I ate a bag lunch, which naturally didn't digest very well, became quite ill and hiked home, which was about a mile and a half from the office. I didn't know whether I had a job the following morning. I was tremendously worried because of family obligations and a very stringent financial situation.

"Fortunately for me, the next morning when I came back somebody else had been given the fairing job and I was asked to design the welded aluminum fuel tanks. This I knew all about. From there on I got along quite well. I was not a designer of a complete airplane. Fuel tanks, fairings, wing fittings, control surfaces, or various parts simply took a good mechanic with an idea of stress and strain and the ability to do calculations." Northrop, with just a high school education, was many rungs down from the MIT-trained Douglas and the MB-1 veterans around him. His entrée into the engineering department was probably his knowledge of drafting, which could be used as a design tool. He was lucky, again, to be on the cusp between the technology's shop-floor past and scientific future.

Northrop remembered Douglas as a "master salesman and a fine engineer, but his primary job during the operation of the company at that time was to contact military or other potential customers." Northrop "never considered him a designer of aircraft." Douglas's friendship with Jerome Hunsaker was the company's lifeline. Yet the firm was still vulnerable even to short gaps between military contracts. "One such time is seared into my memory," Northrop said. "Doug went East to try to get an order which was necessary to keep the house from falling in [mass production of the DT-2]. We were all holding our breath. One morning Harry Wetzel [Douglas's general manager] got a telegram from Doug which said 'let all engineers go except Mankey [the chief draftsman] and Northrop.' I was the one remaining engineer for a period of several weeks, after which things began to pick up. So you can see how things went in those days." Sudden layoffs would become a basic hazard of the industry, only on a far more massive scale than this.

Northrop found the work at Douglas "more or less conventional," meaning boring, and by the latter part of 1926 was spending his time off at home designing "a nice clean little high-wing monoplane." This notion of "cleanness," or low aerodynamic drag, was the core obsession of his peculiar genius, as first expressed in the monocoque S-1 biplane. "It was a radical design," Northrop said of his off-hours project, "far removed from the more conventional types that Douglas was building, and I felt he would not be interested." The Douglas planes, in fact, looked like chunky farm machinery compared to the curvaceous, stripped-down forms that Northrop had in mind. His drawings went instead to an airmail pilot in San Francisco, who used them to build a successful racer. The pilot offered him a job as chief engineer in a new company, but at that moment Allan Loughead reappeared with new angels in hand.

Loughead's path out of the real estate business was provided by a former Army flying teacher, Kenneth Jay, who now made his living as an accountant. On the strength of Northrop's recent freelance work, Jay agreed to write a stock prospectus for a company to produce refined versions of his latest design. Of more interest to Loughead, Jay was also the business manager for a wealthy Los Angeles venture capitalist, Fred Keeler, who had made a fortune selling china, tile, and bricks. Keeler knew the name Lockheed from investing heavily in Malcolm Loughead's successful hydraulic brake enterprise, and so was favorably disposed toward the new prospectus. He agreed to buy 51 percent of the common stock for $2,550 and all of the preferred stock for $20,000. Allan Loughead then put in $2,450 of his own money to acquire the remaining shares. Part of Loughead's secret for salesmanship was in locating, or hustling, rich risk takers like Fred Keeler, Burton Rodman, and Max Mamlock.

Thus, on December 17, 1926, the Lockheed Aircraft Company (Keeler insisted on using the same spelling that appeared on the hydraulic brakes) was incorporated in Nevada. Keeler made himself an absentee president and named his lawyer as executive vice president. The workaday management consisted of Allan Loughead as vice president and general manager, Kenneth Jay as secretary-treasurer, and Jack Northrop as chief engineer. The Loughead shop foreman from Santa Barbara days also quit Douglas to join the new company, bringing with him enough craftsmen to trigger a warning against further raids from Donald Douglas himself.

"If there was any socializing, Allan Loughead did it," Northrop recalled about this arrangement. "He was the hail-fellow-well-met. He knew all about Hollywood, and he would show them around if there was any socializing after business. I was the one and only engineer. I'm a loner anyway. I'm an unsocial person, not antisocial."

At the first official meeting of the corporation on December 20, ownership of Northrop's design work and the monocoque fuselage patent was settled by a further stock issue. Ten thousand shares each went to Allan and Malcolm Loughead, Northrop, and their old foreman, who had been joint applicants on the monocoque fuselage patent. In addition, Keeler's lawyer and Jay received 5,000 and 4,000 shares, respectively, for services rendered. Such private stock issues became one of the hallmarks of the late-1920s boom, often producing astronomical profits even for investors in marginally successful companies.

Over the next six months, spending $19,600 of the company's $25,000 capitalization, the group built Northrop's "clean" new airplane in a rented shop near the corner of Sycamore and Romaine streets in Hollywood. Using the wooden monocoque fuselage technique and a dramatic unbraced cantilever wing introduced during the early 1920s by Dutch aeronautical pioneer Anthony Fokker (his F-IV monoplane, used by the Army for the first successful transcontinental flight in 1923, employed such a wing), the "Vega," as Northrop called it, was one of the era's technological masterpieces—a leap forward that must have astonished Northrop's professional acquaintances, to whom he was a mild towhead who could not eat his lunch if he got upset. Unlike most other airplanes of the day, which still did not look terribly airworthy to an average observer, the Vega's immediate visual impact was of an object that should fly. Perfectly sleek, with no disconcerting tangle of wires and struts, it obviously belonged off the ground.

This was to be Jack Northrop's signature, one that made his creations stand out from those of his rivals even when they were not commercially successful. His design talents matured just at the moment when materials and methods began to be readily available for translating his intuitive vision of airworthiness into physical reality. All Northrop aircraft would be beautiful to behold, perhaps because they seemed to express natural speed and grace rather than a desperate mechanical compromise. The beauty of the Wright brothers' *Flyer* stemmed from its sheer simplicity, a gauzy combination of power and lightness that did not contradict itself. Northrop's Vega was the heir to their aesthetics, setting a new standard for what flying machines should be.

Indeed, some of its features seemed altogether too futuristic to Loughead, who was worried about sales after the S-1 debacle. "Allan kept insisting that we must put some brace struts on, whether they had anything to do or not, because he felt that nobody would buy the airplane unless there was something that could be seen to hold the wing up," Northrop said. "I finally won out." Here was the magical power of engineering at work, an ability to shock, awe, and sometimes frighten the uninformed. The Vega's wings looked like they would fall off without struts, but they did not. In a few more years, the key to such power would pass out of the hands of autodidacts like Jack Northrop, into the sole grasp of a new elite distinguished by university degrees. Northrop himself would soon come to depend on them to substantiate his intuition, for better or worse.

THE FACT THAT the Vega's birth coincided with public euphoria over aviation spawned by Charles Lindbergh's transatlantic solo in May 1927 was of direct consequence to sales that would eventually reach 129 complete aircraft. The Lindbergh celebrity phenomenon is one of the benchmarks of American popular culture, demonstrating for the first time the mass media's power—through saturation coverage by print, cinema, and radio—to turn ordinary mortals into demigods. A shy farm boy from Little Falls, Minnesota, was thus transformed at the age of twenty-five from an airmail pilot into a national icon. The thrust of his fame contributed directly to the explosion of aviation stocks on Wall Street, creating a boom time for all things aeronautical.

For many years, Lindbergh tried to escape the grotesqueries of his public role. When he finally joined movements outside of aeronautics that were congruent with his own instincts—the noninterventionist committees before Pearl Harbor and the conservationist groups of the 1960s—the same public that had worshipped him as a hero dismissed him as a traitor or Luddite. In the late 1920s, however, he could do no wrong, and the aviation industry happily rode his coattails.

His thirty-three-and-a-half-hour nonstop flight from Roosevelt Field on Long Island to Le Bourget Aerodrome near Paris was an ordeal in which fortune figured as prominently as skill, hence the "Lucky Lindy" sobriquet that became a Jazz Age lyric. The airplane he flew was built in two months by the Ryan Aeronautical Company, a San Diego firm, with the backing of nine St. Louis businessmen (Lindbergh had initiated airmail service between St. Louis and Chicago in April 1926). Ryan (unconnected to wartime production official John Ryan) had once owned the Douglas *Cloudster,* extensively modified for passenger service between San Diego and Los Angeles. The aircraft custom-made for Lindbergh was an unremarkable monoplane based on an existing mail transport design whose strongest asset was a reliable 220-horsepower engine manufactured by Wright Aeronautical Corporation, Wright-Martin's successor. Aviation was still a very small circle.

The *Spirit of St. Louis* did not look like an ocean conqueror. Boxy and appearing squashed by its thick wing, it was an ugly duckling compared to Lockheed's sleek Vega. Considering the magnitude of its mission, it was also a tiny aircraft, with a wingspan of forty-six feet and empty weight of just over a ton, beefed up structurally to withstand the stress of flying with a heavy burden of fuel. Indeed, the greatest engineering challenge was loading the plane with enough fuel and then getting this mass off the ground on a controlled course. More than half of the plane's 5,250-pound gross weight at takeoff consisted of gas and oil. The interior was a claustrophobic cell that could barely contain Lindbergh's long limbs. The pilot, who had never made a major water crossing, sat behind a fuel tank that completely blocked his forward view, necessitating the use of a three-by-five-inch periscope mirror. This was a stunt machine, not an aircraft for commerce or war.

The plane carried no radio transmitter or receiver, because of the neanderthal weight of such gear in those days. And Lindbergh decided to rely on dead reckoning instead of a sextant to find his way across the Atlantic, because he doubted the prospect of steady sightings while holding the plane in level flight. Two fishhooks and some twine were the core of his survival kit. Thus, after an understandably sleepless night, when he bounced the *Spirit* up out of the muck of rain-soaked Long Island on the morning of May 20 and headed for Newfoundland, he was perhaps more closely akin to men who floated over Niagara Falls in carefully constructed barrels than to any avatar of advanced technology.

Of course, flying alone across the Atlantic required more stamina than falling off a cliff. Fighting hallucinatory drowsiness, wrestling with the plane's instability, fretting about inexact navigation, he depended on his natural store of perseverance as much as on the equipment around him. It was this aspect of the harrowing flight— the perceived importance of Lindbergh's character in union with the machine that bore him—that enthralled the public. The young pilot not only possessed charisma but exuded a genuine integrity that people craved. After 1,887 miles of brutish weather and physical deprivation, he passed over Ireland's west coast just three miles off course.

"During my entire life I've accepted these gifts of God to men, and not known what was mine until this moment," he wrote. "I know how the dead would feel to live again."

Lindbergh might have still gained fame if he had done the prudent thing and landed in Ireland, but he would not have won the $25,000 prize offered in 1919 by New York hotelier Raymond Orteig for the first nonstop flight between New York and France. So he pressed on. When 100,000 Parisians greeted him at Le Bourget, as delirious as the mobs that used to tear apart the flying machines of fallen bird-men, he was unaware of how his life had changed. "Do you know this Paris hotel; I understand it's quite reasonable?" he asked some-one in the hangar about a name he had brought with him on a scrap of paper.

It was the last time he would ever worry about finding a place to sleep. That night, he was the ambassador's guest in the American embassy. President Coolidge fetched him and his airplane home on a Navy cruiser, then awarded him the Distinguished Flying Cross (the medal's first bestowal). The Congressional Medal of Honor followed. Four million people lined lower Broadway in Manhattan to shower him with ticker tape, sections of which no doubt already recorded the rise in airplane industry stocks.

With financing from Harry Guggenheim, Lindbergh soon launched a nationwide tour to promote aeronautics, covering some 22,000 miles in an impressive demonstration of on-schedule flying that had obvious commercial implications. In December 1927, he flew to Mexico City at the invitation of Dwight Morrow, now the U.S. ambassador there. During his stay, he met Morrow's daughter Anne, whom he would marry in May 1929. Their romance sent the press into crazed combat against the couple's desire for normal privacy, gradually blackening Lindbergh's whole outlook on American life.

For years following the transatlantic solo, at least until the kidnapping and murder of his first-born son in 1932 (when newspaper photographers broke into the morgue to snap pictures of the baby's corpse), Lindbergh was an eager patron saint for the aviation business. Clamored after by commercial interests to lend his name to airline development, he managed to maintain his poise and authority. He would eventually flee with his family to Europe—"refugees from the tyranny of yellow journalism," as Walter Lippmann wrote—but for a crucial period he was the living armature for an entire industry. The Lockheed Vega was just one beneficiary.

IMMEDIATELY AFTER the Lindbergh flight, publishing magnate George Hearst, eldest son of William Randolph Hearst, decided to buy a copy of Lindbergh's plane to enter in an Oakland-to-Honolulu race being promoted by the Dole Pineapple Company. Hearst requisitioned a reporter from the *Los Angeles Examiner* for this task, who quickly learned that the *Spirit of St. Louis* could not be duplicated by Ryan in time for the August 1927 contest. The reporter then heard about the new Lockheed project, and advised Hearst to buy a Vega, powered by the same Wright engine type that Lindbergh had used, before the first one was even finished. The sale price was $12,000—far below cost—but the company was so confident and so elated with the Hearst association that it decided to swallow the heavy loss.

On July 4, 1927, in an Inglewood hayfield now part of Los Angeles International Airport, the Vega, christened *Golden Eagle*, took its first flight. Six weeks later, a two-man crew—one a wartime student of Kenneth Jay, the other a former Douglas engineer—embarked from Oakland for Hawaii against seven competing aircraft. Only two completed the course. Two cracked up while taking off, a common fate for overloaded racers. Two more were forced to return with mechanical problems. And two, including Lockheed's bright orange Vega, were never seen again. Oddly enough, the maelstrom of black publicity that descended upon Dole and the contest's other organizers did not harm the Lockheed Company. The Navy, under orders from President Coolidge, dispatched its Pacific fleet on an air-sea rescue mission of unprecedented scope for the missing crews, but found nothing. Consistent reports from other pilots who had lost their sense of the horizon after dark and drifted into deadly "graveyard spiral" tailspins (there was no instrumentation to help prevent this phenomenon) served to dampen questions about mechanical failures, for which there was no physical evidence anyway. Lindbergh had indeed been lucky.

Though the Hearsts subsequently cancelled an order for a seaplane version of the Vega, luck delivered to Allan Loughead another customer who was not fazed by the Dole disaster. Arctic explorer George Wilkins, who had spotted the *Golden Eagle* from the window of a San Francisco hotel during one of its test flights ("It gave me the thrill that another might experience if he saw his ideal woman in the flesh," he wrote), sought out Loughead and Northrop to build a second Vega for a flight over the North Pole. This time they set the price at $15,000 and promised delivery for January 1928.

In April, after a month of preparations that included hiring Eskimos to dig a mile-long fourteen-foot-wide runway out of the snow, Wilkins and a veteran Alaska bush pilot named Ben Eielson took his new plane on the first trans-Arctic flight—a twenty-hour ordeal from Barrow, Alaska, to Spitzbergen, Norway. They did not find an adoring mob of thousands at the end, however, only a blizzard that took four days to blow out while they camped next to the Vega. Their triumph, which earned the Australian Wilkins a knighthood and Eielson a Distinguished Flying Cross, erased the memory of Lockheed's earlier tragedy and established the Vega as one of the glamour planes of the era.

Orders began to pour in from sportsmen, corporations, and nascent passenger airlines. In the summer of 1928, a New York dealer placed a $250,000-plus order for twenty Vegas. That winter, 150,000 shares of no-par-value Lockheed stock went on the market at ten dollars a share and climbed to $115. The original group of private holders, especially Keeler, reaped a spectacular windfall. It was an honest fortune, as well, based on Jack Northrop's raw talent, Allan Loughead's incorrigible optimism, and a bit of Lindbergh-style panache.

Yet in the middle of this long-sought success, Northrop quit the company. In later years, he would blame one of Allan Loughead's old cronies from Chicago who began to dictate alterations in the Vega design to augment sales. Neither Loughead nor Fred Keeler, the real boss and a canny businessman, was interested in developing new aircraft while the Vega and its variants were so popular (the company built twenty-nine Vegas in 1928 and sixty in 1929). This

made good business sense, given the incalculable risks of experimentation, but it did nothing for Jack Northrop's sense of himself as
a visionary engineer. Keeler was rolling virtually none of his profits
back into the company, content for the moment to squeeze as much
cash out of the Vega bonanza as possible. Even Jay was alienated by
this management attitude and left with Northrop to help raise funds
for their own company.

If Keeler was insensitive to the ego of his most talented employee,
Allan Loughead should not have been. The Vega was the jackpot
he had waited nearly two decades for, but his love of aviation had
long ago shown itself to run deeper than one airplane. Loughead
—and perhaps Keeler, too, after several doses of Loughead's
salesmanship—might have continued to nurture Jack Northrop if
his next goal had sounded reasonable. But Northrop did not want to
develop just any new airplane. He wanted to create a "flying wing,"
a radical concept that would have looked almost as bizarre to his
contemporaries as the Wrights' first machine had seemed in 1903. If
Loughead had misgivings at first about the marketability of the
Vega's unbraced wing, he must have cringed at the notion of a plane
with neither fuselage nor tail.

The flying wing epitomized Northrop's infatuation with "cleanliness." His instinct for streamlining had already produced tangible
improvements in the performance of conventional aircraft, at a time
when an exceptional engineer without university training could still
advance the state of the art. But the quest for clean aerodynamic
forms would become the driving obsession of the rest of his career,
leading him into treacherous technical territory and ultimately to
personal ruin. Along the way he would latch upon whoever could
help him get one step closer, especially if they possessed the educational credentials he lacked. Just who used whom is debatable,
but the flying wing would materialize twice during the next sixty
years as one of the great technological grotesqueries of the century.

FROM MID-1927, when Lindbergh mania was at its peak, to the end
of 1929, when the stock market fell apart, the rising level of investment in aviation reached tidal wave proportions. Such remnants of
postwar pacifism as the Kellogg-Briand Pact of 1928, which renounced war as an instrument of national policy, had no braking
effect on Wall Street's enthusiasm for the warplane industry. During
the last months of frenzy before the Crash, aviation stocks reached

a value of about $1 billion on the New York Stock Exchange, based on earnings of perhaps $9 million. Every company, large or small, was swept up by this manic economy, causing permanent changes in the business.

When the bankers and brokers scrutinized suddenly fashionable companies like Lockheed, Douglas, or Martin, they found firms owned largely by management, with negligible debt (because few people had been willing to lend them any money), that earned their capital several times over every year. Once Lindbergh had broken the apprehension surrounding not just the act of flying but the business behind the act, investors recognized a company like Douglas for what it was: a splendid little cash cow.

"Anything was a cinch for financing," Douglas recalled, "and [the Douglas Company] was good. We had a well-established business, a good record over a number of years—we were making money." Since 1922, all but $40,000 of the company's profits had been plowed back into the business, leaving it with a net worth of $2.53 million by the fall of 1928, about $1.4 million of which was in cash and marketable securities. Net earnings had grown from just over $9,000 in 1924 to $415,000 in 1928. "Who dares to say Romance is dead?" asked a florid Douglas publicity brochure early in 1929. "One would be hard put, indeed, to find among the musty chronicles of legendary heroes a more stirring story of accomplishment than is hidden among the cold, unimaginative figures of the annual financial statement of the Douglas Aircraft Company, Inc."

When quantity orders had first begun to flow in from the Army and Navy, Bill Henry, the *Los Angeles Times* reporter to whom Douglas had given a one-quarter interest, became "frightened" by the fiscal responsibility, according to Douglas, and agreed to let Douglas's father buy out his share for $5,000. Donald and William Douglas, who borrowed the $5,000 from the New York bank that employed him, were thus sole owners in 1928 when they decided to ride the Wall Street bull.

First, Douglas and his father sold the company to a syndicate organized by the brokerage firms of Bancamerica Blair and E. A. Pierce, each already involved in complex webs of aviation financing. A new company was then incorporated in Delaware on November 30 with an authorized capital stock of a million shares. But only 300,000 were initially issued. The Douglases took 200,000 in exchange for the assets of their old company and sold the remainder to the syndicate for a million dollars. At the market's peak in 1929, when a public offering of 60,000 more shares was made, the stock rose to forty-five dollars a share—more than quadrupling the paper wealth of the original insiders.

"Was that warranted by the business you were doing?" a House Naval Affairs Committee member later asked a Douglas company official.

"It was not," he answered.

"It was just pure speculation?"

"I think so."

"And manipulation?"

"Yes, sir, I would think so."

Before federal regulations wiped such schemes out, it was common for the brokerage houses to offer their insider holdings to a "preferred list" of bank officers, corporate directors, and special friends before arranging a public sale. When the Senate Banking and Currency Committee investigated this practice in 1933, it found a "well-conceived plan to excite public interest in the stock so that when it was listed on a public exchange the individuals on the preferred list would be in position to realize a substantial profit." Profits in the Douglas case, while substantial by any measure, were relatively modest compared to some other aviation deals during this breathless period. In one case, the son of Edward Deeds watched a forty-dollar insider investment turn into $5.55 million before the Crash. (In 1926 he bought 200 shares of the new Pratt & Whitney engine company for twenty cents a share; in 1928 a stock dividend turned this into 16,000 shares; in 1929, these were exchanged for 34,720 shares of United Air and Transport, a holding company, which peaked at $160 a share.)

Of the million dollars that Douglas received from the syndicate, $650,000 was placed in the company's cash account. The rest was used to build a new plant in Santa Monica. In the nine years since he had left Glenn Martin in Cleveland and returned to California with several thousand dollars and vague ideas about building transport planes, Donald Douglas had grown a multimillion-dollar enterprise for which the Great Depression would be just a minor annoyance. *Fortune* magazine quipped that Douglas had enjoyed "an aeronautical education for which the U.S. had put up some $19,000,000."

To his family, Douglas was known not to want the company to go public, not to want it to grow much larger than the congenial engineering shop that had built the *Cloudster*. This sentiment was typical of the more technically minded pioneers like Douglas and Northrop, as opposed to mercantilists like Glenn Martin. "After we had to hire the second lawyer, we began to have trouble," Douglas would joke in later years. But neither he nor anyone else could return to the days of cottage industry. Washington and Wall Street would not let them now.

Douglas's nouveau riche neighbor in Burbank, the Lockheed Aircraft Company, also rode the crest of the investment wave. The company was turning out Vegas at a fast clip—twenty-nine in 1928, forty in the first half of 1929—and on May 1, 1929, signed a contract with a millionaire businessman-aviator in Detroit, Edward Schlee, to deliver sixteen planes a month for his distributorship. This deal caught the attention of a group of Detroit investors closely associated with the auto industry who had created a pyramided holding company called Detroit Aircraft, billing themselves as "the General Motors of the Air."

Chairman of the board was Harold H. Emmons, a lawyer who had worked with Howard Coffin during the war as head of motor production for both the Army and Navy. Since the war, Emmons had been an integral member of the clique of Detroit businessmen, including Edward Deeds, who maintained financial stakes in aeronautics. Detroit Aircraft's directors included Charles Kettering and Charles Mott of GM, William Mayo of Ford, Roy Chapin of Hudson, and Ransom Olds. It already controlled seven aviation companies—among them the Ryan Company, builders of the Lindbergh plane—and Detroit's Grosse Isle airport. Capitalization was supposedly $20 million, though the first stock issue had consisted of 2 million shares of no par value.

Lockheed was obviously an attractive target, because of the lucrative Vega and the firm's familial connections through Malcolm Loughead to the auto industry. On May 13, Lockheed board members met to discuss Detroit Aircraft's proposition of a buy-out. For Fred Keeler, who held the controlling interest, there was nothing to debate. Lockheed was clearly at a high-water mark, along with the general economy, and a sale now would bring maximum return on his original investment. From his perspective, the Detroit group was just what a relatively small company needed—deep pockets to support the kind of development work that Jack Northrop had craved.

For Allan Loughead, however, Detroit Aircraft and Harold Emmons et al. were ghosts from the scandals of World War I. Emmons had even co-authored an Army report soon after the Armistice declaring that there were no American airplane builders "who even remotely comprehended the progress that had been made in Europe and were competent to design a complete fighting machine." This was the self-serving rationale behind the auto industry's capture of wartime aircraft production. And so Allan argued hotly against the takeover, predicting that "nothing will come of this Detroit business except grief."

These quaint sentiments meant nothing to Keeler. Allan Loughead was quickly outvoted on the issue at hand. After a two-week hunting trip with a friend from the old days in Santa Barbara who had been one of the pilots from the ill-fated F-1A transcontinental flight, he sent the following telegram to Keeler:

"Accept my resignation to take effect at once. I am sure you realize that I have been fairly dissatisfied with the management of the company for months past. I can no longer remain in false position of condoning mistakes in general policy and operation of company." Loughead had never been a sentimentalist in business affairs, but the company did bear his name, and he was not willing to watch it be absorbed by strangers from Detroit.

At the end of July, the sale was finalized. Detroit Aircraft acquired 87 percent of Lockheed's assets and reorganized the company as a subsidiary. Most of Lockheed's shop workers and technical staff happily accepted the takeover as proof that the company had matured into a respectable concern. Operationally, the Burbank division would be independent and would continue building Vegas and follow-on aircraft, for which there was a lively market. Their optimism seemed justified when Lindbergh ordered a new $22,875 long-distance racer called the *Sirius* in September. He asked for a two-cockpit arrangement to accommodate his bride, Anne Morrow, daughter of the chairman of the 1925 President's Aircraft Board. No one, of course, foresaw what would happen to the economy one month later—except perhaps Fred Keeler, who had made the sweetest deal of his life.

GLENN MARTIN'S BUSINESS fortunes bottomed out in 1924, when he went $96,095 into the red—his first serious loss since 1912. By low-balling Curtiss on the production bid for Navy SC torpedo planes, he put himself back in the black in 1925, turning a profit that year of $297,200 before taxes. The SC work enabled the company to develop improved versions during the next three years, resulting in orders for 223 planes that became standard equipment in the Navy's new carrier-based squadrons. Profits reached the 20 percent level—$499,571 in 1927 and $406,201 in 1928, on sales of $2.64 million and $2.49 million, respectively. "These are very substantial earnings on the Company's capitalization" a Martin brochure proudly noted.

In January 1928, reflecting the high-spirited optimism of the day, Glenn Martin was named one of Cleveland's "thirteen most eligible bachelors" by the *Cleveland Press*. "Sometimes a man goes through life without finding the right girl," he observed. "Most bachelors have justification for remaining single. It happens that I have never proposed."

Though his mother was still his only known social companion, he purchased a splendiferous Stutz "Biarritz" automobile with custom-made snakeskin trim and upholstery. One can only guess the effect on Clevelanders of seeing this odd fellow and his imperious mother cruising through town in such an outrageous car. But it appealed to his Hollywood sense of style and also helped keep him in good standing with one branch of the old-boy network. The president of Stutz was Edgar S. Gorrell, former member of the Bolling Commission and a key officer in the Army's wartime purchasing system who had advocated the role of bombers.

Though he resisted the pressure of most-eligible-bachelorhood, Martin easily succumbed to the merger and reorganizaton fever sweeping the aircraft industry. Since 1926, when his Navy orders solidified, he had considered moving to a larger factory that would be more suitable for the development of seaplanes. By 1927, the Navy was urging him to relocate on the Eastern Seaboard, in order to save the cost of shipping planes to coastal bases. When Cleveland officials balked at buying land for an airport that could accommodate a new Martin plant, he began negotiating with the government of Baltimore, Maryland.

"Glenn L. Martin was driven out of Cleveland because the bankers thought he was a screwball," said an unusually candid Cleveland industrialist. Whether it was his snakeskin-trimmed Biarritz, his live-in mother, or his futuristic faith in airplanes that led to such ostracism was not revealed.

Like many American urban hubs, Baltimore was actively studying sites for a municipal airport, spurred by the Lindbergh boom. In the late 1920s, it was still one of the most charming cities in America, blessed with a perfect harbor on the Chesapeake Bay, thriving blue-collar neighborhoods not quite as hardened by prejudice as towns farther South, middle-class townhouses along tree-lined cobblestone streets, and mansions near Johns Hopkins University owned by those who directed a rich commerce in steel, oil, railroads, and shipping. Baltimore was very much a culture unto itself, with a cuisine drawn from the Bay, a distinct linguistic accent on the lips of natives, and the hallowed *Sun* paper of H. L. Mencken. It was also a tight little fiefdom for its politicians and the merchants or bankers behind them, who were used to getting their way about whatever they thought best for the people of Maryland.

The voters had already approved a $1.5 million bond issue to finance the airport project but were uninformed about attendant cost escalations. Existing plans were discarded when the Democratic mayor who had initiated them was succeeded by a Republican with a different set of political patrons. Through banking circles, Glenn Martin was thus able to interject a proposal to build a large new aircraft factory at a location that had not been among the fourteen previously considered. Situated along the Patapsco River, a Chesapeake estuary with Baltimore at one end and the vast Sparrows Point steel mill at its mouth, the site was ideal for launching and testing military seaplanes. The fact that most of the land that would become a public airport was under eight to twenty feet of Bay water was somehow deemed acceptable by the new mayor.

By the summer of 1928, Martin's relocation had become a full-blown political battle in the state, between powerful urban interests who would profit directly if the airport site were developed and competing interests who would not. Cost estimates for building the airport reached $6 million, though the city possessed only its $1.5 million loan and $700,000 from a pro-business Public Improvement Commission. Surveys showed that construction on the marshy Patapsco shore would require 22 million cubic yards of fill dirt, a $1.3 million bulkhead some 13,000 feet long, and a $300,000 channel for the safe passage of aircraft carriers and freighters. "People think the airport is being built for Glenn Martin and is not to be a municipal airport," asserted one perceptive state senator.

When the public learned that Martin was to acquire a hundred acres of the thousand-acre site for just $50,000, even some local financiers hesitated, if only because of the real estate precedent such a deal might establish. "Ignoring questions of legality, can you and will you explain, in simple language for the benefit of the taxpayer, why the city is justified in selling Glenn Martin one-tenth of all the proposed airport property for a price equal to one-seventieth of its minimum cost, and one-one hundredth of its possible cost to the city?" a prominent investment banker asked the mayor in a letter leaked to the *Sun*. City officials acknowledged this as a subsidy, comparing it to tax exemptions granted to the Baltimore & Ohio Railroad. But editorialists condemned it as a "direct capital contribution."

While the debate grew more acrimonious in Baltimore, Martin arranged to sell his Cleveland plant. In October 1928, with the last big order for Navy planes near completion, a group of Chicago-based bankers and auto executives organized the Great Lakes Aircraft Corporation in Delaware to buy Martin's factory and production licenses for a million dollars in cash. At the end of November—after rejecting an offer from Baltimore to rent land at the new airport instead of buy—he formed a partnership with Louis Chevrolet, who had been part of the auto racing circle that included Caleb Bragg.

Like Bragg, Chevrolet had dallied in aircraft engines. He owned the design for a promising engine, but had few other personal assets besides his fame as a race driver and car builder. Never much interested in business details, he no longer even controlled the commercial rights to his own name, which belonged to General Motors. For $175,000, Martin acquired a 90-percent interest in Chevrolet's small airplane engine company, whose tangible assets at the time were valued at only $12,000. On December 5, 1928, a new Glenn L. Martin Company was incorporated in Maryland with an authorized capital stock of 30,000 no-par shares. Martin took 12,500 of these shares, valued at $1.25 million, for himself as the only stockholder.

On the surface, Martin simply appeared to be taking over Chevrolet's engine business. To this end, he would soon have E. S. Gorrell sign and deliver to the appropriate Army general a letter (composed by Martin) urging favorable consideration of Chevrolet's latest engine. But the fact that the auto inventor immediately became vice president and general manager of the Martin Company, a position for which he was qualified neither by experience nor temperament, made it more likely that Martin was anxious to have an illustrious name on his side for the next gambit. He could not afford to let his "screwball" reputation accompany him to Baltimore.

Early in January 1929, Martin began secretly assembling through dozens of small anonymous purchases a 1,200-acre tract along another Chesapeake inlet near Baltimore, called Middle River. That summer, as construction of a new factory there got under way, he signed a contract with the Baltimore Trust Company, joined by Guaranty Trust and Otis & Company of New York, to buy a portion of his capital stock for $4 million. The contract contained an "out" clause in case of a market crash, indicating how nervous some insiders were becoming about the aviation boom's wild gyrations. On July 1, he secured a $745,253 bid to build nine $50,000 flying boats and an $85,000 experimental torpedo plane for the Navy, but the cash would not be in hand for at least another year. In fact, the company suffered a loss totaling $335,102 on these projects.

Since 1916, Glenn Martin had learned how to manipulate high-stakes financing on his own, but this was not the kind of money needed to pay for the ambitious facility on Middle River, where initial investment alone was estimated at $2.5 million. It was a deluxe edifice by the standards of the day, with individual electric drives for each tool instead of a noisy, dangerous overhead belt system, and floors constructed of expensive wooden blocks. In September, Martin amended the company's charter to increase the authorized capital stock from 30,000 to 1 million no-par shares. The number of outstanding shares—all held personally by Glenn Martin—increased from 12,500 to 475,000 without changing their total value. As in many such watering schemes of the late 1920s, Martin announced plans for a public stock offering but delayed setting a date for the sale. The *Wall Street Journal* reported that Otis & Company would offer 100,000 shares at twenty-four dollars each. Given the giddy excess on Wall Street, there was no reason to believe Martin could not find more than enough takers.

On October 7—a golden Indian Summer day three weeks before the Crash—he and some 250 employees moved into a 298,000-square-foot plant accurately described as the most modern airplane manufacturing hall in the world. On October 19, the company declared and paid a dividend (to Glenn Martin, the only shareholder) of 50,000 shares valued at $250,000, thus bringing his total to 525,000 shares at $1.5 million. He was poised for a killing on the market, if only it would stay put.

Of course, it did not. Within days, some $40 billion worth of investor wealth evaporated, destroying the American economy in one stroke. On November 1, with Wall Street in chaos, Martin turned frantically to his banks and sold $3 million worth of five-year

6 percent convertible gold notes for 91 percent of face value, thus raising $2.73 million in quick cash to save his new plant. The banks forced him to mortgage the entire facility and everything inside, making it unlikely that anyone else would consider him an attractive investment until the notes were retired.

In 1929, the company sold two airplanes. Spare parts accounted for most of that year's gross sales of $625,000. Only in the rarefied mindless binge of the Roaring Twenties could anyone have moved so far, so quickly, with so little.

That Glenn Martin had less than a year's worth of production work in house, had lost $4 million in private financing, and seen his plans for a public stock offering vaporize overnight did not perturb the jubilant officials of Baltimore. They now presided over one of the military aviation capitals of the world.

CHAPTER SIX

DEPRESSION

THE 1930S BEGAN on a note of economic catastrophe and ended with another world war. There is no overstating the consequences of these years on the rest of the American century. Out of the Great Depression came an end to unregulated capitalism and a new consensus about what the government should do for the welfare of individuals. Out of the war came superpower status and a fear of one particular international rival that at times grew into almost a clinical obsession. Fifty years later, these developments are still playing themselves out, while the 1920s have receded into antiquity.

At the start of this period, the names Martin, Douglas, Lockheed, and Northrop were still known mainly by connoisseurs of aviation or finance. But the Second World War and its political aftermath would push them into a limelight all their own, where respect for their tremendous economic power would mix with uneasiness about their merchandise. They would be joined and sometimes overshadowed by new companies, all the while assuming a kind of patriarchal status as links to a relatively innocent past.

By the beginning of the 1930s, the foundation of their industry was solidly military. This status coincided with the War Department's emerging recognition, signified by the appearance in 1930 of the first official plan for a totally mobilized economy, that the armed services were dependent on civilian sources for hardware and that war could not be waged at a rate that overloaded the capacity of those sources. This may seem obvious today, but many professional soldiers still believed that strategy was the principal guiding force in war, not supply. Though the plan did not move much beyond the organizational framework of World War I, the fact that the War Department wrote it meant that the old theory was dead.

Only about one-third of the several thousand airplanes being built every year in the United States were owned by the Army and Navy, but they accounted for two-thirds of the total dollar value. It would soon be virtually impossible to develop a major new design of any type without support somewhere along the line from the armed services. The Crash coincided with the start of an evolutionary leap in technology that greatly increased the cost and complexity of airplanes, especially military designs that emphasized high performance. The transition from wood to tubular metal construction carried a 48 percent increase in engineering costs alone. Monocoque fuselages added another 50 percent. Special alloys, instrumentation, and high-altitude environmental gear piled on more expenses. Between 1928 and 1937, the average price for military planes rose from $15,641 to $39,063. Manufacturers were forced to seek huge amounts of capital, setting back the likelihood that a company could thrive on civilian commerce alone. What business there was became concentrated in fewer, larger firms.

In this regard, the most dramatic transformation of the post-1929 era was Lockheed's. As of December 1928, the company had 138,233 shares of capital stock outstanding, which represented a cash investment of $612,900. After the initial stock market rout, it continued to sell Vegas and derivatives, building a total of forty-five aircraft in 1930—fourteen more than in 1928, but twenty-three fewer than in 1929. They were popular across the spectrum of commercial use, from racers flown by celebrity pilots like Amelia Earhart to workhorse transports in the U.S. and abroad. For a while, Lockheed was the most illustrious example of a small company that had hit the jackpot with an innovative product without turning to the military. This would rarely happen again.

Early in 1930, Lockheed suffered its first Depression blow when the Detroit-based dealership that had contracted to take sixteen Vegas a month defaulted and went into bankruptcy. During March, at a time when a production rate of ten aircraft per month was necessary for profitable operations, Lockheed's income dropped to just $1,500, all from the sale of spare parts. Through September, thirty-four planes were sold, but the company's books showed an alarming loss of $264,000.

Very little of this red ink had been spilled in Burbank, however. Lockheed's parent, Detroit Aircraft, had seen $733,000 of its assets evaporate in the Crash. It had then failed to win a Navy dirigible contract that management was banking on. Most of its divisions were floundering under sales drop-offs even more precipitous than Lockheed's. (Civilian aircraft production dropped from about 5,500 in 1929 to just under 2,000 in 1931, while military rose from about 675 to slightly over 800.) Detroit Aircraft stock that had come onto the pre-Crash market with a par value of fifteen dollars a share would soon level off at twelve and a half cents. In order to keep the organization afloat, the directors decided to leech Lockheed's relatively steady income, mostly to cover salaries and administrative overhead. The Vega was a warm body from which they sucked precious drops of blood.

This myopic tactic merely forestalled the inevitable, as production in Burbank sank below 1928 figures. The corporation's last hope was an Army contract for pursuit (fighter) planes evolved from a Sirius descendant called the "Altair." At the time, however, such low-wing monoplanes represented a radical change in Air Corps buying habits, despite the well-known speed of Lockheed and other types. The Depression, furthermore, was flattening the service's budget for experimental prototypes. Executives in Detroit therefore decided to try a shortcut into the center of the military's procurement maze.

Through the Detroit headquarters, Lockheed managers made contact with an insurance underwriter in New York known as "Colonel Boots," a former Army officer who represented himself as having acquaintances in military circles that could produce business for the company. To the Detroit coterie of World War I procurement czars, this must have seemed a perfectly normal route. Boots was thus given an off-the-books contract and a generous personal expense account. After several months of negotiating with his shadowy sources, the colonel made good by securing one of the Army's coveted experimental contracts. Burbank engineers built a mock-up in March 1931. Assembly of a prototype then took place in Detroit, followed by initial testing that summer.

When the plane, designated XP-900, was handed over to the Army's Dayton, Ohio, test field at the end of September for acceptance trials, it quickly proved to be faster than the Air Corps' current front-line fighters. But Lockheed was an upstart in the tight military market and encountered resistance to the promising new machine from an entrenched ring of contractors and politicians, despite Boots's apparent bribery of Army officials. Detroit Aircraft, moreover, was accurately perceived as a firm on the verge of bankruptcy. The Army's dilemma—a strong aircraft from an economically and politically weak manufacturer—was eliminated during a flight test on October 19, when the plane's landing gear jammed. The pilot was ordered by then-Colonel Hap Arnold to bail out, even though he could have made a safe belly landing. The XP-900 was totally destroyed, along with Detroit Aircraft's chances for salvation.

There is no record of whether Arnold knew about "Colonel Boots." The XP-900 incident is missing from his otherwise detailed autobiography, which notes only that he was relieved from duty at Dayton "in the fall of 1931" and ordered to take command of a base in California on November 26. But the Air Corps was still such a small fraternity that it seems unlikely he could have been ignorant of the affair. An avid proponent of the latest technology who would not be expected to destroy nonchalantly an impressive new machine, Arnold may have acted on the orders of the general in charge of the Army's Matériel Division, who was on the scene that day in October. In any case, the line to Boots was erased.

With a net deficit of $803,127 (of which $800,570 was attributable to Detroit expenses), Lockheed appeared too depleted by this time to cajole the Army or anyone else into building a replacement prototype. On October 24, five days after losing the XP-900, Detroit Aircraft's directors voted to go into receivership. Any remaining cards toppled in November when the Army exercised its prerogative to transfer drawings and production equipment for the fighter to another manufacturer. On Christmas Eve 1931, Lockheed's general manager took a loan against his house and handed out ten-dollar bills as the few remaining Burbank employees went home. Early in 1932, Detroit Aircraft ceased to exist. Allan Loughead's warning had proved to be precisely correct.

THE POISONOUS AFFILIATION of Lockheed with Detroit Aircraft was not substantially different from countless other Depression-era business debacles. What would save the company from oblivion was the association of its name with aviation celebrities like Lindbergh and Earhart who flew custom-made products and, at least for a short time, the demand for spare parts from these influential customers. In this respect Lockheed was rather unusual among the few aircraft companies that survived the Depression, because it carried through on the strength of its reputation in the civilian market, not by becoming a manufacturing branch of the federal government. That would come soon enough, but not just yet.

Lockheed's Vega-derived product line was of special concern to a small California airline company whose name now sounds marvelously picturesque: Varney Speed Lanes. Its flamboyant owner, Walter T. Varney, had bought six $25,000 Orion models, which were passenger versions of Lindbergh's *Sirius*. Varney pitched his service as the fastest link between Los Angeles and San Francisco—he gave his passengers a shiny dime, just like John D. Rockefeller, for each minute a flight was late—and he desperately needed the 200-mile-per-hour Orions to stay ahead. But Lockheed had closed its doors immediately after finishing his sixth plane.

In May 1932, Lockheed's court-appointed receiver placed a notice in Los Angeles newspapers stating that the company's remaining assets—mostly production tooling and raw materials—were appraised at $130,000 and that sealed bids would be taken for a sale of the property on June 6. This was a legal formality; the receiver would hand-pick the new owner beforehand, not wanting to risk a true auction. But it was assumed in local business circles that $60,000 would easily buy the company.

Allan Loughead, who had returned to his real estate ventures and a few unnotable dabblings in aviation after the Detroit Air takeover, thought this figure was an insult and believed he might regain the company for $100,000. He tried to interest an Oklahoma airline owner, but even his vaunted salesmanship could not counter fear of entrepreneurial risk in tenebrous 1932. Varney, however, regarded acquisition of Lockheed as vital to his existence. He laid down $20,000 of his own funds as the magnet for attracting $40,000 from other investors.

Unlike Loughead, Varney was not isolated in seeking scarce financial resources. His Speed Lanes comprised half of a partnership with a respected designer of small commercial planes, Lloyd Stearman, whose successful business in the Midwest had been swallowed by one of the powerful conglomerates of the late 1920s, much as Lockheed had been absorbed by Detroit Air. Stearman-Varney Inc. hoped to manufacture an airliner specialized for markets like Los Angeles–San Francisco. To this end they had developed plans for a ten-passenger all-metal plane that would considerably advance the field. Their key associate in this project was a young venture capitalist named Bobby Gross, the kind of man Allan Loughead always needed but never met.

ROBERT ELLSWORTH GROSS was by birth and education a breed apart from the rough-cut eccentrics who typified the early aviation business—reared in a well-to-do New England mercantile class that ensured its social power by funneling its sons through Harvard and into the great Eastern investment houses. He had been born on May 11, 1897, in Roxbury, Massachusetts, to Robert Haven Gross, whose long career in coal mining culminated as president of the expansive New River Company of Boston and West Virginia; and Mabel Bowman, an independent-minded Yankee who tutored her son at home until he was ten rather than send him to local schools. She was a wrathful and compulsive teacher who forced the left-handed boy to learn how to write with his right hand after he broke his arm in a fight with a neighborhood girl.

Gross once said that he feared his mother for most of his life. But he maintained warm ties with his father and turned to him continuously for business advice or financial help at crucial junctures. The elder Gross had built his own fortune, starting when he hand-packed Quaker Oats cereal in paper cartons and introduced the tasteless mulch to thrifty New England housewives. Robert H. Gross knew how to bargain with cautious strangers, how to use one success to finance something bigger. He was not about to let his son become a homebody who scuffled with girls.

Their formal yet deeply affectionate relationship had its roots in the years when Robert was still not in school, when father would bring the little boy along on business trips to mining interests in the West and South. These were hard excursions—inelegant meetings in railroad cars, small-town hotel rooms, even tents in Montana—intended as an antidote to mother's coddling. They served the purpose well, nurturing a feel for plain talk and handshake deals that later became Robert E. Gross's strongest assets in a business he was not technically trained for.

Gross was never a pilot or an engineer of any stripe. In 1910, at the age of thirteen, he took his first airplane ride with the dashing British birdman Claude Grahame-White during an air show near Boston. The circumstances of this exotic experience were never de-

scribed, but it spoke of a remarkable degree of sink-or-swim abandon on the part of his father—assuming the boy did not run away
to the show by himself. Yet he displayed none of the mechanical
interests that distinguished the teenaged Glenn Martin, Jack
Northrop, and Allan Loughead. The Gross family would have considered garage work the realm of servants. After their rising fortunes
permitted a move to the select suburb of Newton, Robert attended
Newton High School, then spent his senior year at fashionable St.
George's School in Newport, Rhode Island, where redbrick federal
halls stood on a bluff overlooking the ocean. He entered Harvard in
the fall of 1916.

Like so many other privileged young men, Robert Gross arrived at
Harvard because of social imperatives, not academic promise. He
attended lectures in English and economics—typical gentlemen's
fare—and floated through with average marks. He played sports in
which his five-foot-seven-inch stature was not a handicap, captain-
ing the ice hockey and baseball teams. He joined Hasty Pudding and
affected a playboy's passion for foreign automobiles, owning in suc-
cession a Mercedes, Voisin, Panhard-Levassor, and Hispano-Suiza.
These were admired accomplishments in Cambridge, propelling him
to the presidency of the student council.

In 1918, Gross went from ROTC into the Army, where he taught
less fortunate children how to wield the bayonet at a Virginia infantry training camp. His most memorable day in the military was a
dress parade for Secretary of War Newton Baker, when he lost the
reins of his horse upon dismount and spent fifteen minutes chasing
it past unamused dignitaries. After the Armistice, he returned to
baseball at Harvard, where one spring afternoon in the grandstands
his father persuaded a local investment banker to hire the aimless,
though handsome and likable, youth.

Thus in the summer of 1919, Robert E. Gross became an office
boy with the staid firm of Lee, Higginson & Company. He thereby
joined countless numbers of his peers in a profession that offered
lifetime sinecures to the witless or substantial wealth to the ambitious. The 1920s would lead the latter type with special velocity to
the disasters of October 1929.

For the next eight years, Gross worked for Lee, Higginson in
Boston, New York, and London. In October 1920, he married Mary
Bradford Palmer, only daughter of Benjamin Sanborn Palmer, a
partner in the Chase & Sanborn coffee empire. At their wedding
reception, 400 people dined on Timbales of Lobster à la Newburg. In

1925, after Lee, Higginson and Percy Rockefeller acquired a coal company in Colorado and dispatched Robert to find a buyer when it veered toward bankruptcy, he bought it himself for $120,000, then put it in the black with his father's help. Through stock trading in the bull market and the buying and selling of small businesses, he was a millionaire before the age of thirty.

In 1927, Higginson happened to send Gross to evaluate Lloyd Stearman's airplane plant in Wichita, Kansas. When his strong recommendation to back Stearman was turned down by Higginson's conservative directors, he rashly resigned from the firm and used his own $20,000 to buy enough stock to gain a stake in the company's management. The gamble paid off when he arranged for the buyout of Stearman in 1929. He also established with his younger brother, Courtlandt—who had followed Robert's footsteps through Harvard hockey and baseball to Lee, Higginson—the Viking Flying Boat Company in New Haven, a builder of French-designed sport seaplanes, into which they poured some $200,000. Gross was imitating on a small scale the vast, hugely profitable aviation conglomerates of the day. When the Crash decimated his stock portfolio and wiped out Viking's market for luxury aircraft, he fell back on his cushion and waited for sunnier weather.

A telephone call from Lloyd Stearman in 1931 was enough to bring Gross and his family—a daughter, Mary Palmer, had recently been added—to San Francisco, where Robert took a treasurer position with Varney Speed Lanes and put up $25,000 in September to organize Stearman-Varney Inc. As part of the deal, he assumed ownership of an advanced passenger plane design that Stearman hoped would replace Varney's Lockheed Orions. Now, in early 1932, Gross was growing convinced that in spite of the Depression that had drained his personal wealth, the defunct Lockheed plant could be the seed for a lucrative new venture.

"Running as a powerful current through all our plans and operations is the inseparable subject of Lockheed," Robert Gross wrote to Courtlandt, who had remained in New Haven, on January 29, 1932. "We cannot get it out of our minds that somehow or other we must acquire, if not actual, at least working control of Lockheed Aircraft. Although the affairs of the company are so involved that it actually cannot be acquired immediately, still the situation is potentially so interesting, and gives promise of being so valuable that I feel that

you and I must keep each other carefully informed in case an opportunity should come to strike quickly."

Gross's desire to "keep each other carefully informed" was not
based on yearning for his younger brother's advice. He knew that he
might need financial help from his parents to make the Lockheed
deal work. He had already abandoned Depression-crippled Viking
to join Varney and Stearman in San Francisco, and could not afford
to alienate Courtlandt—or, more important, their father—by taking
yet another long-range plunge.

"The company always has operated, and is now operating, at a
profit," Robert continued in his effort to allay Courtlandt's misgivings. "The company has just completed two planes for the Swiss Air
Corps and two planes for the Japanese Army High Speed Bomber
Squad." As for Lockheed's entanglement with Detroit, he gave the
following thumbnail analysis:

"The famous inter-company account between Lockheed and Detroit Aircraft is the big stumbling block which anyone dealing with
Lockheed has first to satisfy. Through the wildest policy of money
matters this account shows that Lockheed owes Detroit $400,000.
We, and the local receiver for Lockheed, feel that this claim can be
reduced by counter items to below $200,000, which, if accom-
plished, would pave the way for an offer to buy the inter-company
account for perhaps twenty cents on the dollar, or around $40,000.
It might be possible to acquire at the same time all of Detroit Aircraft's 89 percent holding of the capital stock of Lockheed for the
same figure, or perhaps a little more. Armed with this Detroit claim
and 89 percent of the stock, a prospective purchaser would be sitting
very much in the driver's seat in making a settlement with the other
local creditors of Lockheed, which amounted on date of receivership
to around $88,000." Gross's lingua franca derived from investment
banking, not stress analysis. He guessed that it would take between
$50,000 and $100,000 to walk away with the company.

He did not see the Stearman-Varney plan and the acquisition of
Lockheed as two unreconcilable efforts. Gauging that it might take
between $100,000 and $200,000 to start building and marketing a
brand-new transport from scratch, he reasoned that if Lockheed's
name, plant, and clientele could be obtained for less, then the advantage was obvious.

Courtlandt was not convinced. "I hope you will appreciate the
fact that our attitude is sympathetic and receptive," he wrote to
Robert on March 16, using the plural pronoun to speak for Viking's

skeleton. "But it is hard for us to form any definite conclusions on the basis of the information which we now have. . . . As I understand it, you wish to buy the Lockheed business because you think that it is profitable, but you also wish to develop a new Stearman plane which would be an improved Lockheed, and you also wish to carry on a transportation business. If you have faith in the new Stearman, what would be the advantage of buying Lockheed?" These were reasonable questions.

But Robert had already made up his mind, dazzled by the gleam of a Lockheed reputation that was brighter on the optimistic West Coast than in the pessimistic East. "It is not a question of our desiring to go into the manufacturing," he wrote to Courtlandt at the end of April. "In fact, we [Stearman-Varney Inc.] have no such desire and would much prefer not to, but it is an absolute necessity to us that we have airplanes and, as the situation is today, if we should lose a ship or two, there is no place that we know of where we could get an airplane. Lockheed has been definitely shut down and is going to be sold within the next few days at a minimum price of $40,000, and somehow we have got to have a place where we can buy airplanes." He was therefore "trying to promote a syndicate here to get Lockheed out of the way and form a new company." Viking could be its Eastern representative, he offered, "so do not think that I am laying down on you." His plan was to make Lloyd Stearman president of the new firm, retain several of the old Lockheed workers to maintain the Orion line, and then make his own bid for control.

Within weeks, it was clear to Gross that he was not going to assemble enough financing to call all the shots—at least not initially. "Walter [Varney] is trying to buy Lockheed himself, after which he proposes to turn it over to me to see what I can do with it," he wrote to his brother on May 24. "Naturally I shall not fuss with it unless I am sure it has a future and unless the returns to me are commensurate with the effort I should have to put into it." Even Glenn Martin would not have used the word "fuss" in quite this way. Here was the phraseology of someone who could take the business or leave it, not of a pilot or engineer infatuated with technology.

After an inspection of the idle Burbank plant, where he was relieved to find that crucial production gear had not already been sold off, Gross proceeded to persuade three independent investors from his social orbit to join Walter Varney: Jacqueline Walker, a wealthy sportswoman who put in $5,000 of her inheritance; her husband's cousin, ex-Army pilot Cyril Chappellet, and his wife, Pat, who pledged $10,000 from their family if Cyril would be hired by the new company; and Thomas Fortune Ryan III, a San Francisco financier, who offered $5,000 to further his interest in organizing airlines. At the bottom of the worst economic disaster in American history, the willingness of these individuals to pour money into a dead-horse airplane company underlined how wealth did not disappear during the Depression. Social credentials were vital in tapping it, however, and Robert Gross's background was elemental to his success. Allan Loughead could never have entered this circle.

"The world was right flat on its face, an all time low," Gross later said. "It couldn't go any lower. Aviation could go only one way: up. And the Lockheed company looked like it could go there with it." Gross shared the blind faith of the Martins and Northrops, if not their humble origins.

His personal contribution was ownership of the all-metal transport design, valued at $8,000, for which he would receive 800 shares of the new company. (Other shares were distributed to Walter Varney, the outside investors, Lloyd Stearman, and to one of Gross's old Harvard classmates, a wartime Army flyer who had introduced him to the Walkers and the Chappellets.) The total cash investment was only $40,000, but there were no other ready sources.

Sometime in May 1932, Gross submitted this bid to the Title Insurance and Trust Company of Los Angeles, Lockheed's receiver. On May 23, the receiver sent a letter to Walter Varney in San Francisco, marked "Attn. Ralph E. Gross," confirming the acceptance of the bid and the subsequent courtroom formality. Whether Robert was aware of this when he wrote to Courtlandt on May 24 is unknown, but the process of acquiring Lockheed was obviously farther along than he had been telling his brother.

At 11 A.M. on June 6, in Los Angeles District Court, a judge was handed two sealed envelopes. Among the spectators were Allan Loughead and Jack Northrop, quietly waiting to see what would become of the company that had taken their youth, their names, their earliest ambitions, but whose fate was now beyond their reach. After examining both bids, one of which was clearly a dummy, the judge announced that they were each for $40,000—one from the Varney-Gross syndicate, the other from a salvage crew called Santa Fe Avenue Machine Brokers and Liquidators. Considering that Varney-Gross intended to continue the business, whereas the other concern would merely dispose of its equipment, he ceremoniously awarded Lockheed Aircraft Corporation to Varney.

"It's yours," the judge said as Robert Gross handed him a down-payment check for $10,000. "I hope you know what you're doing, young man."

JACK NORTHROP'S PRESENCE in the courtroom that auspicious June day was not purely sentimental. Since leaving Lockheed at the pinnacle of its 1928 success, he had been twisted by the Depression's unpredictable vortices. He had stayed on his feet, but was still unsettled enough to have seriously considered teaming yet again with Allan Loughead if Lockheed should somehow come their way. It did not, so his fate continued on a different path.

In the spring of 1929, with backing from the Hearst family, Northrop and Kenneth Jay had formed the Avion Corporation in Burbank. Jay was apparently unable or unwilling to apply the same kind of brake to Northrop's creative ego that Keeler and Loughead had managed for a while, because their first project together was a costly indulgence of Northrop's flying wing fantasy. Even Northrop referred to the idea as a "bug biting me." He said he wanted "to build an airplane where there was nothing but the wing, where everything was included in it—powerplant, passengers, every function that was necessary." Only the giddy economy of early 1929, and sponsorship from Hearsts who wanted flashy racers for the air-show contests, could have made this radical notion seem feasible from a business standpoint.

The first product of Northrop's imagination was a crude approximation of his dream. With a span of thirty feet that was thick enough in the center to contain the pilot and a ninety-horsepower engine, the wing was awkward, underpowered, and dangerous. Northrop did not know how to control such a shape in flight, so he was forced to mount rudders and stabilizers on the end of booms that extended well behind the wing's trailing edge. It was a clever experiment—more at home in an aeronautical research laboratory—but totally impractical.

Needing salable products to pay his bills, Northrop then conceived an all-metal single engine monoplane for carrying passengers or mail, called the "Alpha," which was an evolution of his previous work for Lockheed. As in the past, he leaned heavily on European innovations, but the Alpha made one notable technological advance—extensive use of multicellular sheet-metal panels in its wings, fuselage, and tail. Aluminum alloys were just starting to replace wood in aircraft construction. The prevailing shop practice held that only corrugated structures would not buckle. But the crinkly surfaces that resulted were irritating to Northrop's obsession for cleanliness. He therefore used smooth, thin reinforced slabs—which he had developed for his little flying wing, where stresses had been difficult to predict and he wanted comfortable safety margins—which solved an important manufacturing problem and brought wider industry attention to the otherwise conventional Alpha.

With its gleaming silver skin and streamlined contours, the Alpha seemed to express far more than clever engineering, however. Like the Vega, it looked the way an airplane should look—not like an old-fashioned box kite, but almost like a Brancusi bird. Northrop even enclosed its fixed landing gear in smooth aluminum "trousers" to reduce the drag, completing the visual impact of a metallically feathered hawk. It was a breathtaking machine, radically different from the chunky steam locomotives and automobiles that were still the most familiar products of modern technology. To see the Alpha, let alone fly in it, was to discover a new possibility. This was Northrop's genius and great good fortune, that for a while he was able to translate aesthetic notions about flight into technical accomplishments. Aeronautics would never be as kind to such design instincts again as it was in the late 1920s and early 1930s. Near the end of his life, he rated the Alpha as his most important contribution to aviation, despite its lackluster sales.

In late 1929, with the experimental flying wing ("interesting and cute," Northrop called it) at a dead end and insufficient capital to continue developing the Alpha, Northrop let his company be absorbed by one of the giant aviation holding companies of the period, United Aircraft and Transport Corporation. United was the creation of a coterie of auto industry and National City Bank executives centered around Edward Deeds and his wartime cronies, who had figured in the military procurement scandals of 1917–18. In January 1929, they had begun to merge four highly profitable companies— Boeing Airplane and Transport; Pratt & Whitney, an engine producer controlled by Deeds; Hamilton Aero Products and Standard Steel, two propeller makers consolidated into Hamilton Standard; and Chance Vought Corporation, an airplane builder and Pratt & Whitney customer. Stearman was an additional acquisition. Stock manipulations and pyramiding brought huge profits to the original insiders, in some cases reaching incredible million-percent levels before the Crash. Northrop and Jay received 3,000 shares of United common stock in exchange for Avion.

Between 1929 and 1933, United would account for 48 percent of all aircraft and engine sales to the Navy, 29 percent to the Army, and 48 percent to the commercial market. It was also one of the dominant forces among airmail carriers, which were enjoying heavy subsidies from the federal government. In short, it could not be denied.

The offer to buy out Avion came in the person of William Boeing, who was apparently covetous of the Alpha sheet-metal fabrication technique. Boeing's mail planes were still made of fabric over steel tubing, but on his firm's drawing boards was an all-metal monoplane with a monocoque fuselage and cantilever wing—Northrop specialties. If Jack Northrop was wary about the Deeds clique (there is no record to this effect, and he was never known for political savvy), his misgivings were evidently overcome by Boeing's stature and the proffered deal to establish a Northrop Aircraft Corporation as United's subsidiary. Granting the Boeing engineers access to Alpha structural secrets seemed incidental.

William Boeing may also have been on another kind of fishing expedition for United's directors. They already controlled the northern transcontinental airmail route pioneered by government flyers in the early 1920s. In the spring of 1930, the Post Office Department drew up a new map for airmail, with the Postmaster General personally awarding lucrative transcontinental routes directly to monopolistic conglomerates without competitive bidding. United kept its northern territory. The central route, known as the "Lindbergh Line" because it had been surveyed by the celebrity pilot, went to Transcontinental and Western Air (TWA). TWA had been using Lockheed planes on this route, but that company was now in trouble and an uncertain source for new all-metal aircraft needed for safe flying at night or in bad weather.

Northrop's Alphas had the potential to satisfy such requirements. Moreover, Lindbergh—a TWA technical adviser—knew Jack Northrop as the father of the Vega line. The test pilot of Northrop's experimental flying wing had even been Lindbergh's co-pilot during the mail route survey. This web of familiarities would naturally help sell the Alpha to TWA, thus giving United a stake in another mail line.

As long as United maintained its financial support of his subsidiary, Jack Northrop was secure, or so he thought. But as the Depression strangled the entire commercial airplane market and Alpha sales stagnated as single-engine passenger planes fell from favor, United decided ("very sensibly," Northrop admitted) in September 1931 to consolidate the Northrop operation with Stearman in Wichita. Jack Northrop had no desire to return to a region where "big blue" rivers were muddy ditches, so he exercised a clause in his contract that allowed him to quit United rather than relocate to the bleak Midwest of his early childhood. By the spring of 1932, Stearman in turn abandoned the Alpha.

Thus, as Lockheed slipped into receivership in the fall of 1931 and Allan Loughead began to cast about for some way to get it back, Northrop considered linking his fortunes with the redheaded supersalesman for a third time. Neither could now afford the waiting games of their youth, however. After Robert Gross bought their old company out of bankruptcy court, Loughead disappeared back into real estate, and Northrop finally turned to the military in the guise of Donald Douglas.

"THE DEPRESSION THAT started in 1929 didn't affect us at all," Douglas stated matter-of-factly about his business in Santa Monica. "We weren't doing any civil work."

Like Glenn Martin, Donald Douglas had virtually ignored the civilian marketplace, relying instead on relatively steady appropriations from Washington for military production. His development during the late 1920s of a small luxury "air yacht" flying boat called the "Sinbad" was saved from financial disaster only by later sales of military versions, one of which became the first presidential aircraft. Thus, from 1929 to mid-1933, his company achieved profit levels that seemed miraculous for the period: between 18 and 62 percent on Navy production contracts; between 9 and 32 percent on Army production. He lost only on building experimental prototypes, but so did all the other manufacturers. Between 1930 and 1932, Douglas felt confident enough to pay out some $1.15 million in dividends from net profits of $1.31 million, thereby stabilizing the price of shares that at times dipped as low as five dollars from their pre-Crash high of forty-five.

These spectacular profit levels on government work were not known to the general public, most members of Congress, or anyone else outside the military procurement network. Americans had fled the stock market and were preoccupied with 25 percent unemployment, bank failures, home loan foreclosures, and every other category of economic hell. Early in 1932, a New York investment trust trying to dump its aircraft holdings struck a deal with Douglas to prevent further deterioration of the company's stock. The trust privately sold Douglas 10,000 shares at ten dollars, two dollars below what it was then trading for on Wall Street. With few, if any, other buyers, such a large offering might have triggered a ruinous dive of aviation stocks on the open exchange. The company then held these shares and waited for the market price to rise to eighteen dollars upon the announcement of a large new government production order, thus turning a profit of $80,000.

"And those people who were forced to sell that stock at that time were the losers because of the knowledge that your president [Donald Douglas] had of the potentialities?" a House of Representatives investigator later asked a Douglas officer.

"I would not put it that way, sir."

"You put it your own way."

"This investment trust had asked our president if he would make them an offer for some stock that they wanted to liquidate, which he did. It was put into the treasury account with the idea of holding it there. I know that when we discussed it this was the thought, that it was one way of reducing the amount of outstanding stock, which is always a good thing, because it makes a better deal for the other stockholders."

"It makes a better deal in this instance for Mr. Douglas and his corporation, when they made $80,000 on a turnover of 10,000 shares."

"It happened to be that way."

In the fall of 1931, Jack Northrop and Kenneth Jay approached Donald Douglas about backing them in a new company. It was a logical proposal, since Douglas was known to be cash rich from military production and Northrop was already familiar to him. For Douglas, however, there seemed on the surface few reasons to make such an investment other than as a favor to Jack Northrop. The key to his rather quick acceptance of their plan—Northrop became a full-fledged subsidiary in January 1932, with Douglas holding 51 percent of the common stock—may be found in the ownership of some 86,000 shares of Douglas stock by North American Aviation, another of the period's mammoth conglomerates.

North American—created in 1928 by a former editor of the *Wall*

Street Journal as a holding company—was cross-linked through stock ownership or overlapping boards with a labyrinth of airplane manufacturing and transport companies, including TWA. After a crash on March 31, 1931, into a Kansas cornfield of one of its older wooden-wing passenger planes killed all occupants, including Notre Dame football coach Knute Rockne, TWA was forced to find a new supplier of transports. Boeing was considered the best source, but its output was controlled by United. By the fall of 1931, however, North American managers knew that Jack Northrop had left the United fold. If Northrop operated under Douglas's umbrella as a source of new commercial designs, TWA might tap him through the North American management network.

This is what happened during the latter half of 1932. In August, TWA queried Douglas and four other prominent manufacturers about building new transports to the airline's ambitious technical specifications. At first glance, Douglas was not eager to take on such a demanding project, given the stagnant state of commercial air travel, the competition from Boeing-United, and the healthy condition of his military sales. There was no clear benefit in starting a risky, expensive development program that would entail a shift from the company's traditional line of military business.

"This was quite a departure from what we had been doing," he said. "Up to that time, we were developing [military] planes from planes we'd done before. But this was a departure more than technically. It was a departure from a business standpoint. It was our first major non-government contract." For the first time, Douglas might have to compete like a real commercial entity, and he was not enthusiastic about this prospect.

Strong persuasion was promptly applied in person by the chairman of North American Aviation, Harold E. Talbott Jr. Talbott was the son of one of the original organizers of the scandal-ridden Dayton-Wright Company, a longtime associate of Edward Deeds who had directly profited from Deeds's insider knowledge of Army contracting during the war. He was also a director of TWA. There is no record of Douglas's personal reaction to Talbott's visit, but one may safely conclude that the encounter was stiff. Had the ensuing program not turned out to be the glory of Douglas's career, perhaps his memory would have included mention of Harold Talbott Jr.'s 86,000-share rationale.

Within days, top contracting and engineering staff from Santa Monica were in New York negotiating with their TWA counterparts, led by Charles Lindbergh. On September 20, a $125,000 contract was signed for a prototype called "DC-1" (Douglas Commercial One), plus options for sixty DC-2 production models at $58,000 each. Engines would be supplied by Curtiss-Wright, a 1929 merger (devoid of either Curtiss or Wright family involvement) controlled by North American, despite the availability of superior Pratt & Whitney (United) engines. The design also featured extensive use of Northrop's multicellular metal construction. It was a tidy corporate package, once Donald Douglas saw the light.

It was also the start of a program that came to symbolize not just the Douglas company, but the superiority of American aviation. The DC transports became an icon of the age, so strongly linked in the public's mind with Donald Douglas that the company's role as a primary weapons contractor was obscured. When Shirley Temple sang about the "Good Ship Lollipop," cinema audiences recognized it as a DC-2.

Though Douglas later claimed that the DC-1 cost more than $306,000 to bring to fruition, requiring some $182,000 in company funds above the $125,000 from TWA, financial data surrendered to congressional investigators in early 1934 indicated that the company enjoyed a 24 percent profit margin on commercial work during 1932 and 1933. Since the DC-1 was the only significant commercial activity in house at that time, the question arises whether funds were shifted from military experimental contracts, which showed substantial deficits. Typically, losses on military prototypes were recouped through subsequent War and Navy Department production orders or exports to foreign governments, but this could not be counted on in the commercial world. Surreptitious transfers between public and private accounts were not uncommon; but Congress in this case failed to examine Douglas's books.

"It represented a big risk on our part, not so much on just the one as on the subsequent orders," Donald Douglas said. "We only built the one DC-1." In later years, even after North American was forcibly broken up and its stake in the company removed, he never mentioned the financial interests that considerably buffered the actual risk. Claiming that North American was "largely an investment trust"—a true but absurdly benign characterization—Douglas insisted that he "never had any pressure from anybody" and that "the interlocking directorates caused no problems." His memory was never as creative as Glenn Martin's, but he was capable of the same self-serving interpretations.

Nowhere would this trait be stretched more to the point of delusion than in one very personal instance. Though Douglas was never ostentatious about his wealth, he allowed himself one luxury after his company gained firm footing in the late 1920s—a sailing yacht aptly christened *Cloudster*. The *Cloudster* became an escape both from a business he feared was growing too complicated and a marriage that seldom provided enough solace.

On a summer weekend in 1931, the several company men who crewed on Douglas's sloop brought aboard female companions for their usual cruise. That the women were not their wives, were in fact being paid for their presence, was certainly known to Douglas, though he had not taken part in such entertainment before. On this particular weekend the men arranged for an extra girl named Margaret Tucker—called Peggy—who was to cater to the boss.

When the *Cloudster* berthed for the night and his crew retired with their dates, the shy Donald Douglas compromised with his guilt by just talking with Peggy until dawn. The next morning he gave her ten dollars for her time, a considerable sum in 1931. After returning home, she got a phone call from one of the other women, the one who had invited her along. Asked if she had liked Douglas, she replied that he was nice, but they had spent the night just talking. Asked if she realized who he was—that is, the rich president of Douglas Aircraft—she was awestruck. They soon met again on her initiative, and began a relationship that lasted for the rest of Donald Douglas's life.

In the classic manner of respectable businessmen and their secret mistresses, Douglas at first set Peggy up by buying a hat shop for her to manage in Los Angeles. When she went broke, he gave her a full-fledged dress store. The romance was of course known to the *Cloudster* crew, and thus to a wider circle of company associates, but not to Charlotte Douglas. That something akin to blackmail was in progress was apparently unevident, perhaps because Douglas himself took no steps to sidetrack the affair. Instead, Peggy became an increasingly important factor in his private life, and eventually a major problem in his business world.

Speaking many years later about Hap Arnold, with whom he had begun a close friendship in the mid-1920s, Douglas provided a glimpse into his own contradictory ethics. "If Hap was interested in women, I never saw any of them," Douglas recalled. "I don't think Hap was interested in anyone but his wife. And I know he raised hell with any of his officers that got into family troubles of that kind.

And how! I think Hap was very moral. Hap had no use for anybody who was not moral. I've known officers in the past whose whole careers were wrecked because of that. He never trusted them afterwards.''

Donald Douglas and Hap Arnold remained friends long after the existence of Peggy became known in military-industrial circles. Arnold's son, who married Douglas's only daughter, would claim many years after both men were dead that his father had never been privy to the affair. The issue of who knew what in this instance is of course trivial in the spectrum of twentieth-century moral issues, but it calls attention at least to Douglas's ability to see only what he wanted to see. In his special profession, this may have been a vital trait.

IN THE EARLY 1930s, Glenn Martin's business took a steep Depression slide, worsened by the staggering financial burden of his new plant in Baltimore that had been conceived before the Crash. In desperation over a paucity of engineering talent—the company had been duplicating aircraft designed by the Navy or other manufacturers for years—he tried to hire back Donald Douglas. With considerable charity, Douglas found his old boss to be rather self-centered and oblivious to the fact that by this time he controlled one of the strongest companies in America.

In 1930, the first full year of Baltimore operations, Martin had sales worth $2.2 million, mostly from production of thirty Navy seaplanes on which he turned a profit of at least 15 percent. Only $73,924 stayed in the company's coffers, however, because of extravagant operating expenses and the cost of carrying the gold notes issued immediately after the stock market collapse. This figure stood in stark contrast to the $900,000 income forecasted in a November 1929 publicity paper. The company's net worth fell for the first time since 1924, by nearly $50,000. This was not the surging enterprise Glenn Martin had promised the citizens of Maryland.

Since his earliest days as an arms supplier to Alvaro Obregón, Martin had felt no qualms about exporting the military machines he developed in order to improve his balance sheet. Many Americans, on the other hand, felt in the early 1930s that weapons trading was immoral and contrary to the nation's interests. The last three seaplanes to roll out of the Middle River assembly hall in late October 1930 were shipped, with Navy approval, to the government of Bra-

zil for use as bombers against revolutionary forces there. The price
was $90,000 each, prepaid through an irrevocable letter of credit—
50 percent higher than the $60,000 per plane cost of the preceding
twenty-seven sold to the U.S. Navy. Some of this extra profit may
have been absorbed by commissions to private agents and bribes to
Brazilian functionaries, but such markups were typical for exported
warplanes. In an ironic touch, by the second week of November the
São Paulo government had fallen to the rebels, who took delivery of
the planes for their own purposes.

Earlier that year, the U.S. State Department had quashed Martin's
plan to sell twenty of the same flying boats to the Soviet Union for
about $2 million, even though the only legal arms embargoes at the
time involved China and Latin America. "Russia, according to the
State Department, is following a definite policy of promoting world
revolution," the *Baltimore Sun* explained. "Although the Hoover
Administration does not want to be in a position of selling material
to a country which has an announced policy of world revolution,
especially at a time when two important countries bordering
Russia—India and China—are in revolt, nevertheless it does not
want to take valuable business away from American firms if Russia
can get military supplies elsewhere." Amtorg, the Soviet commercial agency in the United States, denied ever placing the order with
Martin.

In general, Washington viewed the international arms trade as
legitimate business and therefore tolerated gaping loopholes in various embargoes, for the time-honored reason of not wanting to
impede American enterprise. The Navy seemed especially forthcoming, in one instance lending a commander to the government of
Colombia to advise on the purchase of weaponry for use against
Peru. The officer was later caught accepting commissions from at
least two American manufacturers. Airplane builders would turn
increasingly to the world market to bolster their sales as the decade
grew more bellicose.

The precarious condition of Glenn Martin's business was not apparent from his opulent lifestyle. In 1930, he ordered a new Stutz
convertible from E. S. Gorrell, taking the occasion to offer to recommend Gorrell as president of Detroit Aircraft, which Martin felt
was "sufficiently financed and held sufficient future possibilities."
Gorrell politely declined. Martin also thanked Colonel Gorrell for
his "wonderful demonstration of high speed contacts" in bringing
Louis Chevrolet's airplane motor to the attention of Army procure-

ment officials. As with Lockheed's traffic with Colonel Boots, it was all in a day's work.

At the same time that his underlings were struggling to keep the company alive, Martin carried on a self-absorbed correspondence with Stutz about his new car, supplemented by missives from the Martin Company's purchasing agent. The letters read very much like aircraft orders, which Martin needed far more than another luxury automobile:

"I desire one of the new 1931 Convertible Cabriolet Stutz short-wheelbase, which is to be upholstered with M-3 Mission Bedouin grain leather," he wrote. "The body color is to be Valentine Maroon 3128, the striping and wheels color Valentine Tank Gray 7010, the fenders and chassis to be black enamel, all polished parts to be chromium plate. The car should have standard equipment of two spare wire wheels and tires in fender walls, and one Lorrine driving light installed.

"I have requested the Eagle-Ottawa Leather Company, 74 Gold Street, New York, to express two 50-55 sq. ft. perfect hides to you for the specification leather upholstery of this car. In the event the leather is not perfect, please advise immediately."

The special upholstery was inspired by another of Martin's recent purchases: "75 pieces of Macy furniture, consisting of chairs, davenports, etc., all equipped with [Mission Bedouin], which has been very satisfactory so far, and I have decided to have an automobile for my personal use upholstered with this leather."

When the two-year-old Biarritz he already owned began to show some signs of wear that puzzled the manufacturer, Martin had his chief purchasing agent fire off an educational note:

"We do not quite agree with your explanation as to why the Stutz wheels in question have cracked, as the inflation of the tires was carefully checked several times each week by Mr. Martin's chauffeur, and we believe this trouble is due entirely either to the quality of the wheel or the chromium plating; in our experience with chromium plating we have found that if the entire surface is not plated you will always have trouble at that point where the plating stops, and if you will carefully check the other four wheels you will find that they are beginning to show weakness at this point and we would not care to use them further." Martin eventually commanded his own metal workers to replate the wheels.

No one in the company was immune from being called to duty on the leader's personal behalf. When the new Stutz exhibited a few mechanical glitches, Louis Chevrolet himself assumed the role of garage boy, writing a complaint to Gorrell:

"The garage of the Company has done a few adjustments on Mr. Martin's Stutz Coup, and I just tested it for them. Since we replaced the intake manifold, which was leaking, the motor performs O.K., but the transmission of this car is the noisiest one I have ever sat behind, and I know that Mr. Martin has passed the remark several times that he did not like the transmission on account of its noise, but was hoping that it would quiet down after a few thousand miles, but it is just the other way. If you could do anything in regard to that transmission, or replace it by a quiet one, I know Mr. Martin will appreciate it." Such were the duties of the Martin Company's vice president, general manager, and auto pioneer.

During this strained period, the firm's publicity department generated a bizarre history of Glenn Martin's accomplishments, ending with a personal description of the founder that struck an almost surreal tone. The din of carnivals and Los Angeles air shows can be heard between its lines, which must have been approved by Martin himself, if not actually composed. He wanted desperately to be lauded as an industrial hero, even if he had to pen the certificate of merit:

Glenn L. Martin is over six feet tall, and makes a striking appearance. His shell-rimmed glasses and high forehead give the impression of a learned professor, but the skill and dexterity of his long, slender fingers indicate that he is not only a dreamer but also puts his ideas into immediate and accurate execution. He is constantly looking many years into the future in an endeavor to ascertain the future place of his industry in world affairs. He takes especial care to keep himself physically fit and is immaculate in his appearance. His mind is keen and accurate in the analysis of the most difficult problems, all of which he attacks methodically and solves with extreme care and thoroughness. Martin always has absolute control and never becomes confused nor allows himself to be rushed under pressure. His opinions are never biased by personal animosity nor is he subject to flattery. Many of his most important problems are solved as the result of his so-called "hunches" which have proved invariably to be accurate.

Mr. Martin is a born leader of men and inspires in them not only the desire to do their utmost for his company but to make themselves worthy of his friendship. He takes an active part in the design and manufacture of his planes. He does not know the meaning of fear and is ever ready to help others in their perplexities.

As the company continued its descent into 1931, it was Glenn Martin's own perplexities that needed attention. Sales rose to $2.96 million, but this could not prevent a year-end deficit of more than $60,000. The company lost $140,000 on a single flying boat proto-type and $284,000 producing nine Navy seaplanes from another manufacturer's plans. The year might have been even darker with-out an 18.5 percent profit on building twenty-eight seaplanes from a Naval Aircraft Factory design. This work was entirely conventional production that Martin should have been well acquainted with, yet he could not control the spiraling overhead expenses in the new factory. His reputation for running an expensive shop had wounded him a decade earlier during the Army's MB-2 bomber program, and it would continue to plague him for the rest of his career.

Yet he took no measures in his personal life to alter this picture. He and his mother moved into Washington's most chic apartment house, the Kennedy-Warren on Connecticut Avenue. They traveled together to New Orleans for Mardi Gras, and to London, where he delivered the prestigious Wright Memorial Lecture before the Royal Aeronautical Society. The lecture was a description of manufactur-ing innovations at his new factory, which at that moment were dragging him into bankruptcy. In June, he exchanged his Cleveland estate—which had remained on the company's books after the move to Baltimore—as partial payment for a $300,000 twenty-two-room mansion in Hollywood's Hacienda Park. The Los Angeles home was noted in the company's financial records as "used by Mr. Martin as a temporary residence in an effort to dispose of the property."

But an interview conducted by the student newspaper at Kansas Wesleyan College, where Martin had taken bookkeeping courses in 1904, shed sharper light on his bicoastal personality:

Mr. Martin divides his time between Baltimore, where he spends about three and a half days of the week, and Washington, D.C. He works intensely while he is working, and his living quarters are right in the factory at Baltimore. Mr. Martin usually spends two months of the year at Hollywood in his "shack," as he quaintly terms it. His Hollywood home is decidedly a place to rest and recuperate. There is an athletic room, a billiard room, and an entire floor devoted for entertaining purposes. At the back of the house, near the sunken gardens, is a swimming pool. There are large trees around the lawns, which add to the beauty of the attractive place. The interior of the house is beautifully arranged and one of the most interesting rooms is the Egyptian ballroom, furnished in the true Egyptian style, even down to the luxuriant slave couch. There is also a pipe organ in the "shack" and all has been planned in order to insure rest and enjoyment. Mr. Martin succinctly stated that a home should be a place to recuperate with the greatest possible rapidity.

Martin parted the curtains of his private life to a rare degree for this interview, perhaps because of its obscure place of publication. He also opened up his thoughts about his company's products. Surrounded by a culture that had grown suspicious of technology's promise of making war obsolete, he had not aired his vision of warfare for many years. The man whose name was most closely connected with the word "bomber" thus tried to allay the students' fear of what bombing obviously meant.

It is important to note that the new theory with present day bombing is not to bomb the cities of an enemy country. That is, the first operations of bombing are to cut the source of power, communications and transportation. This would be accomplished by bombing railroads and highways not near the city, but in a remote locality between two cities, thereby making it more difficult for man and material to get between the break in transportation. Bombers would revisit and ruin any attempt at repair. Big central power plants in the mountains together with retaining dams and so forth would be destroyed.

Even if he spent only three full days a week at his factory, Martin was much too close to the technology to really believe this peculiarly American fantasy about precision bombing. Like the Air Corps Tactical School officers who had contrived it, Martin relied on the "new theory" to make his profession sound more sophisticated than it actually was. Or perhaps, like Donald Douglas and his commercial transports, such notions were a way to convince himself that life was not largely a matter now of creating instruments for mass destruction.

THE SUBJECT WAS on his mind for good reason. Though Martin's move from Cleveland to Baltimore had been predicated on obtaining better facilities for the production of Navy seaplanes, the company's engineering staff had immediately begun a collaboration with the Army to develop a new bomber that could fly at 200 miles per hour—faster than any existing pursuit plane. Having used World

War I–vintage designs for the past decade, younger Air Corps officers were eager to move into the new generation of monoplanes epitomized by Lindbergh's Lockheed *Sirius*. Though he had not built an Army plane of any type since 1921, Glenn Martin considered high-performance bombers his chosen field. Loss of the old MB-2 had been a terrible blow to his pride.

The Army, for its part, was cautious, at least compared to the days of Billy Mitchell. His 1925 court-martial had stifled proponents of aerial bombardment within the military. And the Depression was shrinking all federal budgets. "It was suggested to us by the [Army] that if we wished to construct a bombing plane entirely at our own expense and with no responsibility on the part of the government other than the consignment to us of instruments, motors, etc., then they would give the plane careful consideration," Martin later recounted to a War Department committee. "If the plane proved satisfactory, they would recommend the purchase of the sample plane at a price of $200,000, which was a little more than one-half the cost of building it. The cost, without profit, was $378,583."

Martin was not telling the whole story, however. The prototype was actually constructed under a negotiated contract that fell within a special category of Air Corps procurement for experimental airplanes. Like many other manufacturers, Martin deliberately bid below cost, hoping to give the Army such a desirable bomber that a quantity order would follow immediately. This is exactly what happened, with the Air Corps first buying ten more bombers for flight tests under the same experimental category, and then large numbers under another regulation that permitted procurement without competition if no other suitable source existed for the product. By the eleventh plane, which reflected myriad refinements and unique shop experience, Martin was indeed a sole source. Thus, the company and the Army both got what they wanted, as all other parties were shut out of the process.

The aircraft itself was a milestone between the old style of lumbering biplane bombers and the fast attack planes of World War II. An all-metal twin-engined monoplane with retractable wheels and completely internal bomb stowage, it looked much like the machines that would devastate Germany ten years later. It was arguably the best plane Martin would ever produce, and was without doubt the linchpin of the company's survival until Munich.

The secret development of this new bomber between 1929 and 1932 may help explain why the Army was not distressed about the loss of Lockheed's XP-900 pursuit plane in the fall of 1931. With a maximum speed of 235 miles per hour, the XP-900 was nearly fifty miles per hour faster than the Air Corps' first metal monoplane bomber, the Boeing B-9, which was flight-tested at Dayton during the summer of 1931. In August, the Army bought five of these at $100,000 apiece. But Lockheed's fighter made the B-9 obsolete overnight. Even more embarrassing from the Army's point of view was that the XP-900 also showed itself to be faster than the 200 miles per hour goal for Martin's prototype, designated B-10, whose flight trials were not slated to begin until summer 1932. If the Depression-strapped Congress had understood this state of affairs, it might reasonably have told the Army to rethink its aircraft development program. But the Army wanted new bombers more than fighters, and that is what it got.

"We are only running about one-third of the business we should do, and that business has to be taken at a loss," Glenn Martin wrote to a family friend in California a few days after Christmas 1931. Since 1919, his company had sold more than $18.5 million worth of airplanes, almost all for military purposes. His net profit was just over $2.7 million, or about 15 percent. It was a remarkable record for the "Flying Dude," but in this bleak holiday season he must have wondered if it would all soon disappear. Had he been in any other business, it most definitely would have.

WHEN JACK NORTHROP re-established his business as a subsidiary of Douglas Aircraft in January 1932, he believed that "the basic idea was to conceive and develop types and pioneer ideas to be fed in for production at Douglas." In other words, he saw himself as the brains behind Douglas's financial brawn. Northrop continued to covet the relatively rarefied atmosphere of prototype design, rather than the workaday discipline of manufacturing. Design engineers were the industry's demigods, and Northrop always ranked himself high among them, though his lack of formal training would soon put cutting-edge aeronautics beyond his ken. Douglas evidently assured him, at least initially, that this would be his role. "We were to run an operation where new types of aircraft could be built and taken through the teething and development stages," Northrop recalled. That Douglas already employed several of the best engineers in the country was somehow beside the point.

Northrop's first project for Douglas was a successor to the Alpha, another single-engine low-wing monoplane called the "Gamma." With a maximum speed of 248 miles per hour, it was very fast for its day, but it was not well matched to any commercial category. The $40,000 Gamma was therefore positioned initially as a customized machine for sport flyers, explorers, and corporations—a rather tenuous market in the depths of the Depression. It nonetheless reached some of the same customers who had bought Northrop's Lockheed models but who now found Lockheed in tatters.

The first Gammas went to Texaco, which called its plane the *Sky Chief* and raced it for the sake of publicity, and to polar explorer Lincoln Ellsworth, who flew his across Antarctica. TWA bought three for carrying mail. Any wider market was wiped out, however, when the Civil Aeronautics Authority placed severe restrictions on single-engine transports for passenger service. Donald Douglas then exercised his majority interest by taking control of the project—Jack Northrop literally transferred the Gamma's blueprints to Douglas by loading them into his two-seat Plymouth roadster.

By the fifth plane, Douglas made sure the Gamma was converted into an attack bomber. Twenty-four were sold to the Chinese government in 1934. A U.S. Army contract followed for 110 modified versions, called A-17s, at $2,047,774. This work ensured the continuance of the Northrop subsidiary, but obliterated its intended role as a design shop. Northrop continued to sell a few custom-made Gammas to sport pilots, and there was a passenger type called the "Delta"—several of which filtered through private buyers into the Spanish Civil War—but after 1934 Northrop was essentially an adjunct to Douglas's traditional military business.

Above all, Donald Douglas made sure that Jack Northrop spent no time on such foolishness as flying wings, which were not compatible with economic depressions.

CHAPTER SEVEN

MERCHANTS OF DEATH

THE YEAR 1934 was something of an anomaly in the history of public awareness of the aviation weapons industry and the men who ran it. The stock market crash five years earlier had radicalized the perceptions of many Americans about Wall Street, Washington, and the responsibility those two centers of power bore for the nation's economic disaster. A substantial number of citizens believed, moreover, that the Depression was partly the fault of inflation from the Great War. As international relations began to deteriorate with Hitler's rise to power in Germany, there was fear of another ineluctable slide into European quicksand. Out of these diverse concerns flowed a broad-based antiwar sentiment that focused on companies and owners who profited from the worldwide munitions trade. For a brief time, the aviation industry looked like an Augean stable that needed thorough mucking.

The movement had accelerated in the fall of 1933, when the Public Works Administration—a New Deal agency intended to create jobs through direct federal subsidies—granted $15 million for airplane construction to the Army and Navy (PWA money also laid the keels of two new aircraft carriers in the spring and summer of 1934). The action merely restored funding already approved by Congress that had been impounded by President Franklin Roosevelt as part of his reordering of budget priorities to fight the Depression. But the controversial grant stirred up long-festering rumors about excessive profits and favoritism in the warplane industry.

This was the peak season for "merchants of death" rhetoric, which was adopted by mainstream groups that had never been polemical about capitalism or the military. A book by that title became a best-seller and Book-of-the-Month Club selection in April 1934. A review in the *New York Times,* always a purveyor of acceptable taste, deemed it "excellent." Given such solid encouragement, Congress launched several investigations into the propriety of the aviation industry's relationships with the U.S. military and foreign governments. These hearings remain notable not so much for what they proved—which was very little—but for the sharp attitude they sometimes took against corporations that Washington had tried hard to keep happy.

The first inquiry was held in February 1934 by the House Naval Affairs Committee's aeronautics subcommittee. Then as now, the congressional military committees existed largely to protect the interests of the services and their industrial suppliers, rarely to search for bad apples. But even Carl Vinson of Georgia, Naval Affairs' implacably pro-military chairman, had apparently been touched by popular suspicions about munitions manufacturers, because he let it be known that he was disturbed by a trading flurry on Wall Street after an announcement of new government orders for airplanes. He was surely aware that this was a most common phenomenon, though the Depression no doubt made it look more vulturous than usual. In any case, he decided to use the occasion for his own ends.

"We want to know what profits these people make and we want to know whether there is favoritism or trickery in obtaining contracts," Vinson intoned, using language that must have startled "these people" whom he had shielded in Congress for many years.

Known as the Delaney hearings after Representative John J. Delaney, the subcommittee chairman from New York, the sessions were chartered to "ascertain profits in the manufacture of airplanes," to discover "if any profiteering has taken place," and to determine whether there had been "any collusion between manufacturers with respect to Government contracts." The members were looking, in today's political parlance, for evidence of waste, fraud, and abuse. Not since the aviation procurement scandals of 1918 had such matters been raised in such raw terms. Though Delaney exonerated the industry in his majority report (only high-ranking Navy officers and top company officials were called to testify), the hearings produced some sensational financial data about the era's major companies.

Among the executives who testified was Glenn Martin. As on similar occasions in the past, Martin appeared to be a stolid businessman with conservative views about the benefits of negotiated (noncompetitive) contracts and increased federal spending on experimental programs (where he always suffered heavy losses). His Egyptian slave couch was nowhere in sight. He revealed, when asked, that his salary had been in the vicinity of $16,000 a year since the early 1920s, that he owned 525,000 shares of his company's capital stock (the only other stockholder being Louis Chevrolet, who held a thousand shares). He did not explain that the company bought his mansions and luxury cars for him. He stated that his overall profit from 1922 to 1933 was just 3.97 percent after taxes. When he told the committee that his last work for the Navy had been completed in 1932 after ten years of business worth a total of $15.4 million, a rather testy exchange took place:

DELANEY: To what do you ascribe, Mr. Martin, the fall of your business with the United States Navy?

MARTIN: I am not familiar with the reasons why I have been unable to secure contracts from the Navy. I have offered designs, but have not been fortunate enough in being awarded a contract, and that is all I know.

DELANEY: Is there any particular reason why you have not, Mr. Martin?

MARTIN: I do not know of any reason, sir.

DELANEY: Are you in competition with other plane companies?

MARTIN: Yes, sir.

DELANEY: And have your designs been accepted by them at any time, and did you get the order?

MARTIN: No, sir; I have not had an acceptable design.

DELANEY: Has the Navy Department made an audit of your books from time to time?

MARTIN: Yes, sir.

DELANEY: And do they find the situation satisfactory?

MARTIN: They have never commented.

Representative Delaney's patronizing air seemed intended to let Martin know that the panel already understood the answers to these questions, namely that Martin had lost his Navy business because of high overhead costs—the same reason he had lost his Army business ten years earlier. A Stutz convertible was difficult to conceal during the Great Depression, and a munitions tycoon who drove one between Washington and Baltimore risked courting disfavor among the more socially unpretentious members of Congress.

The Delaney panel expected to take testimony from Donald Douglas, but heard instead from Douglas's vice president for engineering, J. H. "Dutch" Kindelberger. A man of aggressive ambition who had worked as a designer for Glenn Martin after the Great War, jumped to Douglas in 1925, and would soon rise to the presidency of North American, Kindelberger was a tight-lipped match for the committee's investigators. Of special interest to them was ownership of Douglas stock, which showed the company to be dominated by North American. Of 467,403 shares outstanding, North American controlled 86,116, or 18.4 percent—by far the largest single block. Donald Douglas and his father held the second largest, at 47,820. Delaney pursued the fluctuations of this stock since its issuance in 1928:

DELANEY: [What was the value] at the time the company was organized?

KINDELBERGER: As I remember, the first sales were approximately at 19.

DELANEY: How high did that stock go, Mr. Kindelberger?

KINDELBERGER: It went up to around 45 during the boom.

DELANEY: So the appreciation was a matter of $26 a share?

KINDELBERGER: Well, if you call it appreciation. That is where it went up to at that time.

DELANEY: Was that warranted by the business you were doing?

KINDELBERGER: It was not.

DELANEY: It was just pure speculation?

KINDELBERGER: I think so.

DELANEY: And manipulation?

KINDELBERGER: Yes, sir; I would think so.

DELANEY: What is the stock selling for now?

KINDELBERGER: The last I saw—I just arrived on a boat from Panama last night and I have not seen the quotations—but the last I saw it was around $21.

DELANEY: Then it is just $2 above what it was selling for when it was first issued?

KINDELBERGER: That is correct.

Kindelberger answered, when asked, that Donald Douglas's annual salary was $30,000. He also reported a stunning level of profit on Navy work since 1926: 1927, 25.38 percent; 1928, 30 percent; 1929, 18.15 percent; 1930, 24.98 percent. And on Army business: 1927, 19 percent; 1928, 22.5 percent; 1929, 15.66 percent; 1930, 14.5 percent. In 1931, a loss on Navy contracts of 25.94 percent was offset by Army profits of 19.33 percent. In 1932 the situation reversed, with Navy profits of 14.63 percent and Army losses of 29.98 percent. The aviation industry was known for creative bookkeeping, but exports to foreign governments kept the company in the black during those two Depression years.

Yet when Delaney asked if the company did any business with foreign governments, Kindelberger replied, "Very little—we sold some obsolete types to China and Peru," glossing over the fact that the Chinese government was Douglas's biggest customer after the U.S. War Department. Indeed, the firm maintained an extensive sales network around the world, which explained why Kindelberger had just returned from Panama.

Like Glenn Martin, he argued that friendly agreement with the contractor was the best way for the military to price new airplanes. As for competitive bidding, "I do not think it is fair competition for a company to bid desperately because they want to keep their organization going," he said, putting a curiously un-American spin on the essence of free enterprise. The Army and Navy evidently agreed with this assessment. Records obtained by the committee showed that since the Aircraft Act of 1926, the Army had awarded contracts for new warplanes worth $57.3 million, of which only $3.3 million was on a competitive basis. For the Navy, the total was $53 million, with just $5.9 million open to multiple bids.

In its majority report, the Delaney panel flew its true colors, which were hardly the banner of the "merchants of death" crowd. First, it deemed that industry profits from Navy contracts "have been moderate and reasonable." Navy procurement officials were congratulated for having "endeavored to carefully analyze price quotations and to appraise them in the light of all the technical information available." Second, regarding collusion or favoritism between contractors and military officials, the committee simply found "no evidence." It declared instead that there was "keen competition." It therefore recommended no changes in federal procurement regulations, "since negotiated contracts are necessary until the aeronautical art becomes more stabilized." Finally, it urged "a more liberal use of experimental funds." Once again, the industry had been given everything it asked for, despite some rather harsh words at the outset.

There was a minority report, but it was signed by only one member of the eleven-man committee—Democrat W. D. McFarlane of Texas, the sole member from a state that as yet hosted little aviation business (there were four members from California, two from New York, two from Pennsylvania, and one each from New Jersey and Ohio—all with heavy concentrations of aeronautical activity). McFarlane was far and away the most aggressive inquisitor, constantly raising impolite questions about stock deals, bank connections, interlocking directorates, and other "merchants of death" flashpoints. "How can we determine and declare whether or not the profit made by contractors from the Navy airplane business is moderate and reasonable when the undisputed evidence in the record shows little or no competition?" he wrote. He drew for the record a complex chart showing banking, brokerage, and directorship linkages among the major companies, claiming that it pictured the existence of an "Aircraft Trust." All it really demonstrated, however, was that the aviation industry by now looked much like any other big American business.

In a vain effort to open the hearings to experts unbeholden to the Navy, he placed into the record a statement by Billy Mitchell, who was still a popular figure though long ostracized from the military establishment. Mitchell told his version of the history of military aviation, calling the World War "one of the worst orgies of graft and corruption . . . practically unequaled in our history." The real effect of the much praised 1926 Aircraft Act, he raged, "was to turn over our aeronautical development to what might be termed the aeronautical racketeers." That is, a "small group of people, similar to those who gained control of our aeronautical production during the World War, formed themselves into holding companies into which were taken the factories building aircraft, the factories building motors and engines, and the companies organized to operate the airmail and passenger service." These "manipulators" had been "all-powerful politically," he said, and enjoyed particular success in perpetuating two separate air services in the Army and Navy—"for in that way the functions are overlapping and more money is spent by the Government." It was vintage Mitchell, and nobody paid any attention.

The feisty McFarlane got nowhere with his colleagues, to whom his arguments sounded strident and often inane. Still, he managed to show that the manufacturers that won the lion's share of Army and Navy business were part of the same group that monopolized airmail contracts. And he pointed out that large parent companies could hide the steep profits made by their subsidiaries by filing consolidated tax returns, estimating that the government had thereby lost more than $2 million. But even in 1934, Congress was not a friendly forum for anti-business hypotheses, which were impossible to prove in any legal sense. Besides, Carl Vinson was not about to let the Navy be gouged by old bulls like Billy Mitchell. As a result, the Delaney hearings had little impact on the "merchants of death" debate, serving mostly—as Vinson no doubt intended—to deflect other, sharper attacks against his beloved Navy.

THE ROLE OF CHAMPIONING the movement to confront the arms traders thus fell upon another member of Congress and another special committee. In the same month of February 1934, Republican Senator Gerald P. Nye of North Dakota launched an investigation into "the activities of individuals, firms, associations and of corporations and all other agencies in the United States engaged in the manufacture, sale, distribution, import or export of arms, munitions or other implements of war." Somewhat like Representative McFarlane's, Nye's political compass was not aligned by isolationism or pacifism, but by being a defender of rural farmers against big-city business monopolies. In the 1920s, he had presided over crucial phases of the Teapot Dome scandal investigation and more recently chaired a special panel on senatorial campaign expenditures. He was thus inclined by experience to take the "merchants of death" literature as something more than chic intellectualizing.

The Nye committee was far more powerful than Delaney's. Consisting of seven senators appointed by the Vice President, it could subpoena witnesses and documents. In May, when Roosevelt called for ratification of the 1925 Geneva Arms Traffic Convention (which merely provided that each signatory would supervise arms deals by its own nationals according to its own laws), the President expressed gratification that the Senate would be scrutinizing the munitions trade. "The peoples of many countries are being taxed to the point of poverty and starvation in order to enable governments to engage in a mad race in armaments which, if permitted to continue, may well result in war," FDR said. "This grave menace to the peace of the world is due in no small measure to the uncontrolled activities of the manufacturers and merchants of engines of destruction, and it must be met by the concerted action of the people of all nations." This was not the pap of a best-seller but the imprimatur of the White House.

The Nye investigation therefore enjoyed considerable clout, at least during its early stages (it dragged on into 1936). For a while it even discouraged the Army from increasing its budget requests. Most significantly, it helped clear the way for neutrality laws later in the decade. But like the Delaney hearings, it had no deep effect on the weapons business, gaining historical value instead through the records it left that otherwise would have remained locked in corporate files and eventually been destroyed.

Among the many companies that came under its purview, the committee included Lockheed and Douglas as examples of aircraft firms with onerous export programs. Testimony about the two airplane builders was not taken until 1936, well after the media spotlight had turned away from the hearings and the whole "merchants of death" movement. But what the Nye panelists found most disturbing well after the fact were efforts by each company to skirt various arms embargoes, plus an eagerness to do business with two countries that were looking increasingly like enemies of America: Germany and Japan.

The chief foreign sales agent for both companies was Anthony Fokker, the brilliant Dutch engineer who had designed for Germany the best single-seat fighter of World War I, as well as the first interrupter gear mechanism that made it possible to fire a machine gun through a rotating propeller. Typically, Fokker was sold the technical data for American planes and could then do with them as he wished, arranging for exports or licensed production abroad. (In the case of the DC transports, Douglas received a royalty between $2,000 and $3,000 on each plane, plus about $100,000 for manufacturing rights.) Lockheed and Douglas gained a salesman with enormous influence throughout Europe, as well as a degree of legal detachment from the darker side of international arms deals.

The committee found, for example, letters between Fokker and Donald Douglas that referred freely to the use of bribes to obtain foreign sales, to Fokker's hope of doing business directly with the German government rather than with transnational companies there, and to the fact that Germany would rather purchase technical information than a manufacturing license for certain Douglas planes. "The fact that under the Treaty of Versailles Germany was under restrictions regarding the manufacture of and acquisition of military, naval, and aviation equipment was of no import to Fokker, apparently," Nye investigators charged. They also discovered a letter in which Donald Douglas had urged Fokker to evade the problem of getting a State Department permit to ship warplanes directly to the Cantonese government in China, since Washington only recognized the Nationalist government in Nanking.

At Lockheed, the committee found that Robert Gross had been negotiating to grant a license to a Czech subsidiary of Skoda Works, a giant European munitions conglomerate reputed to be active in the rearming of Nazi Germany. "Correspondence on this matter in the Lockheed files clearly indicated that the Skoda people were interested primarily in the military version [of a new Lockheed transport]," the panel noted. Gross also authorized a commercial agent in Berlin to try selling XP-900s to Germany, Bulgaria, and Ethiopia. Though the fighters were considered obsolete by the U.S. Army, the transaction emphasized how lightly the arms embargo against Germany was taken by American companies.

"The Committee took cognizance of information from an official source, not disclosed on the record, which gave further and very definite information regarding German activity in the military aviation field in 1933 and early 1934," the Nye members added, naming Douglas among four prominent American firms involved in the activity. "It is apparent that American aviation companies did their part to assist Germany's air armament. It seems apparent, also, that there was not an adequate check on the foreign shipments of these companies by the appropriate branches of the Government—the War and Navy Departments."

The committee also made note of Lockheed's sales to Japan, which although arranged by a commercial agent for ostensibly civilian customers such as the *Mainichi Shimbun* newspaper in Osaka, were clearly funneled to the Japanese military—as Robert Gross had mentioned as early as January 1932 to his brother.

Courtlandt Gross appeared before the Nye committee as Lockheed's assistant secretary, a sinecure apparently created for him so that he could testify as an officer of the company. "I have been an officer for a very short time and for a specific reason only," he told the committee, whose chief counsel wondered if it was "customary" for officers to earn commissions on sales. Gross claimed that Lockheed's effort to sell military transports to Skoda was "temporarily in abeyance," that he did not know whether Skoda was working for the German government. He insisted that the company had "never built a military airplane," that those offered for export were "commercial airplanes with military modifications."

Douglas took much the same oxymoronic approach in dodging the committee's questions, sending its Washington representative, John Rogers, to testify. Rogers, a Cornell-educated engineer with long experience in the aircraft business, maintained that he was "not familiar in any detail" with the letters and documents obtained by the committee, except for those that bore his own signature. The company's DC-2 transport had "no military features," he said. "Any airplane can be converted to military type, but those planes furnished abroad were no different than these airplanes supplied to our airline operatives." Nye's chief counsel then entered into the record a letter written to Donald Douglas by Anthony Fokker, whom the counsel identified as "the man that kept Germany in the air during the war." Referring to his efforts to sell DC-2s, Fokker had made the following comments:

> Selling the license to the Swiss factory of Dornier at Alten Rhein means disclosing all the technical information to the Dornier Company in Germany, and we hope to do a better business with the German Government direct, or with one of the German firms with which we are negotiating.
>
> Germany probably will not buy officially the license on the Douglas, but unofficially may buy the technical information.
>
> Spain, after my personal visit at Madrid, intends to buy two airplanes for troop transportation. . . . This business will not be closed for at least half a year, as it takes this time to obtain the necessary funds and pass all the Government departments. Besides that, many people have to be "seen."
>
> Sweden: The Government was interested to acquire the license for the state factory, but the Geneva Treaty Convention prevented them from acquiring any large bombing equipment, and they are now negotiating for smaller. Maybe we can place the Lockheed.
>
> As to the price situation, we must make a considerable margin for official and secret commissions.

Here was the kind of incendiary document that would never escape the office shredding machines of a future epoch. "He puts this 'seen' in quotation marks," the committee counsel pointed out to Rogers. "Do you take that to be a general comment on how things are done over in Europe?" Yes, Rogers replied, "It seems to reflect something of the sort." Asked if he knew anything about Germany's interest in buying DC-2 technical data, Rogers said, "I do not; and if anything did happen, it would have been a transaction between Fokker and the German Government, not with the Douglas Company."

"He had all the information?" the counsel asked.

"He had everything," Rogers admitted.

"He could sell it if he wished?"

"Exactly."

Referring to a bomber version of the DC-2, the counsel quoted a letter from Donald Douglas in which he implored Fokker to "use this data judiciously and not show it anywhere outside of England; the ship has been flying for the last month and is now at Dayton, where it is undergoing tests by the Army." But Rogers maintained that the aircraft "was developed by the Douglas Company at its own expense," that the "United States Government had no control over it and would not have any control over it until such time as they purchased it."

"So [Douglas] could let Fokker have the specifications?" the counsel asked. "Then why this care about it?"

"Frankly, I do not know what [Donald Douglas] had in mind when he sent it over there," Rogers answered.

"Do you know the situation, Mr. Rogers?" the counsel replied incredulously. "Mr. Fokker has contacts all over. Does not the company put itself very much in his hands by this? Suppose the [U.S.] Army had bought that plane. Did it, by the way?"

"Yes, it did."

"All right. Then, before the Army has the planes, the specifications are over there, and Fokker—or somebody, perhaps, whom you trust less than Fokker—could peddle them around to whomever he wanted to, could he not?"

"That is quite possible."

The committee put into the record several letters that revealed the premium prices placed on exported aircraft. For example, a DC-2 worth $58,000 to TWA was sold to the Soviet government for $101,480, along with a $40,000 Northrop Gamma outfitted as an attack bomber for $76,100. Japan's Nakajima Aircraft Company—a primary builder of military airplanes—bought a license to manufacture DC-2s for $80,000, along with the right to sell the planes to the Manchukuo government in China, which the United States did not recognize. For an additional $20,000 a year, Douglas agreed to furnish Nakajima with engineering data for all changes developed in the DC-2 design.

To shed light on at least part of the reason for these premium prices, the committee produced a memorandum from an unnamed U.S. government official summarizing a conversation between the official and Rogers: "It had been a custom for years for [the Douglas company] to pay 'generous commissions' to its agents abroad in connection with sales to foreign governments, and [Rogers] had never doubted that a large part of such commissions found its way into politicians' pockets."

The committee also reviewed correspondence (not placed in the record) "which indicated that in return for certain favors performed for the War and Navy Departments by one of Douglas's foreign agents, the Douglas company expected definite favors in return from the War and Navy Departments on matters connected with some foreign sales transactions." That is, in return for these unspecified "favors," American military officials would turn a blind eye to federal restrictions on weapons trading.

Anthony Fokker, who might have provided explosive evidence to the committee, was never questioned in public. In September 1935, Nye dispatched a special investigator to serve a subpoena on Fokker, who was living in Alpine, New Jersey. With the help of the U.S. Coast Guard, the investigator boarded Fokker's yacht *Helga* off Montauk, Long Island, where the famous warplane designer had slipped away for a few days of bluefishing. Fokker was not aboard, but was cornered later at the Montauk Yacht Club. Nye then announced that it would be unnecessary for Fokker to testify in Washington if he would make a satisfactory deposition to the committee.

Fokker's affidavit was not released until after the hearings were over, and then only because the press accused the panel of suppressing evidence. In October 1936, an article in the trade magazine *Aero Digest* alluded to Fokker's allegation that he had once contracted with Elliott Roosevelt, the President's son, to sell military planes to the Soviet Union. Fokker said that Roosevelt received $5,000 in cash for the deal, which did not ultimately go through. Nye decided unilaterally to release the deposition, causing a brief storm of criticism from Democrats just one month before national elections. But the affidavit was quickly forgotten, or, perhaps more accurately, left untouched as a too-hot political potato. Democrats did not want to hurt Roosevelt, and Republicans dared not open the issue of exports.

By the time the Nye committee released its full voluminous report and document collection—which presented a fragmentary picture, at best, of an emerging military-industrial network—national debate had turned away from the "merchants of death" angle to the potent political issues of neutrality and isolationism. Though later scorned by President Harry S. Truman for "pure demagoguery in the guise of a congressional investigating committee" that had been "backed by isolationists and 'America Firsters,' " the Nye panelists were not motivated by isolationism, a pejorative term for pacifism. In an authoritative account of the committee's two-year existence, historian John E. Wiltz concluded that "the majority hoped to persuade Americans that the United States should avoid economic ties with countries at war, that it should not become an arsenal for belligerents."

Through a glass darkly, as Wiltz put it, the Nye committee saw that modern warfare created ominous problems for a liberal democracy. But its proffered solutions, such as nationalization of the munitions industry and radical limitations on wartime profits, were not taken seriously. Close association between the military services and munitions manufacturers "constitutes an unhealthy alliance in that it brings into being a self-interested political power which operates in the name of patriotism and satisfies interests which are, in large part, purely selfish," the panel warned. Such associations were "an inevitable part of militarism" and should be "avoided in peacetime at all costs."

Unless the arms trade were restricted, many Americans believed in the early 1930s, disarmament was a farce. But after Congress passed the first neutrality law in August 1935, the context of debate changed drastically. The business practices of individual companies seemed trivial compared to events in Spain, East Africa, and eventually all of Europe. By 1938, "merchants of death" was a quaint anachronism. The darksome records of Lockheed, Douglas, and other weapons manufacturers were conveniently forgotten.

FOR GLENN MARTIN, 1934 was the year when his financial rope ran out. Though the new bomber his engineers had developed with the Army was a technical success, winning the 1932 Collier Trophy for "the greatest achievement in aviation in America during the preceding year," the company lost $633,379 building forty-eight of the aircraft for $2.44 million. At the end of 1931, the company had been

$167,709 in the red; by the end of 1933, the figure climbed to $335,374. Assets were valued at $2.4 million, but liabilities stood at $4.3 million. Without another contract from the Army for eighty-one more bombers worth $3.75 million, the operating deficit would approach a staggering million dollars in 1934.

In short, Martin was broke and had no choice but to go through Section 77B of the Federal Bankruptcy Act. The company was otherwise unable to retire $2.8 million worth of five-year gold notes that had been issued to raise cash immediately after the 1929 Crash. While the notes were outstanding, a stock dividend could not be paid, thus making shares unmarketable on the public exchanges. In April 1934, Glenn Martin allowed 150,000 of his 525,000 shares of company stock to be "retired," or eliminated, thus reducing the number of outstanding shares to 376,000. Of these, 325,000 were offered to the public in a desperate attempt to raise capital, but only 20,000 were sold (to a single investor in New York named Charles R. Crane, apparently a champion risk taker) at $11.50—not enough to cover the company's 1933 deficit, let alone what was projected for 1934.

In an effort to paint over the failure, Martin's broker announced that the sale would be deferred pending the letting of military contracts for the new fiscal year, which began in June. But the consensus of industry insiders was that the asking price was too high and that no amount of new work would justify it, given the torpid condition of the company.

In order to qualify for the new Army order, which the federal government would not approve if the company defaulted on its gold notes, Martin was forced by August to petition the holders to refund them for new paper that would mature in 1939. Slightly over $1 million of the notes were held by Martin's principal bank, the Baltimore Trust Company, which was itself bankrupt. (Since 1929, Baltimore Trust had been one of the primary investors in Martin's new plant. When it went into receivership in March 1933, the Martin Company lost access to $318,143 in deposits. Glenn Martin's personal savings were also impounded, and he became liable as a shareholder in the bank to its creditors.) Baltimore Trust's receiver agreed to refund its notes. But the National Aviation Corporation of New York, a large holding company that owned $500,000 of the notes, objected violently to the plan, finally forcing one of its directors onto the Martin board. All in all, it was a confused, desolate picture.

There remained the matter of the snowballing operating deficit. In order to climb out of this deep morass, Martin turned—like so many other of the era's stalwart but broke Republican businessmen—in November 1934 to the New Deal's Reconstruction Finance Corporation. Given the fact that the company employed nearly 2,000 people and was perceived as critical to the nation's defense, there was little difficulty in obtaining a $1.5 million loan from the RFC and the Federal Reserve that would have been impossible to secure from commercial banks. Martin swallowed his pride and mortgaged his plant to the Democrats.

Throughout this cloudy period, Martin generated silver-lining publicity from a contract with Pan American Airways to build three giant "China Clipper" flying boats for transoceanic routes. Pan Am originally called for planes that could carry mail between the United States and Europe—nonstop, with midair refueling—but this idea proved too ambitious. Instead, the plan was modified for passenger travel across the Pacific, using Hawaii, the Philippines, and Guam as stepping-stones. It was the kind of work Glenn Martin loved most— exotic, luxurious, exclusive. The four-engined planes carried forty-eight passengers for distances up to 3,200 miles. Their crew of six was the most elite in air travel, required to have knowledge of the Clipper's technology as well as of the culture of its destinations. Both companies made much of the fact that Lindbergh was chairman of the technical committee that determined the Clipper's specifications, though he was becoming more and more reclusive, a famous man with a valuable name much like Orville Wright had been twenty years earlier. Newspaper accounts put the contract's value at $800,000, but failed to explain that it merely granted Pan Am the option to buy $800,000 worth of airplanes under stipulated terms.

In fact, Martin lost heavily on the three Clippers: $230,000 in 1934 and $270,000 in 1935. The company planned to amortize these sums over a future production contract, but learned in June 1936 that Pan Am would not go into quantity production because the planes were too expensive to operate. The project's fate was sealed when one of the Clippers disappeared without a trace near Hawaii, raising the ghosts of the 1927 Dole disaster and shattering the illusion that seaplanes could ditch safely in the ocean and wait for rescue like disabled passenger ships. Still, the glamorous airliners—with their haute cuisine and sophisticated service—became as closely associated with Glenn Martin in the public's mind as his bombers, symbolizing civil aviation in the 1930s as vividly as Douglas's DC transports. But Martin never built commercial flying boats again or another passenger airliner of any kind until after the Second World War.

Martin's financial distress during these mid-decade years was best illustrated—along with his eccentricity—after the death of his father at age seventy-eight in June 1935. Neither Glenn nor his mother had kept any notable contact with Clarence Martin after they left the West Coast in 1918. He maintained several feeble businesses around Santa Ana, including a wrecking company and a small orange grove, joining the legions of sunburnt tawdry men who reached benign dead ends in California or Florida. There was never a divorce, just many years of estrangement during which the former traveling hardware salesman from Iowa remained trapped in poverty while Araminta and son scaled the heights of American capitalism. Glenn did not keep Clarence totally hidden the way sister Della was—he admitted having a father if asked—but there was no sign whatsoever of kindly emotion for the decrepit old man.

The only evidence was of remarkable callousness. Six months after his father's funeral, Glenn sent a letter to the undertaker in Santa Ana, a family friend since childhood. "I happened to inquire of a funeral director friend in the East about the merits of casket construction and costs of services," he wrote officiously. "Since there seems to be a marked difference in costs, I would like to ask you to review your statement with the thought that you might like to make a substantial reduction." The bill for the funeral had amounted to $986.88.

"My dear Glenn," the undertaker replied, "I am at a loss to know how to answer you concerning your account. As you probably know, I have been engaged in this business, in this town, for over forty years, and yours is the first case where fair dealing has been questioned, and the one case where it was least expected. We have decided to make you this offer. If you will send us your check for $36.88, being $25.00 for music and $11.88 for state sales tax, we will cancel all other charges as we value a clear record above Dollars."

"I appreciate very much your frankness and the generosity in leaving the matter entirely up to me," Martin wrote back. "Mother and I at the present time are a long way from being well-to-do and when I requested that you give us consideration it was not intended to reflect on your honesty in dealing, but was merely intended to ask for a split in profits." He enclosed a check for $750.

Thus the world famous warplane magnate shaved $286.88 off the cost of his father's burial by haggling with a small-town undertaker. The taxpayers of America might have erected a statue in his honor had he ever paid such parsimonious attention to the multimillion-dollar business of building bombers for the U.S. Army.

ON JUNE 28,1932, the front page of the *Burbank* (California) *Review* carried stories of momentous change. "Roosevelt Forces Win Important Victory" was the headline on the lead story from the Democratic convention in Chicago. A short wire dispatch from Berlin near the bottom of the page said that "the alarming spread of political chaos, street fighting and rioting throughout Germany indicated today the government might have to resort to martial law to prevent more serious disorders." And there was, of course, some of the depressing detritus from a wrecked economy: a deputy county clerk who lost $70,667.07 playing the stock market between June 1930 and December 1931, on trial for embezzling $75,000.

But in the top left-hand corner of page one was a rather upbeat story for the era. "Lockheed Co. Ready to Take Care Orders" the headline said. The newly reorganized company "is now ready to take care of all orders for airplanes as fast as they come in," the lead paragraph trumpeted. "However, until new orders begin to come in the local plant will not resume production on an extensive basis in order to avoid an unwarranted rush of men to the factory seeking employment."

Like Donald Douglas ten years earlier, Robert Gross was in the awkward position of having a physical plant and a company name but no capital to finance production. Unlike Douglas, he did not have a government order to use as bait. His most valuable asset in this regard was the word "Lockheed," now a piece of corporate property unconnected to the pioneering brothers who had been born with it.

When the *Burbank Review* article appeared, the books of Lockheed Aircraft Corporation showed a balance of $864.82. By the end of July, the sale of spare parts for various old Lockheed planes had brought in another $1,623, including $156 from "A. Earhart." During the same period, the company spent $1,538, mostly on overhead expenses. This left a balance of $950, but there were also

$3,790 worth of unpaid bills, including $2,541 in accrued payroll. It was obviously a small-time operation, but the judge's parting words must have been haunting Robert Gross's sleep.

In early August, San Francisco broker and Lockheed investor Thomas F. Ryan III went East with Gross's blessing to seek further financing for the company. On August 13, Ryan wired Gross that if Lockheed were reincorporated in Delaware, then a stock purchase could be arranged with a Washington, D.C., brokerage house that would bring in $60,000 in three payments between October 1 and November 1, with an option for $40,000 more within a year. Given the harrowing business conditions of that summer, Gross and the other original Lockheed investors quickly authorized him to pursue the deal.

"The credit for this goes to young Mr. Ryan," Gross wrote happily to his father on September 2, "who went back East and induced these people to come in." Gross felt that the deal would produce "the money we need to work with, without giving up control of the company." Meanwhile, Lockheed would continue to limp forward on its "good looks" and spare parts business.

The next problem that Gross had to confront was his brother Courtlandt, who journeyed out from New Haven to see with his own eyes what had come to pass. The Viking Flying Boat Company was a dead fish more obviously now than ever. Robert therefore formalized an agreement whereby Courtlandt would act as Lockheed's Eastern representative and sales agent. It was a rather weak palliative, given that the company had nothing yet to sell, but second-born Courtlandt was in no position to decline.

"This is the day before election," Robert wrote to his father on November 7, reporting the arrangement with Courtlandt, "and I am looking at the outcome with mingled feelings. I feel that Mr. Hoover deserves another chance if he wants it, but I also feel that it might be a kindness to him to turn the job over to a Democrat who, regardless of qualifications or ability, can probably be depended upon to get congressional co-operation." Such were the ambivalent sentiments of the Republican aristocracy about FDR, which Gross would soon have to swallow in much the same fashion as Glenn Martin.

In the last month of 1932, after news of Lockheed's rebirth had settled through financial circles, Gross came face-to-face with the same distant forces that had derailed Martin in 1916. But Gross was infinitely more sophisticated than the "Flying Dude," and was able to keep his wits. The encounter was with J. Cheever Cowdin of Bancamerica-Blair, a director of twenty-three companies, including TWA, North American, United, Curtiss-Wright, and Douglas. Cowdin was the apotheosis of pyramided holding companies, a human switchboard of financial entanglements. In 1930, he had been among the handful of representatives from large transport companies who had helped the Postmaster General fix major airmail routes, thus preserving lucrative subsidies for themselves. Like Harold E. Talbott Jr., Cowdin could not be ignored.

"Two days ago I received a phone call from Mr. Cowdin, requesting an appointment," Robert Gross wrote to Courtlandt on December 16, showing deference with the use of Mr. "The upshot of this whole affair was that he looked into our figures and our plant in some detail and finally told us that he was talking for any of three companies: North American Aviation, National Aviation, and General Motors." Cowdin asked Gross if he was interested in selling control of Lockheed to one of these groups. "He was somewhat surprised and, I think, secretly chagrined to find out that we had around $100,000 in the business and that . . . it would take at least $120,000 to purchase 51 percent."

This level of investment apparently moved Cowdin to select just one of the cards in his hand. "The more we talked with Mr. Cowdin the more he tried to give us the impression he was talking for General Motors . . . to the exclusion of the other two trusts," Gross continued. "He also tried to give us the impression that our old theory was still in force, i.e., that in order to get business from TWA we would have to hook up in some way. Cheever tried to rush us into the deal in typical Wall Street fashion by giving us a big dinner at The Town House and having his lawyer in the next room; but of course we did not fall for just travelling expenses and told him we would think it over." Here was the kind of back-room action Representative W. D. McFarlane of Texas sorely needed to know about two years later but never uncovered.

"I have told the boys here that, as far as I am concerned, at some future date, if the General Motors company were to make us a decent proposition, even involving control of the business, but giving us personally the management of the business with enough scope to make some money, I would hesitate a long time before turning it down," Gross concluded.

A week later, he had lunch with Harry Wetzel, Douglas's general manager, whom he referred to as "the active man in the Douglas company." Wetzel believed that Cowdin had come to California specifically to promote a deal between Lockheed and General Motors. "He went on to say that in times past [Douglas] had been propositioned by [General Motors] and that the General Motors Company had tried to move in on them from all angles, so that he and Donald were practically dizzy trying to figure out how many scouts General Motors had out for the same prey," Gross remembered. "I asked Mr. Wetzel what, in his opinion, could be [General Motors'] idea, in view of the fact that they had a whole bag of troubles of their own; he said that they wanted to capitalize on the name 'Lockheed,' which he thinks is the best in the business next, of course, to Douglas."

Gross badly needed Cowdin's largesse, though he was not yet ready to admit it. In efforts to garner airmail business from the government, he had been stymied by the Postmaster General's public statement before a congressional committee that he would never allow "single-motored, fast-landing" planes to carry passengers. This was widely assumed to be a direct blast at Lockheed's Orions. And in his attempt to sell equipment to the major airlines, he was handicapped by the company's independent status. "We have succeeded in getting together enough cash to keep the wolf from the door," he wrote to his father at year's end. "The difficulty seems to lie in the fact that the four big airlines who in total consume 75 percent of all the transport planes, have close affiliations either through ownership or influence with airplane manufacturing companies. Naturally, this works to the advantage of the airplane manufacturer who happens to be tied into a big airline. At the same time it has certain disadvantages, since it tends to brand a given airplane as private property of a certain airline." Here was the essence of "Trust," in workaday terms.

By the first month of 1933, it became clear to Gross that Lockheed faced difficulties far deeper than he had foreseen even a month earlier. He learned that TWA was going to buy a Northrop Gamma, which, while not strictly competitive with the Lockheed Orion, ruined a sale he had been counting on. "I have tried to emphasize to the group here that there is no need for us to get panicky because we lose two or three sales," he wrote to Courtlandt on January 18, "and it is not a good time to be stampeded into making some wild-eyed move which calmer reasoning would cause us to steer away from." There seemed to be "only two places to drive at," he said. "One, the organization ourselves of transport lines, and secondly, the possible getting of government contracts for war machines." The latter was by now a well-worn path for aviation businessmen, whether part of a "Trust" or not.

His problems were compounded when the underwriting deal arranged by Thomas Ryan fell far short of expectations, damaging the company's tender image on Wall Street. The Washington, D.C.– based brokerage house with whom Ryan arranged the purchase of 60,000 shares (his brother worked for the firm) subsequently resold the right to market these shares to another firm in New York. The new broker failed to find any substantial buyers, forcing Gross and his fellow officers to buy 41,00 shares with their own money to keep the company afloat.

At the end of January, a shaken Gross queried a New York investment banker about the possibility of selling the company off. "I wish to point out that between the time that we purchased the company in June [1932] and the latter part of November we had practically no business whatsoever," he forthrightly admitted. "We had not yet received sufficient capital in the business to be in a position to satisfactorily execute an order." Since then, the company had won three orders for Orion types, two of which were ready for delivery. Orions cost about $16,000 to build, Gross explained, and sold for $20,000; Vegas cost the company $14,000 and sold for $16,000. He estimated the cost of bringing out a brand new transport at between $60,00 and $70,000.

"It would seem entirely probable that a moderate-sized, carefully administered company acquiring the manufacturing facilities and the technical development of Lockheed . . . should be able to improve its position under every advantage, and would be able to occupy a prominent place in a steadily expanding market," he concluded optimistically, using the same rationale that he had found so convincing himself a year earlier.

Besides such quick-killing strategies, Gross was preoccupied with the headaches that were plaguing all small businessmen in the midst of a depression. An Orion ordered by the Air Express Corporation, a feeder airline, had been sitting finished in the Burbank plant for ten days, unpaid for. "We have wired, written, telephoned, spoken to and, in short, done everything we could to advise the company that the plane is ready and that we would like our money, but to date have had absolutely no answer from them," he complained, telling Courtlandt to have a "frank talk" with the airline's owner in

New York. Cash was a rare commodity, and a $20,000 airplane could languish for a long time in its assembly bay.

On March 2, like unexpected crossfire, the governor of California ordered a complete closing of all banks in the state to forestall mass withdrawals, leaving Gross with "a few cents over five dollars in my pocket" and "a twenty dollar piece in the vault." Lockheed had about $28,000 in cash at the time—not a rosy picture even when the banks were open. A panic was avoided when Washington guaranteed new deposits, but the effect of the episode on general finance was grim.

Obviously wishing that Cheever Cowdin would reappear, Gross considered turning to the big trusts to finance development of the new Lockheed transport, which was undergoing wind-tunnel refinement at the University of Michigan. In mid-March, however, his desperation eased somewhat when TWA ordered three Orions instead of the Northrop planes it had first favored. TWA's reversal may have been based partly on the immediate availability of one Orion built for Air Express but still unpaid for, as well as the deeply discounted offering price of $13,800 per plane. The decision could only have been an ironic blow to Jack Northrop, whose TWA Alphas would now be replaced by a Lockheed product he had grandfathered. For Robert Gross, it was precisely the foot in the door he had been waiting for. "It is worth almost any sacrifice on our part to get a plane which the post office is willing to put its stamp of approval on," he said of the TWA Orions, which would be used on a night mail flight between Los Angeles and Kansas City.

At the end of March, one of Walter Varney's seven Orions crashed in a freak rainstorm, putting further pressure on a business that was already foundering from a slump in ticket sales. Varney asked for help from Gross, who demurred. Their relationship had soured since the acquisition of Lockheed, because of Varney's dissatisfaction with the company's shaky start and Gross's perception that Varney was not doing enough to "carry the Lockheed flag." Varney was never a force in the Lockheed management, and would grow increasingly distant from its affairs.

After the threat of bank closures was resolved with deposit guarantees, Gross returned to the problem of selling the company's stock. Through the Walker family, he secured the services of a San Francisco brokerage house to sell 325,000 shares at $1.40, from which Lockheed would net $1.12. In order to sell directly to customers

without going through the New York Stock Exchange, the brokerage established a shell corporation with its own license and salesmen. At least $50,000 worth would have to be sold by June 15 for
the deal to remain in effect. "Generally speaking, getting money for
the company has been a hard task," he wrote after signing the
agreement in April, "and I am not sure we have gotten it by any
means, as there is no definite commitment. However, it begins to
look as though we would surely get some money and I rather think
that if we can continue to do some business here we will get quite
a substantial amount."

The Burbank plant was actually quite busy, with seven planes in
the work bays—three for TWA and four under repair. Progress on
the new model was at a standstill, however, until financing was in
hand. "While I am enthusiastic about the type, I feel that we have
made several mistakes in it," Gross said, "but because I know so
little about airplanes I am deferring to the judgment of the other
boys." Gross had originally envisioned a $30,000 wooden aircraft
that could carry from eight to ten passengers with "very brilliant
performance." He guessed that this would fill a niche "not touched
by Boeing, Douglas, Northrop or General Aviation [General Mo-
tors]." But the plan now was for a $35,000 all-metal plane of larger
capacity that "looks slow to me," he worried.

During these months, as Robert Gross constantly pressed his
brother to circulate among the financial elite in Manhattan on Lockheed's behalf, he wrote letters that would have sealed the arguments of the McFarlanes and Nyes in Congress who suspected the
existence of a powerful network of aviation money men, but who
could never quite prove it. Gross may have been merely prospecting
for gold around his social circle, but paragraphs like the following
from a letter to Courtlandt on May 11, 1933, suggest something
more intricate:

> I had a very nice talk with Mr. Hanshue [Harris M. Hanshue, pres-
> ident of profitable Western Air Express, forced by the Postmaster Gen-
> eral to merge with unprofitable Transcontinental Air Transport in 1930
> to form TWA] the other day. He is very friendly toward Lockheed and
> would like to do everything he could to help us. He says quite frankly
> that he thinks the General Motors group have gotten control of West-
> ern Air and feels that the business will be gobbled up and lost in the
> scuffle. When it becomes no longer a question of doubt on this point
> he is going to make up his mind what he will do. I think it would be

a relatively easy matter to get him actively interested in Lockheed, but I am proceeding rather slowly as I do not want to act hastily.

I had a long talk with Mr. Cord Saturday afternoon [E. L. Cord, who in 1932 gained control of the Aviation Corporation, another of the major holding companies, in a savage proxy battle against Lehman Brothers and W. Averell Harriman Co.; Cord was known to the public for his namesake luxury automobiles, spectacular stock manipulations, and for having offered to carry mail at half the prevailing rate, then slashing pilots' pay by 40 percent, triggering a strike; he also invested heavily in naval armaments; was enjoined by the Securities and Exchange Commission in 1937]. I guess I was with him for four or five hours. He came out to the plant, went all over it, climbed all through the Orions and talked breezily and glibly about an order for twenty. I think he was just talking. He told me some interesting things about the airplane developments, one of which was that he expects to buy some Boeing planes in return for certain concessions from the United group. I think the concessions are political. He also said that in his opinion the men who could help Lockheed more than anybody are Richard Hoyt [one of the original members of the Wright-Martin syndicate; a director of TWA, North American, United, Aviation Corporation, Pan Am; and chairman of the board of Curtiss and National Aviation] and Guy Vaughan [another Wright-Martin alumnus and Curtiss director]. He said that he and Mr. Hoyt had struggled over business for a long time, but that he had always found Mr. Hoyt pleasant to deal with and considered him the keenest, shrewdest man he had met in this business. Why not . . . see if you can't get an introduction to Mr. Hoyt and tell him about our business, and try to get his ideas and help? If he tries to just turn you over to Cheever Cowdin I would be inclined to slow up on it. I also think it would do no harm to tell our story to Hank Dupont [Henry B. Dupont, officer of General Motors]—I see now that he is a director of North American.

Here was a road map for the "merchants of death" theorists, but they were of course never privy to such private communications. Congress would force the dissolution of certain interlocking directorates and stock ownerships in 1933, but the social web of financial power would remain. Gross was also quite explicit in his correspondence about the export business, some of which would eventually come under the Nye committee's gaze. In discussing with Courtlandt a possible sale to the government of Brazil, whose battles with revolutionaries and arms dealings with embargoed neighbors made it the second-largest importer of American warplanes next to China, Robert made the following observations:

"As for military airplanes, you know that the Lockheed company has only developed two studies and it of course would be very much of a job to build up an entire new fighting plane on the strength of an order which, on the face of it, would be subject to political caprice. Would it be possible, do you think, to submit the standard Orion with camera and bomb adaptation? This would be a relatively simple matter and I think we could tackle it.

"A guiding principle in dealing with [exports] would seem to me to be that, before the Lockheed company could do any building, it would have to have a substantial down payment in money and an absolutely irrevocable evidence of full payment where the business is admittedly of an unusual nature; it seems that we ought to be positive that our money could not be cancelled out on us just because somebody did not get paid off enough down below." Lockheed would soon learn the export arts as thoroughly as any of its competitors.

Gross left no stone unturned in his search for business. Near the end of May, he journeyed with Cyril Chappellet and Elliott Roosevelt to Mexico for a private audience with General Plutarco Elías Calles. Calles, whom Gross called the "real ruler of Mexico" and "the man who makes all the presidents and tells everybody what to do," had employed Malcolm Loughead in 1914 as chief engineer of his primitive air force. The Lockheed contingent petitioned the strongman for a mail contract between San Diego and Mexico City. Elliott Roosevelt was evidently untroubled by the political implications of his presence—the Nye committee never learned of this particular meeting.

In June, though the sale of Lockheed stock was still insufficient to fund full development of the new all-metal transport, Northwest Airlines signed a contract for one of the aircraft—now officially called the "Electra"—at $35,000, plus an option for two more. Gross made note of the fact that "both the Douglas and Northrop factory are busy on their new experimental planes [DC-1 and Gamma] for TWA and we understand confidentially are having the dickens of time with both of them." The Gamma was "not so hot," he had heard on the grapevine. Near the end of the month, Gross's Saturday afternoon with E. L. Cord paid off when American Airways ordered six Orions at $19,200 each. "This is extraordinary news for Lockheed," he exulted.

Buoyed by this slight upturn, Gross returned to the matter of exports. Though the company had yet to sell any planes abroad, it maintained a contract with the export firm of Miranda Bros. in New York. Another agent had recently presented Lockheed with a definite plan for selling fighter planes to Brazil, however, bringing the complexities of international arms trading to Gross's direct attention. "I would be more enthusiastic about the kind of business which [the agent] proposes, but on the face of it, where it is admittedly of a grafting nature, it seems to me it raises certain objections," he told Courtlandt, who was in favor of the proposed deal. "For instance, if the Lockheed company became involved in a sale of goods to a South American country and our agent was shown to have been bribing the officials of that country, might we not ourselves come in for certain international litigation with some chance of recovery being had from us? I confess I do not know anything about international law."

Gross's hesitation was thus based not on moral qualms, but on concern about Lockheed's liability. Congress had recently passed a joint resolution prohibiting "the sale of arms or munitions of war in the United States under certain conditions," singling out Bolivia and Paraguay for embargo. (Aircraft were not specifically mentioned as "munitions," but the government would later prosecute Curtiss-Wright for shipping four planes to Peru that Washington charged were actually meant for Bolivia. The case was not settled until 1940, when Curtiss-Wright officers pleaded guilty.) The way around this problem, of course, was to sell the airplanes or the relevant engineering data to an independent agent—not directly to the foreign government—who would carry on his business one step removed from Lockheed. This would all become clear to Gross in due time.

Raising capital at home continued to be his first concern. The company had stumbled on the New York Stock Exchange, and the attempt to organize a shell corporation in San Francisco to sell shares had now fallen through as well. Gross therefore decided to bring the effort back to Los Angeles and try to take advantage of the company's hometown reputation. Even then, he found that "we could not do anything with the [brokerage] companies of the highest standing . . . and we must turn to one of the smaller houses of a decent and respectable background, and above all, not members of the New York Stock Exchange." His strategy was to try selling first in California, then nationwide if the first offerings were successful.

To set up this sale, Gross reorganized Lockheed yet again. He folded its Delaware incarnation and established a holding company called Southern California Aviation to keep all of Lockheed's outstanding stock. Southern California Aviation was in turn controlled by Gross, Chappellet and the Walkers. Walter Varney was no longer a factor, having relinquished his interest to concentrate on his troubled airline.

These tortuous maneuvers (compared to the giant pyramid companies like North American and United, Lockheed's were actually quite simple) seemed to have no other purpose than keeping dust from settling on the company's stock. Slight name alterations and changes of incorporation venue allowed Gross to present a fresh face to investors. If he could not sell in New York, he switched to San Francisco. If not San Francisco, then Los Angeles. His finesse in this regard was crucial to the company's survival. The days when Allan Loughead would comb Hollywood for rich angels were as long gone as homemade biplanes.

Work on the Electra proceeded slowly and at great expense. By mid-August, production jigs were finished and aluminum was starting to be cut. It was beginning to look like a worthy successor to the Lockheed Vega, a plane both beautiful and technically sweet. There was something futuristic about its distinctive triple tail, yet an overall sturdiness in its compact dimensions. Hollywood would love the Electra in later years, making it the getaway airliner in the last scene of *Casablanca*. Sale of Lockheed stock from Los Angeles took hold, bringing in $31,178 by the end of the month. "My main concern during the past year has been to get capital," Gross said wearily, but he now felt that his "immediate worries about capital are over" and that "the money is and will be forthcoming." His original conception, that sales of established Lockheed products would tide the company through development of a new transport, seemed for the moment to be proving out.

During these months, Gross—a deeply conservative Republican—complained bitterly about New Deal programs that he felt were hindering his ability to run a profitable operation. The National Recovery Act's labor provisions were especially galling to him, because they set a forty-hour-week standard for work time in an effort to boost employment. "For a business constituted as ours is, we get no immediate relief through a general increase in pay rolls and shortening of hours," he said. "In fact, it is a distinct sacrifice . . . because we are not doing business freely with the consuming public, and hence freer circulation of money is not necessarily reflected in greater sales." The NRA "makes it very difficult for us to get [the Electra] done on time," he fretted. "We have received permission from the N.R.A. Compliance Board to work overtime and extra shifts, and while we are not working the same men both days and nights, we do have two [eight-hour] shifts."

Gross's resentment of what he saw as federal meddling in his private business affairs would grow over the years into a personal political crusade. Already he was thinking beyond Burbank. "With no thought to detract from the great struggle our president has made against economic ills," he said sarcastically, "I view with consistently increasing alarm the unmistakeable march toward socialization of industry and nationalization of business."

By the end of summer, stock sales had brought in $55,000. The company employed about 300 people, turning a profit of $1,700 in August and $13,700 for September (due to the simultaneous delivery of five airplanes). "We actually did about $4,000 better than this," Gross said, "but because the airlines are howling about the profits we are making we decided to be conservative and tuck a little of it away in development." In a letter to his father, he wrote a capsule summary of the air transport business's complex horse race:

"The aircraft field is narrowing down now to a very interesting struggle between two or three groups. The United group, whose Pratt & Whitney engine is still the best product, are on one side of the fence; and the General Motors–North American Aviation–Wright Aeronautical group are lined up against them, with Cord sitting in the middle not knowing which way he is going to jump. United are somewhat concerned over the fact that the General Motors–TWA group are probably going to discontinue the use of their engines and use Wrights, due to company alignments rather than merit of the product. This makes Pratt & Whitney and United plenty upset and I look for some fireworks. We are playing pretty close to Pratt & Whitney since we are using their engines exclusively."

Most of this action occurred far over Gross's head, but it colored his workaday world. Until the Electra entered the market, he was obliged to keep one foot in the 1920s. "The Orion and Vega business holds up pretty well," he marveled, "but every order we get we always say it is the last, and still they come." Nonetheless, the old designs were very profitable for the company. When orders came in and there were no construction difficulties, "we can whack them out like hot cakes," Gross boasted.

Though the Electra was still unfinished, interest was growing in important quarters. In November, a U.S. Navy commander visited the Burbank plant to order one of the aircraft for the Secretary of the Navy "as a sight-seeing and general purpose transport plane for congressional investigations and political journeys," or junkets, according to Gross. And Pan American decided to take three for use on an Alaska route. With the successful sale of 150,000 shares of stock, from which the company took $168,000, the project seemed on firm ground at last. Sales in 1933 would total $355,989, for a pretax profit of $36,354. The Electra, whose development cost had soared past the $100,000 mark, was scheduled for a maiden flight in early 1934. To Gross, the only "flies in the ointment" were the NRA and unionization of Lockheed workers by the American Federation of Labor.

In order to squeeze every last drop from the company's stock, Gross asked his father to "initiate a little buying" on the Los Angeles Stock Exchange. The "least outside interest" in Lockheed shares "has very immediate effects," he noted, pointing out how a recent order for fewer than a thousand shares had pushed the stock from $1.50 to $1.75. "Frankly, it looks to me as though a nice little turn could be made in it without tieing up any capital to speak of." Laws against market manipulation and insider trading were still in a fetal stage, so Gross was free to throw his father's weight around.

It was in January 1934 that Gross first made the acquaintance of Anthony Fokker—coincidental with the rise of the "merchants of death" movement. Fokker came to Los Angeles primarily to seek a license for sale of Douglas DC transports abroad. When that was accomplished, he sought out Gross in Burbank about a similar deal for the Lockheed Electra. Word was obviously spreading quickly that Lockheed had produced another hot new aircraft, and the world was starting to gather at Gross's doorstep.

"At the outset I was a little cool to the idea because I was not so enthusiastic about giving our case into the hands of a competitive airplane manufacturer," Gross recalled about the offer from Fokker, who also built his own planes in Holland. Fokker wanted a two-year exclusive sales agreement for Europe, Great Britain, and Russia, with an option to manufacture the Electra abroad after purchasing one for evaluation. The deal would give Lockheed $25,000 in cash, $1,500 per Electra manufactured, and $12,000 a year until a ceiling of $70,000 was reached. "I felt a little chagrined about letting him have the Electra for Russia," Gross said, "but frankly I felt that if he exercised the option we would get the thing we are all looking for, namely money, and with no risk to ourselves." As usual, Gross cautioned his brother in a letter outlining the arrangement, "Fokker wants this matter given no publicity."

On February 23, the Electra prototype flew for the first time from the company's airstrip. Gross wired his brother that it "handled splendidly, lands beautifully and is plenty fast." The projected cost of building the plane in quantity had risen to $38,000 each, which meant it would have to sell for $45,000, according to Gross's calculations. He feared this price was too high and predicted that it would be difficult to generate business. But Fokker soon delivered an order for six, and Doris Duke ordered one for her personal use, as did Frank Phillips of Phillips Petroleum. Engelbert Dollfuss, the reactionary chancellor of Austria who was in league with Mussolini, wanted one, too (just before he was murdered by Austrian Nazis during a coup attempt). This was hardly a flood, but a meaningful trickle that at least indicated a resumption of the kind of carriage trade Lockheed had thrived on in the late 1920s.

The Electra's long-awaited debut happened to coincide with Roosevelt's cancellation of domestic airmail contracts that had heavily subsidized the major aviation holding companies since 1930. Combined with revelations about excessive profits and lack of competition coming from various congressional panels, this action heightened public perceptions of trouble in the aviation industry. But before the public could comprehend the forces at work, a gruesome string of fatal crashes among Army pilots who had suddenly been handed the job of flying the mail routes diverted attention from underlying issues. Twelve dead aviators and sixty-six crashes or forced landings seemed to confirm serious inadequacies in the Air Corps, at the same time strengthening the hand of those politicians who wanted to clean up the commercial aviation business. This crisis hastened the separation of transportation and manufacturing, helping independent companies like Lockheed to enter larger markets—though this was not yet clear in the trenches.

"The outlook at the present time is extremely confused," Robert Gross wrote at the beginning of May. He gauged the company's strategic position to be strong but felt that the entire background for the development of aviation was being clouded by changed attitudes of the federal government, whose "contractual obligations with the mail lines and with the builders of military equipment I consider unreliable, if not downright wrong."

The outlook for exports, on the other hand, Gross deemed excellent, "there being little question but that a major conflict is in the making in the Balkan states and probably in the Orient." Here, in a single sentence, was the central fear of the "merchants of death" writers—that strapped arms manufacturers would greedily feed the flames of foreign wars.

As the full cost of developing the Electra and bringing it onto the market became clearer to Gross, he began to consider more seriously than ever the necessity of military sales. "I have been casting around in my mind for some time for the answer [to rising costs], and, while I make the admission with some reluctance and misgiving, still I feel that it is about the only thing to do: I refer to getting a big contract for the Electra or some modification of it from the military services.

"I have been loath to go into the military thing at all, feeling that it was a lot nicer not to have to depend on the light and shadow of politics, but the job of putting out the Electra has proved such an undertaking that some way or other we have got to scare up a lot of business to support it."

Whatever Gross's "reluctance and misgiving" derived from, it was not due to guilt about being an arms merchant. In June, the company received its first order for an Electra from Japan. "We assume the plane is really for the Japanese army but the order actually is passed through a commercial firm," he noted. Lockheed's factory payroll alone was now running $45,000 a month, with just $180,000 in cash on hand—a tenuous situation undoubtedly driving Gross to take all offers. There were twenty-five affirmed orders for Electras, but Gross was "still generally in the dark about costs." The company was carrying monthly losses of more than $5,000. Moreover, the first Electra had stability problems that needed to be fixed before the Department of Commerce would fully approve the design for mass production.

Gross found some solace in the fact that his closest competitor for small transport sales, Northrop, was also having trouble. "One of the hardest breaks in years happened to the Northrop company last week," he wrote on July 30, "when TWA rejected all of the Northrop transports [Gammas] that they had previously contracted for. The rumor is that TWA, having lost out on their mail appropriations, made up their mind that they simply could not use the Northrops and as a result they found a way to break the contract. At any rate, Northrop has all the planes and is pretty despondent over them. Northrop company also had a body blow when a wing fluttered off one of their newest planes [the Delta] flying in Europe. Nothing is being said about it here and we are very careful not to talk about it, but it just goes to show the risks in this business." It was at this juncture that Donald Douglas pressed Jack Northrop into pursuing military sales to the U.S. and Chinese governments.

In the first weeks of August, Robert Gross and Lockheed began to suffer their own body blows from the fates. The romance of starting a little airplane company was quickly dashed as the firm entered the real world of building and selling complex new products in a weak market. Federal approval of the Electra prototype was delayed by unforeseen problems with the aircraft's stability in certain load configurations. During one test flight, it was forced to make a one-wheel landing because a landing gear was faulty. The matter was corrected, but it piled more development costs onto an already stretched budget. Then, on August 7, the second Electra crashed on a night takeoff from Milwaukee airport after both engines stalled. All six passengers were injured. "This has naturally been a trying experience because such marvelous publicity has been given out about our plane and to have this nip us in the bud just when we are beginning to bloom a little is a mighty hard break," Gross said after working hard to keep the company's name out of press reports.

Loss of this plane, which was operated by Northwest Airlines, disrupted the company's delivery schedule of more Electras for Northwest, Pan American, and Fokker's export customers. Northwest now needed two more to complete the set of three required for its Midwest mail route, and Pan Am expected six to be delivered on time for transport routes in Alaska. Fokker was waiting for seven, including two for Poland, two for Holland, one for France, and one for Spain. It might have been a workable problem for a mature assembly line, but not for an infant organization trying to learn as it stumbled forward. Even with a perfect schedule of deliveries and payments, Lockheed was on thin ice.

To make matters worse, a strike of aluminum workers in mid-August choked off the company's supply of a vital material. Gross, who had just calculated that the company lost $48,000 during the first six months of the year, immediately laid off 180 men—half his work force. The strike served at least one useful purpose, introducing the Grosses to the family of a local aluminum dealer whose son would later marry their daughter.

"The last week has been the hardest one I have ever had in this company," Gross wrote to Jacqueline Walker's husband, Randolph, on September 10. "Our position is somewhat strained and is due largely to the fact that we have $75,000 worth of airplanes on the floor and cannot hope to get the money in for them for two or three weeks.

"We are trying to be as economical as we can, having cut our overhead expenses considerably and are being as frugal as we know how. In spite of this we are going to have mighty close sailing.

"I do not want you to think . . . that I have lost my nerve, but I have had some bitter disappointments these last two weeks and am facing facts."

Three days later, Gross replied to a query from Courtlandt about a military model of the Electra for a foreign interest through Fokker. Names and nationality were later scissored from both letters to conceal the customer's identity, the only surviving clue being that the representatives in New York did not speak English. The desire was for a bomber, with special concern about whether the Electra could be significantly overloaded past its nominal 10,500-pound limit, "assuming [the user] had no particular concern about the life and limb of the operator." In the absence of any firm proof, it seems most plausible that the customer was Germany, the only European government controversial enough at this time to warrant such censorship. Robert sent Courtlandt a document entitled "Electra Specifications—Military" to be forwarded to the representatives.

On the same day, September 13, the first Electra to Pan American suffered a landing gear malfunction during a test flight. It was of a different nature than the prototype's mishap, but it also forced a one-wheel landing. Damage to the airplane was negligible; the company's already precarious production cash flow was mortally wounded. Gross immediately cut the factory payroll to just eighty men, and even the precious engineering staff from thirty to six. "We have to have financial help," he wrote bluntly to Cyril Chappellet, who was in Mexico City pursuing sales. Like Glenn Martin, he saw no other recourse than applying to the Reconstruction Finance Corporation for a loan, "using the argument that the government wants to do something for aviation, that the Post Office realizes this airplane is a national asset, and that we could give employment to many men." It had a definite socialist ring.

At the end of September, Gross was compelled to look beyond the turbulent horizon of his own business by a telegram from Senator Nye "questioning [Lockheed's] policy in selling airplanes abroad." Nye's inquiry, Gross assumed, was based on a concern that "the

planes might be for military usage." Gross's response was to pre-
varicate. "With the help of an attorney I answered the telegram as
completely and as forcefully as I could, refuting, of course, all implications. I do not think there is anything to be concerned about at
the moment."

Within a week, Gross received the following letter from a Wall
Street broker heavily invested in Lockheed stock:

"A very disturbing rumor has just reached us to the effect that
Senator Nye has asked the Lockheed Company not to sell any planes
abroad unless it can guarantee that such planes will never be diverted to military use. The rumor goes on to say that the Lockheed
company is seriously disturbed, and because it has no way of guaranteeing what Mr. Fokker will do with planes sold to him, is considering some method of securing cancellation of the Fokker
orders."

The broker suggested that Gross forestall Nye's "government by
inquisition" by "releasing full information on the subject to the
press." This might also head off any attempt by the senator to block
Lockheed's RFC loan application, he added. In addition, "we have
been somewhat concerned here because of information we have received with regard to landing gear troubles which seem to have
caused a considerable amount of delay in production and partial closing of plants, resulting in some labor trouble."

Though Gross brushed off the broker ("generally speaking, these
people get in my hair"), he fretted to Courtlandt that "it is obvious
he is getting some rather confidential information." The real concern was not so much Nye, who could and would be finessed, but
the Wall Street gossip mill, which was spreading unflattering news
about Lockheed.

"I am trying to live from day to day," Gross said, "and take one
problem at a time, solving that and then going to another. The first
one was how to get around the first three planes for Pan American.
Barring any unforeseen accidents, I think that matter will now take
care of itself. The next problem is Fokker, who, I admit, is something
of an enigma."

Fokker soon solved the problem himself by nullifying his order
with Lockheed. Although the company had proceeded with the
construction of his seven planes, delivering the first one well beyond
the time specified in the contract, official tests showed that it fell
short of guarantees on maximum and cruising speeds. Fokker responded by cancelling the entire batch. Northwest Airlines imme-

diately accepted the first Electra built for Fokker, but Lockheed was in limbo regarding some $70,000 that had been invested in parts for the next six. Despite yet another financial setback, Gross confessed to being "tremendously relieved to have certain aspects of the matter cleared up," namely the nature of his relationship with Fokker.

Nonetheless, he was wary of Fokker's power and influence, and so instructed his colleagues that any publicity on the episode "indicate that Mr. Fokker cancelled his contract with us and that we did not cancel it with him." If asked for an explanation, "I would say 'that the political situation in France during the past two or three weeks has been extremely electric, that this first plane was sold to the French as his client, and that with the delays which the French caused Mr. Fokker in actually taking acceptance, it was decided to release this plane for sale domestically to meet a large and insistent demand.' " This was nonsense, but it served to protect the company's fragile reputation.

Despite Fokker's cancellation, he continued to be a factor in Lockheed's overseas business. In late October and early November, the subject of trade with Nazi Germany came up directly in correspondence between the Gross brothers, who were considering another sales agent for that country. "Mr. Fokker is unpopular in certain quarters in Germany," Courtlandt noted. "The fact that he made money there during the war and left abruptly has contributed to this feeling of resentment, but more important now is the fact that Mr. Fokker is a serious competitor for all the German manufacturers and by producing good airplanes has made the German airlines pay substantial prices for their purchases. He is on good terms, however, with Heinkel and has always worked with Heinkel on a very cooperative and close basis in Germany, in Russia, in Scandinavia and elsewhere. These two firms have always interchanged technical data and the two men are considered to be close personal as well as business friends." Here, once again, was the kind of network that Nye and the "merchants of death" theorists suspected but never firmly established. That Hitler had been chancellor since January 30, 1933, and had just withdrawn from the League of Nations make such business-minded comments as Courtlandt's all the more disturbing in retrospect.

In the financial community, Lockheed's measure was constantly taken by standing it next to its illustrious neighbor, Douglas Aircraft. As the company's representative in New York, Courtlandt bore the brunt of defense, insisting to bankers and brokers that "Douglas had lost a substantial amount of money this year and that all of the manufacturers that were changing over from wood or fabric to metal were having difficulties and that the only significant difference between [Lockheed] and the others was that we did not have access in 1929 to a lot of easy money from the public which would still carry us." His analysis was correct, but did not foster much sympathy among the Depression-wizened investors.

Robert Gross knew that the company must soon receive a transfusion of cash to survive. Besides the RFC loan, which might have to go as high as $300,000 (a substantial fraction of the firm's current paid-in capital), Gross considered selling the Electra design and production equipment to Curtiss; getting a sweetheart loan from National Aviation, which held a large block of Lockheed stock; or merging with other aircraft companies, even perhaps Martin. In any case, the price of an Electra would have to rise drastically again, to at least $50,000—far above his original notion of a $35,000 transport.

As of November 30, the company's balance sheet showed a loss of $161,806 on sales of $466,335. To fortify his power for the maneuvers he saw looming ahead, Gross forced Lloyd Stearman out as president of the company. Stearman had insufficient financial leverage against such a move, and evidently no strong allies among the other officers. "Although the ordeal was harrowing and quite a wrench," Gross wrote to his father on December 21, "I had to do it and feel it was best for all concerned. Lloyd will always remain in my mind an honorable boy of a great deal of ability, but whose best work can be done in an organization where there is a great deal of experimentation and development work, which is not the case with us." In other words, Stearman the design engineer had outlived his usefulness. With Stearman and Walter Varney now gone, Gross had burned his bridges and was alone at the top. "The boys made me president as well as treasurer and here I am," he said coldly.

The letter to his father in Boston had a far more serious purpose than announcing ascendancy to the company's top job. Gross had just learned that the RFC had approved a life-saving loan, contingent on the participation of Lockheed's local deposit bank. When the bank refused to lend him $20,000—Lockheed already owed it $40,000—much to Gross's embarrassment ("I realize that it is not an encouraging feature that the bank turned me down"), the RFC officer in Los Angeles made it clear that the company's management would have to demonstrate their faith in the firm by personally putting up the sum. In short, this meant that Robert Gross would have to use his own money, or else watch Lockheed sink into bankruptcy court again. Since he had no such funds, he turned to his father, "my ace in the hole as usual."

> Naturally I do not like to have to ask you to do this any more than you like to have me, but I feel it is our chance to really put the thing over and that, having had all our tough luck at the beginning, with the aviation picture brightening up so fast, it would really make a very attractive thing out of Lockheed—particularly if the government would swing in for funds necessary to put us over the hump.
> So, I felt justified in putting the matter up to you . . .
> This is not much of a Christmas letter, as the cheer and hope are all on your side. By this I mean you are putting out all the cheer and I am spending it for you. I do believe, however, that this company's reputation is becoming world-wide and I feel that in another year or two we will have a real proposition for Courtlandt and myself if we can just keep the old boat sailing over this hump.

The terms of the $150,000 RFC loan were humbling from every angle—economic, political, and personal. Gross was required to mortgage the whole factory, including warehouse receipts. He had to pledge that the money would be applied only to operating expenses. The sub-loan from his father was to be unsecured. And the ongoing deal with Anthony Fokker would have to be without financial liabilities of any kind. These were bitter pills for any business person to swallow, let alone a man of Robert Gross's stature.

THE FRENETIC, headlong rush of Lockheed through the Depression's worst months illustrated several factors that distinguished the company—and the industry as a whole—from others that did not survive. Leaders such as Robert Gross were businessmen first, political consciences second. They were ready to cast off their bitterness about the New Deal in order to take advantage of benefits like RFC loans. And they were ready to ignore popular antiwar sentiments in order to use export deals to prop up faltering domestic sales. The basically apolitical, profit-oriented motivation for many of their decisions may seem unnotable today, but it was a crucial characteristic in the 1930s.

By mid-decade, international political turmoil was beginning to obliterate the economic concerns of the five years following the

Crash. In a sense, 1934—with its congressional investigations and "merchants of death" movement—marked a dividing line between the old aviation industry and a new one that would soon grow to unimaginable proportions. The business would never again be small enough for the Martins, Douglases, Lougheads, Northrops, or even Grosses to enter by the seat of their pants. And it would grow less and less tolerant of the political ambiguities these men embodied, if only for the sake of commercial success.

CHAPTER EIGHT

EXPORTS

FROM 1935 UNTIL the outbreak of the second global war, the American aviation business transmuted into an international munitions industry, a status it has kept ever since. The financial scope and physical size of companies that had started out in backyard garages, barbershops, or pottery sheds changed by orders of magnitude until they were unrecognizable to their founders. Whether they were well managed or not was important primarily to the extent that output stayed abreast of astronomical demand. By 1939 the flow of cash from military customers at home and abroad was a thunderous wave that washed away all weaknesses, past and present.

Until the Munich crisis of 1938, however, Washington remained wary of foreign entanglements in general and the role of air power in particular. With his background as a Navy Department administrator, Roosevelt was first and foremost a Navy man, unconvinced of the necessity for vast air forces. The military establishment, for its part, still regarded the Air Corps as a parvenu. And many Americans continued to view the potential for bombing cities with distaste, though what debate there was on this issue was riddled with contradictions. Indiscriminate bombing was something the Fascists or Japanese did, so people told themselves. Yet in reality it would be the United States and Great Britain that stretched the concept of strategic bombardment to its limits in the firestorms over Berlin and Tokyo.

There was also the amorphous issue of profiteering. Revelations about excessive earnings on warplane contracts during the Depression led Carl Vinson to co-sponsor a 1934 bill that limited annual profits to 10 percent on work for the Navy (there was no Army limitation until 1939, when orders from both services were limited to a 12 percent average profit over five years). At the same time, he pushed through a five-year expansion program authorizing the Navy to buy 1,200 new planes. This seemed like a large number, but it was child's play compared to what lay ahead.

Even Billy Mitchell soft-sold the level of force that might be needed for attacking an enemy from the air, though more out of bravado than conscience. In the summer of 1934, when a special War Department committee headed by former Secretary of War Newton Baker recommended an increase in the strength of the Army Air Corps to 2,320 planes by 1940 (a figure derived from the Army's "worst-case" studies of defending against a coalition attack by Britain and Japan), Mitchell claimed—or bragged—that 400 would be quite enough. "The only reason to build 2,300 airplanes is to feed hungry contractors," he said with characteristic impolitesse, being no fan of Baker, whose fanciful promises about American aircraft production in 1917 had accomplished little besides spurning Germany to redouble its output. After the hero of St. Mihiel died of a heart ailment in New York in 1936, surely much to the relief of the military and industrial hierarchy, no one took his place as a true believer in American air power.

Some contractors were hungry, indeed. When E. S. Gorrell, a member of the Baker committee, asked Glenn Martin whether the "basic principles" of the airplane business needed changing, Martin told his friend and loyal Stutz dealer, "It has not been a business, Colonel Gorrell." Federal spending "does not provide an opportunity for profit that will interest investment," he explained. "We cannot induce capital to come in on the basis of the [Air Corps Act of 1926] and the treatment that the aircraft manufacturers have received since that time. The money invested in aircraft, Colonel, is invested on the thought or hope that somewhere in the future it will be a great business." What he dared not elucidate in public was that aircraft stocks were held by speculators, not investors, who were waiting for the industry's only true mass market: war.

This "great business" was not as far in the future as Martin thought, of course. His crystal ball may have been clouded by strong popular sentiment to stay out of Europe's problems, as embodied by the Neutrality Act of August 1935. Enacted largely in reaction to Italian dictator Benito Mussolini's threats against Ethiopia, the law called on the President to define munitions that could not be sold or shipped to belligerents abroad. FDR responded by including "aircraft, assembled or dismantled, both heavier and lighter than air, which are designed, adapted or intended for aerial combat."

Despite this attempt at inclusivity, the act was weakened by the fact that it did not apply to civil wars, such as Spain's (which Congress specially embargoed in January 1937—six months after General Francisco Franco revolted against the Spanish government—thereby cutting off Loyalist forces), or armed conflicts that the President did not wish to recognize as wars, such as the Sino-Japanese "Incident" of 1937–38. Aircraft exports from American manufacturers surged through these gaping loopholes, oiled by the contrived dilemma about whether transports such as Douglas DCs and Lockheed Electras were civilian or military machines.

Between 1935 and 1938, U.S. exports of aviation products to Europe totaled nearly $42.5 million, or 43 percent of all U.S. exports to the area—astounding figures, given that even by 1939 the aircraft industry accounted for only 0.6 percent of American manufacturing employment. Buyers were led by the Soviet Union with $9.7 million, the Netherlands (a prime intermediary for politically controversial customers) with $9.1 million, the United Kingdom with $6.5 million, Nazi Germany with $2.2 million, and Fascist Italy with $2.08 million. Exports to Asia were even higher at $58.7 million, led by China with $20.1 million, Japan with $15.5 million, and the Netherlands East Indies (an even more active middleman than its mother country) with $10.5 million. Exports worth $24.6 million to South America were dominated by Argentina, Brazil, and Peru, each of which funneled aircraft into regional conflicts that might otherwise have been off limits to American arms producers.

While it is easy to understand why countries in Asia and South America, with relatively primitive technical capabilities, might turn to the United States for the latest aircraft, it is less obvious why this would also be the case in Europe, which had so outclassed American aviation during World War I and continued to produce innovations through the 1920s. By the 1930s, the United States was turning out university-trained engineers equal to any in the world, but there was also the straightforward factor of money. Airplane manufacturing was a luxury business that could be afforded only by the richest societies, or at least those willing and able to divert public funds into a rather narrow slot. A wealthy country like the United States that was also physically vast supported an active domestic market for transports. And a military establishment that did not have to worry about invasion procured long-range bombers and torpedo planes rather than interceptors. These three types accounted for the bulk of American exports abroad. It is significant that the best fighters at the start of World War II were British, German, and Japanese.

In 1934, noting the penetration of American airplane companies into the European market, the Swiss trade journal *Inter Avia* observed that Europe's passenger fleets were "quite obsolete." European companies had "not kept pace with American progress" and were "in no wise prepared to deliver, and even less to export aircraft for the needs of smaller European countries." The editors singled out Anthony Fokker's success in selling the best American planes as "quite logical," given that they were two or three years ahead of European airliners in performance and comfort. "The aircraft industries of Europe are now standing on the defensive," the journal warned. "The struggle will not be an easy one, for if last year the American invader was yet encamped outside the gates, he now has a secure foothold within, and he shall show what he can really do."

As a fraction of all U.S. aircraft production, exports accounted for just 10 percent in 1929, but grew to 40 percent in 1934. After a slight dip in 1935 while the Neutrality Act was fresh, the dollar value of aircraft exports nearly tripled by 1937, though they declined to about 34 percent of total production as that figure itself started to rise sharply. Even at just one-third of all output, exports brought in half of the industry's profits. It was easy sustenance, and the warplane builders turned to it like starved children.

GLENN MARTIN WAS saved from oblivion by the export market. After securing a Reconstruction Finance Corporation loan at the end of 1934 that permitted him to survive bankruptcy and proceed with a second order of Army bombers, the company built 115 B-10s for the U.S. government, with the last delivered on August 8, 1936. The Army badly wanted these aircraft, which were its first of the new all-metal monoplane generation and answered, too, the Baker committee's emphasis on planes that could support ground forces rather than strike distant cities. Of 115 planes of all types purchased in fiscal 1934, eighty-eight were B-10s.

But the experience of nursing Martin through bankruptcy helped put the Army off from doing further business with the firm. A new generation of officers came to the same conclusions about Glenn Martin as their predecessors had in the early 1920s. After Army planners circulated a requirement in May 1934—during the depths of Martin's financial crisis—for bombers with twice the B-10's payload and range, Martin lost out to a Douglas modification of the DC-2. The company did not build another warplane for the Air Corps until the exigencies of World War II erased the memory of past difficulties.

For a second time, then, Martin found himself estranged from the Army in a field that he felt should be his alone. The B-10 was a far more original breakthrough than the MB-2, and might reasonably have been expected to father a long line of advanced Martin bombers. Since the early 1930s, the Air Corps had been leaning toward ever larger bombers of longer range (apparently as ineluctable as ever faster, higher-flying fighters), which would have satisfied Glenn Martin's greatest ambitions. Financial turmoil was a pivotal factor in frustrating his quest—the company could not afford to shoulder the rapidly rising costs of new designs. But Martin's reputation as a "strange chap," as Donald Douglas politely called him, could only have hurt in a straitlaced business that was still transacted on a personal basis between top company managers and military officers.

In 1935, for example, a newspaper writer observed that Glenn Martin wore unusual suit jackets, which were "longer than the popular vogue" and had fancy lines of braid sewn around the sleeve cuffs. "Because of the slope of my shoulders, I have an especially wide lapel on my clothes," Martin informed the reporter. "My tailor in New York likes to see me come in because I like things to be exactly right and he knows how to make them attractive." When the influential *Saturday Evening Post* profiled him in 1937, it quipped that "Jimmy Walker in his prime never approached the braided splendor of these workday jackets." In military circles, any kind of sartorial extremism would have been received almost as negatively as the Martin Company's financial report.

From August 1936 to the end of the decade, Martin functioned primarily as an exporter of military aircraft. The B-10 assembly line continued to roll, now for the benefit of foreign governments eager for weapons that would revolutionize the destructiveness of warfare. As early as January 1936, when the company was still delivering three B-10s a week to the U.S. Army—but had just learned that Douglas would build the successor—Glenn Martin informed stockholder Charles Crane that exports would be a rapidly growing outlet. "We have concluded a contract with Construcciones Aeronauticas to reproduce the bomber in Spain," he wrote to Crane. "The Spanish Government has placed an order with this firm for 42 bombing planes, Spanish built, and the Spanish Government has approved the purchase of eight identical planes to be built by the Martin Company. The license contract with Construcciones Aeronauticas provides that they will pay us $100,000 in advance on account of the 5 percent royalty as provided in the contract."

In February 1938, when the press reported that seven Loyalist B-10s had been shot down in Spain, Martin claimed that these were copies of a single bomber shipped to the Soviet Union in September 1936 (it had been ordered in November 1935 for $116,718). He also stated that the eight planes ordered by Madrid directly from the Martin plant had been cancelled after Congress enacted the special embargo of January 1937. But he said nothing about the license contract with Construcciones Aeronauticas, which would have entailed the transfer of complete drawings and technical data. The question of whether the Soviets could reverse-engineer a sample B-10 and produce eight combat-ready copies in barely sixteen months was not answered, nor was Martin asked whether Construcciones Aeronauticas had exercised its manufacturing rights. In May 1939, when an Italian Air Ministry report revived the subject of B-10 losses during the Civil War, Martin suggested that the planes may have been built "from photos."

Nineteen thirty-six, Glenn Martin's fiftieth year, marked a turning point in the company's fortunes, due entirely to export sales. In July, six B-10s were sold to China for $924,093, followed by three more after the Japanese invaded the mainland. In August, three were sold to the Kingdom of Siam for $313,097, followed by another three for $305,216. After Argentine pilots tried a demonstrator in competition against German and Italian planes (one of the rare occasions when a U.S. warplane was actually flown to its foreign customer, the usual course being by freighter), the Argentine Army ordered twenty-two at $2.748 million and the Argentine Navy took twelve at $1.4 million. Turkey ordered twenty for $2.2 million. And the rapacious traders of the Netherlands East Indies bought twenty-six for $2.84 million, eventually taking a total of 112 bombers—second only to the U.S. Army. Profits on these sales were high, given the basic B-10 cost of about $72,000 per plane. The year was capped off with the sale of a single Clipper flying boat, with complete technical data, to the Soviet Union for $1.05 million.

With this tremendous surge of business, Glenn Martin assumed a more public role as industry elder and sage. At fifty, the tall, skinny, bespectacled youth of the war era had grown heavier, with a girth that stretched the waistline of his custom-made double-breasted suits. He was threatened with a double chin, and gray strands threaded through the black hair that he parted straight down the middle of his head. "When he walks around the factory, there is something of the schoolmarm's benevolent despotism in his stride," a local newspaper reporter wrote.

His spacious office at the factory was decorated with Oriental rugs, leather chairs, and oil paintings of Martin aircraft. On a fireplace hearth were andirons shaped like the B-10's distinctive bulbous nose. Next to the office was a private apartment where he sometimes spent the night rather than be chauffeured back to the home in Baltimore that he shared with his mother. When he felt the need for a few days off, he liked to take the train to New York, where he shopped along Fifth Avenue and went to the theater. All in all, these were the best few years of his life.

In the fall of 1936, press reports verified that Britain would speed up munitions purchases in response to German aircraft production, with the Martin Company a likely recipient of orders because it still had excess plant capacity at Middle River. Hitler had already renounced the Treaty of Versailles and instituted a policy of rearmament. In October, the Italian and German governments established the Rome-Berlin Axis. By January 1937, Martin paid off its RFC loan, wrote off the original $600,000 development cost of the B-10, declared that the year just ended had been its best ever (showing a $700,000 profit compared to 1935's $318,364 deficit), and made application to sell 350,000 new shares of one-dollar-par-value stock on the New York Stock Exchange. The "great business" was just beginning.

Though Glenn Martin was known for having an iron stomach during times of stress, the Depression years finally took their toll on his health in late 1936 and early 1937. He spent twenty days at the University of Pennsylvania Hospital in Philadelphia with a hemorrhaging duodenal ulcer. Doctors warned him against "too much mental strain." The years since leaving Cleveland had been "a real life battle," he wrote with uncharacteristic honesty about his business problems. But he believed the worst was over when the new stock offering was taken at thirteen dollars a share, bringing $4.55 million.

Indeed, the financial community began to notice the Martin company's recovery when those shares soared to twenty-nine dollars. In April, the *Wall Street Journal* reported that the company's first quarter profit margin was 20 percent, "unusually high for an aircraft manufacturer," thanks to foreign orders. By summer, investment advisers were rating the company's financial position as "strong," while noting that "operations were unprofitable during most of the last decade and maintenance of satisfactory earnings apparently depends on the volume of buying by the United States and foreign governments." Indebtedness to holders of gold notes and dependence on military business "place the stock among the more speculative of the aviation equities," one financial periodical concluded. This warning may have been prompted by the fact that the company's 850,000 outstanding shares had leveled off at around fourteen dollars after the initial peak. About 42 percent were still owned personally by Glenn Martin.

Total employment at the Middle River plant surged to 2,300, the highest in the company's history. A $2.5 million expansion that would double the factory's capacity was scheduled for completion by November. "Business is moving very nicely for us and I believe we are on the eve of an expansion period in aeronautics around the world that we have all been waiting for," Martin wrote to a member of the Irvine family in California on April 21, 1937. Five days later, Nazi planes bombed Guernica.

But Glenn Martin's thoughts did not dwell on what he termed the coming "expansion period." Early in May, he traveled to Los Angeles with Minta to celebrate the twenty-fifth anniversary of his 1912 flight from Newport Beach to Catalina Island, a feat whose innocence must have seemed as remote as selling kites to Salina first-graders. He was in a buoyant, generous mood—taking time before he left Baltimore to dictate a letter to the manager of the Lord Baltimore Hotel, offering to "pick up some lovely oddities and attractive things on the West Coast" to help decorate the hotel's cocktail lounge. The commemorative flight took place in a magnificently outfitted China Clipper instead of a rickety biplane. Congratulatory messages came from many of the era's aviation leaders, including Eddie Rickenbacker, Howard Hughes ("Just a note to congratulate you," scrawled on a shred of brown paper), as well as such interested observers as the Soviet ambassador and the chairman of the

RFC. Los Angeles newspapers headlined the occasion as in days of old, recounting Martin's career with all of the triumphs and none of the disasters. Here was the adulation he and his mother yearned for. It would seldom be so sweet again.

Just after arriving on the West Coast, Martin learned from one of his vice presidents that the company had bid advantageously for a $5.3 million contract to build twenty-one Navy flying boats. Martin's position in the late 1920s as a leading manufacturer of Navy airplanes had been eroded by competitors, especially Consolidated Aircraft Corporation—founded in 1923 by Major Reuben H. Fleet, a former Army pilot who had commanded the first airmail flights between New York and Washington, D.C., in 1918. Fleet's most important early acquisition was the design rights to planes owned by the defunct Dayton-Wright company. He also hired Dayton-Wright engineer Virginius Clark, the Bolling Commission officer who had been Donald Douglas's superior in the Army's aeronautical division during the war. The threads of competition between Martin and Consolidated were thus tightly interwoven, forcing Martin to obtain a restraining order in February 1936 after Fleet hired away forty-one of his workers.

The two rivals were the only bidders for the new Navy seaplanes. "I am decidedly optimistic," Martin's vice president wrote to Martin at the St. Francis Hotel in San Francisco, where Glenn and his mother were staying before going down to Los Angeles. "You are a damned good guessor on prices and I stand willing to buy you the best dinner you can find at any time you wish to have it." With the official contract award later that year, Martin returned to the Navy fold and was finally able to take advantage of the Chesapeake Bay site he had paid so dearly for in 1929.

Though it was not evident from any aspect of his business, Martin still tried whenever possible to pull his public persona away from the image of a weapons magnate. "You should make clear my real attitude towards this industry and the particular phase of it which has brought me most prominence," he lectured a journalist in June 1937. "My interest is not primarily in warfare nor in bombardment, except that they offer another field in which to demonstrate the utility of large airplanes. Large airplanes have always been my prime interest."

The only large airplanes Martin had ever built in quantity were bombers, of course. The B-10 production line, paid for by the federal government, had pulled him out of bankruptcy and—once the U.S.

Army was finished with it—generated huge private profits from export sales. The company's financial turnaround between the end of 1933, when production of the first lot of B-10s was getting under way, and 1937, when exports were in full swing, must have seemed miraculous to outsiders. On December 31, 1933, Martin's deficit stood at $335,374, with sales that year of only $495,796. At the end of the first quarter of 1937, the company's surplus was $6.036 million. By the end of the year, there was a $17.6 million backlog, with exports accounting for 62 percent of it. Net profit for 1937 was $1,144,858, driving up the price of Martin stock, which had been worthless just three years earlier. The B-10 made Glenn Martin the richest of the surviving aviation pioneers, a status he knew how to enjoy better than any of the others.

In November 1937, with his fame as an aeronautical business genius widening, he accepted an invitation to become a member of the newly formed Lilienthal Society for Aeronautical Research in Berlin. Named after glider pioneer Otto Lilienthal, the group was ostensibly organized as an academy similar to the American Institute of the Aeronautical Sciences. But its charter specifically envisioned it as "a means for furthering relations between our friends abroad and the German aircraft industry." In October, Charles Lindbergh had attended the society's annual meeting in Munich on an invitation arranged by the United States military attaché, Major Truman Smith. Smith was intent on improving the quality of intelligence about German aircraft developments, but whether his project included Glenn Martin is not known. Corresponding members like Martin were encouraged by the society to supply technical reports for review, which was obviously not the direction of information flow that Smith desired. Whether Martin actually sent any is unknown, but he was evidently not put off either by the fact that the society existed at the pleasure of the Nazi Air Minister or by Hitler's hell-bent rearmament program.

"The year just closing is the best in our history," Martin wrote a few days after Christmas 1937, "but we already have the assurance of 1938 being a greater year not only in volume but also in profits."

Soon after the contract for Navy seaplanes was finalized, Martin acquired a 106-foot eighty-five-ton motor yacht from Laurence Fisher, the president of the Fisher Body Company in Cleveland. Purchased and maintained with company funds, the vessel was supposedly intended as an observation post for seaplane trials on Chesapeake Bay. In reality, it was Glenn Martin's private cruiser, luxuriously outfitted for the comfort and convenience of himself, his mother, and their guests. His annual salary was listed in stock registration statements as $25,000, but he padded it considerably by having the company buy his boat and his cars. Christened the *Glenmar*, the vessel was anything but a work boat, as indicated by the following description from company files:

> Living accommodations include one double and two single staterooms, both exquisitely appointed, with exceptionally comfortable beds; a sun room, forward and aft cockpits and a spacious after deck with permanent awning. In mild weather two can sleep on this deck in comfort, as installations include two built-in sofas with heavy coil springs and deep mattresses.
>
> A large, high-ceilinged, paneled dining room seats ten persons and is complete with silver and china services. Chairs are tapestry upholstered. Serving unit and china-silver storage are built in.
>
> Adjoining the staterooms are large toilets and showers. Each stateroom has ample closets. Additional closets are scattered throughout the vessel. A powder room is readily accessible from all places where guests gather. A built-in bar has plastic top and all the necessary glasses, mixers and other equipment.
>
> Finishes in all staterooms, dining room and lounges is natural wood of various types. Chairs, lounges and settees are variously covered in fabric, leather or tapestry. Telephones connect the staterooms and lounges with bridge, galley and engine room. An automatic record player is included in the furnishings.
>
> A motor launch and two row dinghies go with the Glenmar.

Here was a pleasure craft fit for a head of state. Nonetheless, for years Martin kept up the ruse that it was little more than a tender for seaplanes maneuvering around the Bay. He used a similar half-truth to camouflage the company's purchase of a seven-passenger sixteen-cylinder Cadillac touring sedan for his personal use. In a bizarre letter to the Governor of Maryland—in which he referred to himself with royal plural pronouns—requesting a special license tag number for the car, Martin said that a number "ending in three ciphers" would "make us feel a great deal like a lieutenant in your Administration."

"As you may know, we are equipped with two-way radio in our

offices while testing large aircraft, as well as equipping our boats in the Chesapeake in the same manner, enabling us to telephone between the clipper ships while flying and the boat standing by in the Chesapeake, and the offices of the company.

"My new car will soon be equipped with the latest in two-way radio, enabling us, while on the road, to talk to a plane under test or our boats in the Chesapeake, and later it may be possible for us to arrange with your State Police Radio to be privileged to radio your stations, if, while driving in the City or State I should witness either a serious accident or lawlessness and would be governed by the wishes of the Commander or your own good offices."

Martin's first utilization of the "Caddy," as he called it, was to transport his mother to the Breakers Hotel in Palm Beach, Florida. When the car was brought back to Baltimore two months later, he found time to dictate to his secretary several complaints and instructions for mechanics at the downtown Cadillac dealership:

1. When the car is operated rather continuously at cross-country cruising speeds there is an excessive amount of hot air coming into the car in a strong breeze up past the sides of the front seat, as well as under the seat, and considerable warmth coming in the forward portion of the front seat.
2. The right-hand windshield wiper operates higher than the left-hand wiper and I believe both wipers are pounding the window sills too much on the downward stroke.
3. The radio should be tested for loose connections, as I am not sure that we are receiving the service we should from the installation.
4. The finish of the instrument panel appears to have been injured during assembly by a scribe or marker overrunning the cut-out hole for one of the instruments.
5. The rear heater fan motor appears to be much more noisy than the front heater fan motor.
6. I would like to have the carburetor tested for correct jet setting for best service conditions and also whether or not it is a little rich.
7. All the brakes should be adjusted.
8. Service the car for all squeaks and rattles, and in this connection, the speedometer shaft appears to be noisy at low speed.

9. There is also a rattle at times in the center of the instrument panel at certain speeds when driving moderately slow over a brick pavement or washboard type roadway.
10. The car has just been oiled and should not need oil service.

Louis Chevrolet was no longer with the company, so Glenn Martin signed the letter—yours very truly—himself. He had started out, after all, as a garage hand.

DONALD DOUGLAS EXPERIENCED neither the fiscal cataclysms of Glenn Martin nor the teething pains of Robert Gross. During the middle to late 1930s, stock in his company routinely sold at twice the price of Martin shares and four times that of Lockheed's. His annual salary in 1935 was $30,000, and he personally held 13,100 shares of the company's capital stock, which paid a dividend that year of seventy-five cents per share. A steady flow of military work through his Santa Monica plant cushioned the development cost of the DC line of transports, which were inspired and partly subsidized by domestic airlines. With a solid reputation in military circles and the phenomenal success of the DC-3, of which 10,654 were built starting in 1935, the enterprise seemed invulnerable. As of December 1936, unfilled orders totaled $24.5 million. Douglas was truly "the Ford of the aviation industry."

Perhaps because of this nearly seamless fortune, compounded by the antipathy toward New Deal policies that Donald Douglas shared with his fellow executives throughout the industry, it was his fate to preside over one of the decade's ugliest suppressions of organized labor. "The coming of the New Deal gave us lots of problems," he later recalled. "The NRA was really tough. We had a lot of trouble adjusting ourselves to all these new ideas."

The National Industrial Recovery Act, which established the pro-business NRA to develop codes of fair competition and declared that workers had the right to bargain collectively with employers through representatives of their own choosing, became law in June 1933. Because of the fractured nature of the airplane business—with airframe, engine, and component manufacturers seldom working as a unified trade—such codes had not yet been applied to the aviation industry when the Supreme Court found the NRA unconstitutional in May 1935. Two months later, the National Labor Relations Act—also called the Wagner Act, after the senator who championed it—was passed, which encouraged collective bargaining under the auspices of a federal board.

During the 1936 presidential campaign, FDR promised to continue his effort to improve working conditions, to increase wages and reduce hours, by seeking at least a partial re-establishment of the NIRA. As the Depression eased—especially in the aircraft business, where profits rose 60 percent during the 1937–38 recession, while general industry earnings fell by about 70 percent—labor leaders fought to maintain the momentum of their movement. The Supreme Court did not rule favorably on the Wagner Act until March 1937, creating a vacuum in which many employers ignored the demands of unionists.

In industries with particular relevance to the military, unions were also frustrated by consistent exclusion from government planning for wartime mobilization. Of direct consequence to airplane manufacturers was the ascendancy of John L. Lewis, who in November 1935 created the Committee for Industrial Organization, or CIO, after failing to persuade the more conservative American Federation of Labor (AFL) to organize along industrial lines rather than traditional craft or trade groups.

In the wake of a successful CIO sit-down strike at General Motors, the nation was swept by similar actions in early 1937. The sit-down tactic, in which workers brought production to a sudden halt and refused to leave the plant until management at least agreed to talk with their leaders, was especially galling to company founders like Donald Douglas, who regarded the factories as personal property. At the Santa Monica plant, where labor relations had been sour for more than a year, about 60 percent of the employees were affiliated with the Aircraft Division of the CIO's United Auto Workers, though the company did not recognize it. Most of the rest belonged to the Douglas Employees Association, a so-called company union that had been set up by management "largely because of the threat of what was coming from the N.R.A.," in Donald Douglas's own words. The group's president freely acknowledged that it was a pawn of the company.

Encouraged by the strike at General Motors, about 500 of 5,600 Douglas workers launched a sit-down strike on Tuesday morning, February 23, 1937, after the company fired three CIO organizers and quashed attempts to hold an election. They demanded a flat wage increase of fifteen cents an hour (averaging nineteen dollars a week), time and a half for overtime and double time for Sundays and holidays, plus recognition of the UAW as collective bargaining agent. The standard assembly line wage for experienced workers in the aircraft industry was then less than fifty cents an hour.

The strike began "with a demonstration like that of college football fans between halves of a game," the press reported. The main power switch was pulled, bringing the world's largest aircraft factory to a standstill. Plant managers ordered "loyal" employees, including all female office workers, to go home. Donald Douglas issued a statement charging that the strikers were trespassing, emphasizing that $19 million of the plant's current production was for the U.S. government.

Douglas maintained the same hard-line attitude against the unionists that had helped percolate the crisis, refusing to join a peace conference called by the director of the Federal Regional Labor Board. "We will hold no meeting with [the FRLB] as long as our plant is illegally occupied," he announced. Police then ordered the sit-downers out of the factory, warning that force would be used if "the powers that be" so desired. The strikers responded by barricading their shops.

On Thursday morning, February 25, a Los Angeles grand jury indicted the men for "conspiracy to commit forceable trespass," a felony charge. When this news reached the Northrop subsidiary in Inglewood, about seventy-five of its 1,100 employees began a sit-down strike of their own. Meanwhile, a force of 350 police officers and deputies, heavily armed with machine guns and tear gas, surrounded the Douglas plant by midafternoon. A Red Cross field station was set up nearby, while police squads infiltrated factory buildings with pistols, clubs, and gas canisters.

In the face of this threat, a spokesman for the workers shouted that "we'd rather die fighting than leave this plant under arrest." Realizing that they were isolated, however, the strikers sent a telegram to President Roosevelt pleading for his help: "Sheriff's squads armed with machine guns mobilizing at Douglas plant preparing to enter and oust sit-down strikers. Government inspectors are in plant and know that no damage has been done to government property and government planes. We appeal to you to intervene." FDR did not reply.

In a last-minute attempt at mediation, a local representative of the National Labor Relations Board met with Donald Douglas, the mayor of Los Angeles, and the county sheriff. But he was only able to promise the strikers prompt hearings and caution them about the overwhelming police force. Enraged by this palliative, some of the workers armed themselves with fire extinguishers and hand tools, even wheeling three DC transports into line so that their propeller blasts would blow tear gas back through doors and windows. But the majority were intimidated into capitulation, letting the police commander serve the felony warrants and arrest them as they filed out of the plant under machine-gun cover.

"The sit-down strikers in the Douglas plant were ready to defend the plant at all cost," a CIO attorney said after 343 men were locked up that night in the Los Angeles County Jail, "and agreed to permit the Sheriff's army to peacefully enter the plant only when [the NLRB representative] informed us that the cases against the Douglas company for violating the Wagner law in discharging union men and refusing to recognize the unions or conduct an election under the auspices of the government would be heard immediately.

"We depend upon the support of all decent public opinion, for the strike will continue until we force Douglas to abide by the law in recognizing our union and granting our just demands."

Douglas management delivered its answer through the district attorney: "The principle of collective bargaining is not involved. What is involved is the principle of whether or not a citizen is to be dispossessed of his own property by striking or resigning employees. Likewise, the protection of millions of dollars' worth of government property is involved.

"This office serves notice that every resource at its command will be employed to prevent dispossession of the rightful owners of property through the so-called sit-down strike method. Los Angeles County is still functioning under constitutional government. We will keep it so."

When the Santa Monica plant reopened on Monday, March 1, Donald Douglas claimed that 3,130 workers out of the normal day shift of 3,800 had returned to their jobs. Strike leaders countered that 3,500 employees were now on the union's rolls. Some production resumed, but Douglas conceded that it was concentrated in Army and Navy projects. He continued his tough anti-labor stance by barring the indicted strikers from re-employment. More than a hundred men remained in jail after the UAW posted a $118,000 bond. In order to separate organizers from rank and file, bail was refused to nonresidents of Los Angeles County, men "whose attitude is not such as to warrant release,'" and anyone who could not give "acceptable" references. A few pickets walked around the plant, guarded by officers with guns and tear gas who limited them to "dirty looks" and "no dirty words," as one reporter wrote.

When Donald Douglas addressed his employees at midweek, he lambasted the strike as the mischief of foreign influences—a common insinuation at a time when many union organizers had non-Anglo-Saxon surnames. "The plant will be operated under the American plan," he said, "under which any able-bodied man can get a job regardless of his labor affiliation. No one need fear harm in coming to work here. The sheriff and the Santa Monica police chief have promised plenty of men to protect you. In addition, a war chest is being raised by manufacturers in Los Angeles to assure protection of workers under any conditions and to aid them in any other ways. This does not mean that the management of this plant will refuse to deal with workers groups or their representatives. It has been said I refuse to meet committees of workmen, but this statement is ridiculous. I have never refused to meet anyone nor to discuss any subject."

Douglas then unilaterally granted a five-cents-an-hour pay raise. The *New York Times* reported that workers "waved their lunch pails and tossed sandwiches high into the air as he spoke," but did not explain whether these were friendly or derisive gestures. Within weeks, under federal mandate, the UAW was certified in a plant-wide election.

On April 5, 1937, Douglas bought out Jack Northrop's 49 percent share in the five-year-old subsidiary. His decision stemmed in large part from War Department hesitation to buy any more Northrop planes until labor problems there were resolved. Even before the labor crisis, however, relations between Douglas and Northrop had become strained because of Douglas's belt-tightening at the Inglewood plant, spawning rumors that Jack Northrop was "beginning to stroll around the countryside looking for new pastures." A major Wall Street speculator was said to have offered Douglas $1.8 million for the Northrop operation, a very generous figure, considering that it earned only a few hundred thousand dollars during its existence. But the buyer backed out when Douglas asked for $1.9 million.

While the Douglas plant in Santa Monica settled down to occasional nonviolent confrontations, further turmoil at the Northrop facility—where Jack Northrop was even more hostile about New Deal workplace innovations—led to a UAW sit-down strike on September 2. Disputed firings under a seniority agreement, not wages, were the issue. Exercising his now total control over the subsidiary,

Donald Douglas closed the factory on September 7, putting 1,393 people out of work. "Effective today, the Northrop Division of the Douglas Aircraft Company no longer exists as a separate industrial entity,". he announced. "All production at its Inglewood plant has been stopped and no future activity is contemplated there."

Out of a total Douglas backlog of $34 million, contracts at the Northrop subsidiary were valued at $4.168 million, including Army, Navy, and foreign orders. Work on sixteen Army aircraft, worth $604,000, was in progress—the last of an order for 100 A-17 attack planes. After the strike began, Army and union officials arranged for completion of five planes that were in final stages of assembly. Though a shutdown had been rumored, it was widely assumed to be a bluff on Donald Douglas's part. "No structure can offer shelter and security to labor and capital when its very foundations are continually destroyed by industrial termites whose main objective is strife and dissension," he fumed after the plant's doors were locked, obviously not interested in diplomacy.

Industry analysts saw reasons other than labor strife for why Douglas would take such drastic action. The Northrop plant had proved expensive to operate, for example, and was tied to a property lease considered too costly for airplane manufacturing. Consolidating Douglas and Northrop work under one roof would make a more efficient system. On the other hand, Douglas was already running at full capacity and could not absorb Northrop's contracts without shop expansion.

On September 22, the War Department notified Douglas that it would not accept further deliveries of A-17s because of suspicions that the planes were being sabotaged. Whether this was predicated on fear of inferior workmanship or actual evidence of intentional damage was not made clear. In any case, the move undercut the union, persuading many of the financially strapped workers to consider a deal with Douglas management that would reopen the plant. By October 21, over 500 of them had signed a company contract appended to re-employment applications that read: "I agree that I will not go on strike, or seize company property, or occupy company property without authority from the company, and will perform my duties in an efficient manner and not indulge, singly or jointly, with others in 'slow-down' or 'pace-making.' " It was humiliating language, but few were in a position not to sign.

The plant reopened on October 23, with union leaders powerless to do anything other than file a protest with the NLRB against what they called a "yellow dog" agreement (five months later, an NLRB examiner held the Douglas no-strike contract to be illegal, found the company guilty of unfair labor practices, and ordered the reinstatement of one former employee). Production of A-17s resumed. On January 1, 1938, Jack Northrop officially resigned as vice president of Douglas Aircraft and general manager of the division that bore his name. He said only that he would take a six-week vacation, "get back in shape," and not re-enter the aircraft industry. Many years afterward, his description of being bought out by Douglas was icily simple: "They were in a position to say 'we'll give you so much money for your 49 percent,' and there wasn't anything we could say except 'thank you.' " It was not a cash transaction, but he received enough Douglas shares to earn the man whose family had once lived in a tent on the Santa Barbara beach a comfortable income.

Donald Douglas retained highly refined memories of these milestone labor conflicts in the Southern California aviation industry, which in early 1937 employed 40 percent of the nation's 30,000 aircraft workers and was well on its way to becoming the business's epicenter. "The Wagner Act didn't create problems for us at once, but of course we could see what was coming," he recounted in later life. "The Employees Association was formed, a company union. I think we encouraged them to do it—we didn't dominate them, but we encouraged them, because we could see the writing on the wall—and we dealt with them very satisfactorily until the sit-down strike [of February 1937].

"They sat down within the plant and promised to blow the plant up if we tried to get them out. That strike was started by ten people. None of us knew it was coming. We came in one morning, and the police came in and said 'there's a bunch of guys sitting down out there that say they ain't going to work, and a lot of others milling around.'

"They threatened a lot of things, but when the chips were down, the boys didn't do anything but go along with the sheriff. There was no destruction. It was started by a few chaps who didn't have any real backing behind them. This was after the Detroit [General Motors] affair, so it sounded like a good idea to a lot of them.

"They hadn't come to us with grievances. Nothing. It came out of a clear sky. They'd made no overtures or threats. There was nothing to strike about. I guess these fellows just saw a chance to make big heroes of themselves. They practically told me that afterwards. Most all of the people we evicted came back to work."

Douglas was able to afford meeting the workers' demands without financial sacrifice. Between 1936 and 1938, the company's net sales and income nearly tripled, to $28.3 million and $2.15 million, respectively—both the highest in its history. The sit-down strike was a signal, in fact, that the enterprise had finally grown up. It was no longer Donald Douglas's intimate little shop of engineers and craftsmen pioneering an eccentric new field. The founder no doubt still perceived it as such, which may help explain his reaction—and especially Jack Northrop's—to the labor crisis as an insult perpetrated by foreigners. The company was simply growing larger than he had ever conceived. Since reorganizing in 1928, his payroll had climbed from 264 employees to 7,197.

Exports helped fuel this tremendous surge. Ironically, they also helped advance the technological prowess of the country that would soon trigger a metamorphosis of the American aviation industry: Japan. At least five Northrop or Douglas designs were transferred to Japanese manufacturers. In 1935 and 1936, two Northrop-5 attack bombers—closely related to the U.S. Army's Gamma-derived A-17s—were shipped for testing by the Japanese Navy. They were then handed over to Nakajima and Mitsubishi, Japan's premier warplane companies, for engineering analysis. Of special interest was the compact Northrop retractable landing gear, which had been designed for another Northrop product, a single-seat fighter called the "3A." The 3A prototype was lost in a crash, but a version built by the Chance Vought company was eventually sold to Mitsubishi, which, according to Chance Vought's president, "turned it into the Zero"—the Japanese Navy's deadly fighter of World War II fame. The Zero's pedigree was not quite so all-American, but the Japanese were obviously not just collecting U.S. aircraft as modern art.

In 1936 and 1937, Douglas shipped two flying boats, originally intended for Pan American, to Greater Japan Air Lines, which turned them directly over to the Japanese Navy for evaluation and eventual production of military models. In 1938, a Japanese trading company acquired thirteen fully assembled DC-3s, nine partly or totally unassembled DC-3s, plus the license rights for manufacturing the coveted transport. The DC-3 thus became the Japanese Navy's standard cargo carrier during World War II, with more than 500 constructed by Showa and Nakajima.

Finally, in late 1939, only months before Japan joined the Axis, a prototype DC-4—Douglas's revolutionary four-engine transport with twice the capacity of the DC-3, under development since 1936 but spurned by the major domestic airlines as too expensive to operate—was shipped partly assembled (under the Neutrality Law, many aeronautical parts did not need an export license) to Greater Japan Air Lines, which again acted as a front for the Imperial Navy. The sale price was reported to be at least $725,000, perhaps more than a million including plans and jigs, which helped recoup development costs estimated at $2 million. A Douglas spokesman said that both War and State Department officials favored the transfer, which in any case "would not injure national defense" because the DC-4's construction placed control machinery under the cabin floor, where bomb bays would ordinarily be located.

Technically, the contract with the Japanese had been signed on March 1, 1938 (three months before the plane's maiden flight), thus antedating State's July 1, 1938, letter to manufacturers that discouraged exports to countries "which are attacking civilian populations," as Japan was then doing in China. Douglas evidently never took this missive seriously, because DC-3 transports were known to be coming off the assembly line in September 1938, bound for Japan. Indeed, Douglas employees put the DC-4 together in Japan, where it was soon reported as having crashed in Tokyo Bay. In fact, Nakajima had secretly dismantled it for use as the model for a four-engine long-range bomber.

Donald Douglas's blatant disregard both for Japan's feverish militarism and the U.S. government's efforts to curtail the flow of arms abroad defies all but the most cynical explanations. One possibility—purely speculative, since there is no record of his thinking—is that he acted at least in part to recover losses suffered not only during the DC-4 program but also in a secret project to develop an intercontinental-range bomber, designated XB-19. Conceived in 1935 over some domestic political protests that it was a purely offensive weapon, the prototype, then the largest American aircraft ever built, encountered such technical problems and construction delays that the company asked the Army to cancel it in mid-1938. The Army refused, and Douglas eventually lost more than $2.5 million on the program.

In later years, Donald Douglas would claim that he did business with Tokyo only "before things started hotting up." But the DC-4 deal refutes this statement. Besides, Japan was unfurling its true colors as early as 1936, when it signed the Anti-Comintern Pact with Hitler. The sale of military hardware to Japan "created problems with the Federal government, very much so," he admitted. But he added nonchalantly that "we had to go into Washington and see some of the people in the Roosevelt administration and get the thing cleared." When the State Department balked at granting an export license for the DC-4, "we had to work the political game again, so we could ship the plane."

Since his earliest days as Jerome Hunsaker's protégé, Donald Douglas had always worked the political game in Washington to utmost personal advantage. In the late 1930s, he made money by helping to arm both the greatest military dictatorship in Asia and the country that would soon have to fight it to the death.

WITH THE INFUSION of a $20,000 personal loan from his father and the first disbursement of a $150,000 loan from the RFC, Robert Gross began 1935 on relatively firm ground. During his first business trip of the year, he conferred with the Navy in Washington and the Army in Dayton, Ohio, about entering design competitions with military versions of the Electra transport. He also met in New York with several major investors, including Walter Chrysler and Harold Talbott, who apparently looked upon Lockheed as an attractive target much as Gross had three years earlier. Gross judged their interest to be "on terms that we could not afford to take," and the matter was dropped. Hostile takeovers were not yet part of Wall Street's regular fare.

In February, Lockheed shipped the second of two Altair monoplanes—from which the ill-fated XP-900 fighter had been derived in 1931—with retractable landing gear to the *Mainichi Shimbun* newspaper in Osaka, Japan. Supposedly used for newsgathering, the 221-mile-per-hour aircraft was of obvious interest to the Japanese Navy, especially regarding its undercarriage, which was the first to fold flush with the wing. And though the airframe itself was considered obsolete, it came equipped with an up-to-date 550-horsepower Pratt & Whitney engine. Japanese interest in the Altair had perhaps been renewed by the fact that an improved version of the old XP-900, built by Consolidated, had just become the U.S. Army's only operational two-seat monoplane fighter.

Nearing completion when the Altair left the plant was the company's first Electra export sale, also to Japan. Here there was no ruse about the ultimate customer. A Japanese Navy inspector "is here watching their plane," Gross noted. Electras were in no sense obsolete, but were untouched by Washington's various export restrictions because of their categorization as commercial transports.

"The word has gone out around the trade that we are out of the woods," Gross wrote to his father. "People have stopped calling me up to ask when I can pay them a little money on their accounts and generally the air seems very much cleared." He added that "there is a heavy inquiry from abroad for planes."

Orders for Electras continued to come from domestic airlines as well. Congress had reacted to the airmail turmoil of February 1934 by passing new legislation that curtailed subsidies to carriers, which fell from $19.4 million in 1933 to $8.8 million in 1935. Airlines thus had fresh interest in high-performance planes like the Electra as they scrambled to cut costs and attract passengers. Speed, especially, had sales appeal, and the Electra was known to be fast.

But Gross understood by now the absolute necessity of doing business with the military. At the beginning of March, he formally notified the chief of the Air Corps that Lockheed was "desirous of entering the military aircraft field and in this connection has done considerable preliminary work along lines of all-metal, high-performance, multi-engined designs." In particular, "we would be prepared to build for the Army a 'triple threat,' multi-engined, all-metal monoplane, which could be used as a light bomber, long-distance observation or ground attack and would have a top speed of 270 miles an hour." Stripped of military jargon, this was essentially the Electra with more powerful engines than those placed on "civilian" models like the one sold to the Japanese Navy.

In a rather poignant closing sentence, Gross admitted that Lockheed had "very little experience or knowledge of the procedure with regard to procurement of military aircraft by the Army Air Corps and we would greatly appreciate your furnishing us with proper and adequate information." Out of either naiveté or inadvertent honesty, Gross was asking for something that Congress, the War Department, and the oldest aircraft manufacturers in America had been unable to crystallize since 1918.

"I am secretly much interested in this bomber project," Gross told his brother Courtlandt. The company's chief engineer felt that there was "no question but that we could achieve 270 mph with it and you can imagine our excitement when two of the Pratt & Whitney representatives told us that within a month they could furnish the engines supercharged to 18,000 feet, which would work out to give us a top speed of 310 mph. Some buggy!" No doubt the Japanese were making similar projections.

Gross's new enthusiasm for military work collided immediately, however, with the realities of Army procurement. He could not sell an Electra to the Air Corps in the way he could to Pan American. "We have received an indication from the Army Air Corps Materiel Division at Wright Field [Ohio] that they would permit us to send an engineer to Dayton to converse on the subject of the latest installations of bomb racks, armament and military accessories, if a design competition should be announced that would be suitable to the class of planes that we build," he explained to Randolph Walker. "A design competition is now in force for bombardment airplanes. The specifications in the [Army's] circular proposal for this, however, require an airplane considerably larger than any adaptation of our Electra. . . . We could forward some preliminary data, with the result that, though our design would be thrown out, it might be examined sufficiently to awaken the interest of the bombardment board to such an extent that a second design competition might be announced with specifications that would let us in."

Though a neophyte, Gross was already attuned to the military's method for acquiring particular aircraft from certain manufacturers while skirting regulations that called for open competition on any purchase. If the Air Corps wanted an Electra bomber, it would simply circulate specifications among qualified manufacturers that could only be satisfied by Lockheed's plane.

"The only way we can be sure of getting military business is through a design competition," he told Walker. "As the four design competitions now in force are for aircraft out of our class, we shall have to await a design competition being announced after June 30, or commit the 'social error' of putting an airplane that does not fit into one of the present competitions. I know this all sounds very involved and confused, but I thought you would like to realize how ridiculous the terms of this new Procurement Act are."

In fact, the terms that Gross found ridiculous had been adopted to answer demands from members of Congress and the General Accounting Office that competition be re-emphasized over negotiated experimental contracts that fostered favoritism. Military officers, on the other hand, felt that design competitions were unworkable. By soliciting bids on the basis of fixed specifications, they froze the design process and risked quick obsolescence. The choice between competition and negotiation, between low price and superior quality, was never clear-cut.

The Army made clear to Lockheed that the Electra was of potential interest, but pointed out that there were only three ways to get the business: first, by building one on speculation and turning it over for tests; second, by winning a design competition; third, by selling one to the government under the "Extraordinary Military Secrets" section of federal procurement regulations (which had covered the prototype Martin B-10). "The much-coveted section is, of course, the most desireable from our standpoint," Gross told Randolph Walker at the end of April, "and means that we would get an outright development contract with money guaranteed for the sample plane. This is of course the apple of the aircraft manufacturer's eye and everybody tries to get a contract under this section." He added that the Army had sharply limited such awards and "showed no disposition whatsoever" to give Lockheed one.

For the first six months of 1935, the company earned a profit of $152,000, still leaving at least $30,000 in Electra development costs unamortized. "The problem is how to sustain the volume business," Gross fretted, reciting the industry's basic quandary. "I have reached the conclusion that between the military possibilities and a newer and smaller airline plane lies the answer." Recent legislation that sharply reduced federal subsidies to airmail carriers was forcing the airlines to seek higher-performance transports like the Electra, hence Gross's desire for an even faster version for feeder routes.

As the Electra's reputation for quick, efficient service spread through aviation circles, military buyers moved to acquire the plane as-is for VIP transport. "Military authorities are becoming Lockheed conscious," Gross observed in October. So, evidently, were Nazi authorities in Berlin. The Electra's chief designer, Hall Hibbard, traveled to Germany to deliver a technical paper before the German Aeronautical Scientists Board. Neither Hitler's announced policy of rearmament nor the official persecution of German Jewish scientists in 1935 acted to cancel Hibbard's trip.

Japanese authorities were also pulling at Gross's coat sleeve. "Coming to the matter of a license price for Mitsubishi," he reported to Walker in early December, "I feel that a sound proposition would be to sell them the license, including one sample airplane, for $100,000." (The price of an Electra to domestic airlines was then about $50,000.) Gross confessed that he felt "a little awkward" about dealing with Mitsubishi through a sales agent, because "we are in direct communication with Mitsubishi already on the Lockheed XP-900 fighting airplane." Nonetheless, he wanted "to do the business and make the money for the Lockheed Corporation."

At the end of December 1935, Gross was able to compose a far different report to his stockholders than a year earlier. Initial Electra development costs of $151,000 had been entirely written off, and nearly $30,000 had been spent on evolving a new model. Working capital rose by $414,000 to $476,000. Accounts receivable increased from $72,000 to $415,000. A profit of $218,000 had been earned on sales of $2 million. "In the relatively short period since the present Lockheed Aircraft Corporation was formed," he stated, "such important changes have occurred that the builder of aeroplanes has been hard taxed to keep abreast of developments—particularly on a profit basis." As for military sales, "the building and marketing of military and commercial aircraft are two very distinct and different businesses, and it does not follow that because a company is successful in one field it will be automatically successful in the other," he observed. "With no criticism whatsoever of our Military or Naval authorities, the matter being entirely beyond their control, the laws governing the procurement of service aircraft are so complicated and the process necessarily so long that no aeroplane company can permanently invade the military market without financial resources which this company has never possessed." Lockheed had the technical ability and the plant facilities to develop modern warplanes, he said, but "entrance to this field, now being seriously attempted, must be on a business basis." Glenn Martin could not have put it better.

As the Douglas DC transports began to dominate major air routes, Lockheed plowed its financial resources into matching the Electra more closely to remaining segments of the commercial market. In 1936, the company developed scaled-up and scaled-down versions of the plane, both of which found more success with foreign airlines and the military than with their intended domestic customers. Riding the export wave, Lockheed sold six-passenger Electra "Juniors" to three Indian maharajahs, the Argentine Army, the Brazilian Ministry of Aeronautics, and the Netherlands East Indies (the largest single operator). Although considered a commercial aircraft, almost half of the 130 Electra Juniors built between 1936 and 1942 were sold to military forces in four countries.

To compete directly against the DC-3, Lockheed produced a

twelve-passenger "Super" Electra and tried to sell its higher cruising speed against the DC-3's 50 percent greater capacity. To offset landing and take-off speeds that were too high for a commercial aircraft, the company acquired confidential information from the Martin B-10 bomber program. A new type of flap, or air brake, had been installed experimentally on a B-10 with positive results, whereupon Lockheed found the engineer who invented the mechanism and bought the rights to it.

Despite this and other technical innovations, the $75,000-plus Super Electra never seriously challenged the massive popularity of the $100,000-plus Douglas transport. Exports again saved the day, especially to Japan, where the Tachikawa airplane company acquired a production license in 1938. ("Our cash position has been materially strengthened by the Okura deposit," Gross observed with satisfaction, referring to the Japanese sales agency.) Tachikawa soon developed a military version of the Super Electra for the Japanese Army Air Force, sharing production of 119 planes with Kawasaki Aircraft Engineering Company between 1940 and 1942.

Lockheed turned to the export market with special urgency when Northwest Airlines' Super Electras began to crash with alarming frequency, due to a design flaw in the tail structure. Northwest had originally asked Lockheed to develop a transport on the scale of the DC-3, but at Robert Gross's urging had bought Super Electras even before they were approved for passenger service by the Department of Commerce. Three out of the airline's eleven Supers were lost starting in January 1938, forcing Northwest to abandon the aircraft in favor of the DC-3. In a confidential agreement of January 1939, Northwest dropped all litigation against Lockheed stemming from the first crash, and Lockheed promised to resell or buy back all of Northwest's Super Electras. The grim episode wiped out the domestic airline market for the plane.

"I feel Europe is 'hotting up' fast," Gross told his brother just before sending him on a trip to England to capitalize on burgeoning aircraft orders there in the fall of 1936. "Looking down the line for the Lockheed Company, I want to be braced to give the stock holders of this company the best run for their money." Between 1936 and 1939, Lockheed's foreign sales organization would grow from one man to 130.

New projects were pushing the work force in Burbank toward 2,000 employees, inevitably undermining the shop-floor camaraderie of a small business. But the stockholders, of which he was one of the largest, were always foremost in Gross's mind. When sit-down strikers closed Douglas's plant in nearby Santa Monica in February 1937, Gross applauded the police action that jailed the workmen. "The way the Douglas Company rallied the local authorities to their cause and the courage of the District Attorney in taking a firm stand and throwing the sit-downers out is one of the bright spots in the country at the present time," he wrote to Courtlandt. "Our own problem has never been as acute as the Douglas Company's for the reason that we are smaller and that there is a much better basis of understanding between the Lockheed management and the employee. However, feeling ran extremely high here in the middle of the week; and with the Douglas Company out cold and Northrop following two days later, the schedule among the labor agitators was to take on Lockheed next."

Gross took two deliberate steps to prevent a strike at Lockheed. First, he stopped hiring new men, on the assumption that many of the Douglas strikers would try to find jobs at Lockheed. "I think you and your associates are familiar with the method which radicals and subversive aliens use in applying for positions," he wrote to the vice president of Northwest Airlines, whose first Super Electras were behind schedule at the plant for many reasons other than labor unrest. "They go under assumed names, forged references, etc.; and we decided that in order to be absolutely sure that no undesirables get on here, we must shut down the bars on all hiring."

Second, he played the AFL and the CIO against each other in their rival attempts to organize Lockheed, recognizing the International Association of Machinists, an AFL affiliate, before the more radical UAW could gain a foothold. "This move was made as a protective move only," he stated. "The great majority of Lockheed people do not really want any union, but they realize that if they do not have something with which to combat this infectious spread of radicalism which is sweeping the country, they are about to have unorthodox labor tactics by a militant few." Gross's estimation of unionism's appeal was somewhat short, given that the Machinists signed up eight or nine hundred of Lockheed's employees.

In a further effort to "head off agitation," he then granted a six-cent wage increase, with time and a half for overtime. This was a liberal raise for the era, one that created the highest wage rate in the industry (including a forty-eight-cents-an-hour hiring rate, compared with forty cents at Douglas and Martin). His tactics were effective from a management perspective, at least to the extent that he was able to douse the sit-down threat and fend off a union shop. Lockheed was the only major aircraft company in the Los Angeles area that did not experience a strike in early 1937.

For several months during the labor crisis, Gross seriously considered making an offer to Douglas for the deeply troubled Northrop subsidiary. After discussing the subject with a Douglas director, "I got the impression that a decent offer that would save Douglas's face before his board and stockholders would buy the Northrop Company," he wrote to Courtlandt in May 1937, when the sit-down threat had receded somewhat. "What that amount is is a question, but I think myself that somewhere between $1.5 million and $2 million would do it. I do not know whether the company is worth this much or more, but the Douglas Company itself recently established an evaluation of about $1.6 million on the company when they bought the minority interest in Northrop in exchange for 9900 shares of Douglas."

Gross never followed through—Lockheed did not yet have the financial depth to expand in this fashion—but musings of a takeover indicated that his ambitions were unsatisfied by the Electra's success. "It has always been my dream to found a strong, well rounded empire in aviation," he confessed after one of the most powerful Wall Street investors offered to put him in charge of a conglomerate that would include Lockheed and Northrop. "This particular deal would give me a chance to work out our own destiny; and if I could keep my head, would give me a chance to make a lot of the boys in the business rich." But Lockheed alone kept his attention.

On June 2, 1937—just five years after he plucked the company from bankruptcy court—Gross learned that Lockheed had won an Army design competition for a new twin-engine fighter. It was a milestone achievement for the company, technically and politically, given that the other entrants—Douglas, Boeing, Curtiss, Consolidated, and Vultee—were the strongest in the business. The aircraft, called the P-38, was a radical design, looking like two airplanes joined together with a twin-boom tail (that would inspire the first Cadillac tailfins of the 1950s) and top speed of 400 miles per hour, necessitating use of the same air brakes snatched from the Martin B-10 and applied to the Super Electra. Lockheed received $157,500 for the prototype, without engines and accessories, which was the first of more than 10,000 P-38s in eighteen versions manufactured through the Second World War. Gross could not know it at the time, of course, but the P-38 marked the beginning of a golden age beyond his imagination.

As the export markets continued to grow, Gross further refined his attitude about selling production licenses abroad. "I still am not interested in licensing a country which might become a serious competitor in the markets we are now developing," he instructed his European sales agent based in Poland. "On the other hand, I have no particular objection to licensing some minor country or selling a license where I am positive it is never going to become a competitive factor and take bread from our mouths. Japan is a good example of this. I am not going to try to sell a license to Japan, but if they ask me for one, I would quote them on the theory that they never have had anything for export in our game and I do not believe they ever will." Not a trace of political wariness seemed to enter these ruminations.

At home, the company was being reminded of its still adolescent technical and financial stature by problems with the Super Electra. Northwest Airlines was plagued by difficulties along the whole spectrum from bad locks on baggage doors to structural failures in the aircraft's tail assembly. By the end of November, engineering rework on the Super had brought all other projects—including the P-38—to a standstill.

"All this has its reflection in one set of conditions which hitherto have been fairly uniform in our business," Gross informed Randolph Walker, "namely that we could bring out a new model and get the cost to 'normal' by say ten or twelve units. It is now apparent to me that the bigger and more complicated the airplane the longer it is going to take to bring cost to normal, and it would not surprise me at all if we failed to achieve normal cost before twenty units are built. . . . This is the experience which Douglas has gone through and they have learned it to their disappointment as well."

The direct implication was that new designs could not be undertaken without a guarantee of quantity production—a fact of life that pushed all manufacturers ever deeper into the military market.

By 1938, AIRPLANE builders did not have to experience the subtleties of technological change to feel the pull of the military. As in the final months of peace preceding the Great War twenty-five years earlier, the entire industry was now under the gravitational influence of

Europe. When Prime Minister Neville Chamberlain of Great Britain stepped from his personal Lockheed Super Electra in Munich on September 15, the world turned a dark corner. For Martin, Douglas, Northrop, Lockheed, and every other embodiment of American aviation, it was the start of a geometric expansion that ranks as one of the greatest industrial wonders of all time.

"War clouds in Europe have had a silver lining for aircraft and engine manufacturers in the United States," crowed the *New York Journal and American* in May 1938. "Exports of aircraft products from the United States during the first three months of 1938 totaled $14,320,273, a 99.3 percent increase over the corresponding period in 1937." The *Wall Street Journal* noted that export contracts carried a profit of at least 18 percent.

At home, total federal expenditures for military aircraft reached $116.8 million in 1938, nearly twice the 1935 outlay. In the spring of 1938, Douglas stock was trading at forty-five dollars a share and Martin at twenty-three. Immediately after the "Peace of Munich," they climbed to sixty-seven and thirty-two, respectively. Lockheed stock began 1938 at ten dollars and reached thirty-three a year later.

Although the official line of the Roosevelt administration remained strict neutrality, the President constantly undermined it by liberalizing the War Department's policy on release of military aircraft to foreign customers, which itself had been stretched in response to industry proddings. His most blatant action in this regard was the decision not to cut off the flow of arms to China and Japan in 1937. He also approved numerous instances of letting French and British military officials inspect the latest American warplanes. The neutrality laws passed by Congress were not mere wallpaper, but FDR used his discretionary power to skirt them whenever they stood in the way of his foreign policy objectives.

In February 1938, Robert Gross stated in a letter to his brother that a "proposal from the U.S. Army Air Corps for an attack bomber to be submitted on January 8 of next year for a quantity order running as high as 500 articles offers a great temptation to rework the [Super Electra]. It is almost exactly the airplane they call for as far as size and performance are concerned. There is also a strong possibility that the Dutch and other governments would be interested in the [Super Electra] as a bomber, and we are looking into this energetically."

Aircraft designed according to U.S. military specifications were not supposed to be exported until as much as one year after the start of full production. This usually meant several years after leaving the design stage. But in April, after a mere five days of work that revealed just how absurd the arguments were about whether airplanes were civilian or military, Lockheed engineers put together a full-scale wooden mock-up of a Super Electra bomber for the arrival of a British commission searching for American warplane sources. (When the weary Brits arrived in Burbank, those officers who liked golf were put up at the Riviera Country Club—a deft bit of salesmanship on Lockheed's part.) Underscoring the airplane's split personality, the bomber version retained the row of seven windows on each cabin side, a rather cushy feature for combat missions.

By the end of June, after intensive technical and financial discussions in London between Lockheed managers and the Royal Air Force, the company signed a $25 million contract with the Air Ministry for 200 "Hudson" bombers, plus as many more as could be delivered by December 1939 up to a maximum of 250—then the largest single order ever received by an American aircraft builder. (The 250th Hudson was finished seven and a half weeks ahead of schedule. When Congress enacted a neutrality law in September 1939 that forbade American citizens to deliver weapons to European belligerents, Lockheed bought an airstrip in North Dakota that straddled the Canadian border. Hudsons were flown in from California, then hauled by mule teams across the line. The procedure was blessed by the Roosevelt administration.) It more than compensated for the Super Electra's failure in domestic airline markets. In one swift stroke, Lockheed left the ranks of small business forever and became a major power in the weapons industry.

The deal also marked a rise in stature within the company for Courtlandt Gross, who carried his brother's authority to London. Courtlandt had recently married Philadelphia socialite Alexandra Van Rensselaer Devereaux—great-granddaughter of Anthony J. Drexel, founder of the Drexel, Morgan & Company banking firm— soon after her divorce from Rodman Wanamaker II, the department store heir. In an industry still considered somewhat declassé by the Eastern establishment, these were magical names. His career at Lockheed, which until now had looked like a tagalong to Robert's, took a sharp upturn.

Robert Gross later claimed that Washington had feared sending the British commission to a company already close to the U.S. military, like Douglas. Lockheed's reputation was as a builder of commercial transports, so "planes from us wouldn't interfere with

American military production." If true, then this was clear evidence that the Roosevelt administration was pursuing a policy of bolstering U.S. warplane production capacity through foreign orders.

Yet speedy delivery of Hudsons to the British while production of P-38s fell far behind schedule led the Army to suspect that Lockheed was favoring its more profitable foreign contracts. Thirteen test models of the P-38 ordered in April 1939 were not delivered until June 1941. By November of that year, P-38 production was ramping up to expected numbers, but the delays were costly to the Army. The Lockheed plane was at that time the only American fighter in the same league as the famed German Messerschmitt 109 or British Spitfire.

Whether because of the secretiveness of the Hudson negotiations or the technicality that the deal did not involve a warplane owned by the U.S. military—the Super Electra was a commercial item, and its bomber version was developed directly for the RAF—there was no public outcry about breaching political neutrality. It fell upon the Douglas company to trigger a national debate about Roosevelt's ambiguous stance. In January 1939, a French warplane purchasing commission stopped first at the Martin factory and then traveled west to visit Douglas. During a demonstration of a prototype DB-7 attack bomber, a French captain survived a spectacular crash near the Los Angeles municipal airport that killed the plane's test pilot.

Douglas tried to conceal the officer's identity by telling reporters that he was a company mechanic. But when the truth leaked out, along with details that the American ambassador to France had wangled the plant visits through pressure on the Army from Secretary of the Treasury Henry Morgenthau and FDR himself, antiwar congressmen demanded to know the President's real intentions vis-à-vis arms transfers to Europe. Senator Nye, for example, charged that the incident bespoke a "military alliance" with France. The debate unleashed strong support for aiding Britain and France against the Axis dictatorships, however, and served to open the floodgates for tens of millions of dollars in further orders to American manufacturers.

In February 1939, France bought a hundred DB-7s from Douglas, the first of 7,478 of the durable bombers and derivatives built until October 1944. After the defeat of France in the spring of 1940, the British government took over delivery of the planes. Still later, the governments of Australia, Brazil, Canada, Holland, the Soviet Union, and South Africa made large purchases. Before American rearmament accelerated in July 1940, these military exports—plus Douglas's commercial sales—amply covered heavy losses in the company's business with the U.S. government. In August 1941, Donald Douglas testified before Congress that in 1940 his company lost a total of $1.14 million on contracts with the Army and Navy. Yet net income for that year reached $10.8 million, almost four times the figure for 1939.

The Martin Company was another immediate beneficiary, with an $11 million order from France for 115 B-10 derivatives called the A-22, originally designed to compete with the DB-7 as a new Army attack bomber. Glenn Martin was "thoroughly angry," the press later reported, about losing the Army contract, which would have been his first since 1932. But the French purchase nearly equaled the company's 1938 sales of $12.4 million, which had yielded an 18.9 percent profit. Martin stock climbed to thirty-nine dollars, topping the list of ten most actively traded shares for six sessions in a row.

"I would prefer to get into the domestic market, which is our ultimate future," Glenn Martin wrote in the midst of this incredible acceleration of business, "but we have now practically closed the door to all commercial inquiries and are preparing for 100 percent military production." Rapid plant expansions—an assembly hall for the French bombers was erected in seventy-seven days at a cost of $1.8 million—and massive new hiring put over 10,000 people to work in round-the-clock shifts on twenty-five acres of factory floor space. Easily half of the $7 million net worth of Martin's facility was paid for by export sales.

Fortune magazine, signaling the approval of the Wall Street establishment, took enthusiastic note of the company's financial resurrection:

> Push through the door of Building C, the latest addition to the Glenn L. Martin Company's plant at Middle River, Maryland, and the hum of power-driven machinery is drowned in a bedlam of riveting now low and deadly like the spray of machine-gun bullets, now rising to a high-pitched scream. Across some 240,000 square feet of floor space, which by night is flooded in the blue-white light of mercury-vapor overheads, there stretches an array of workbenches, of great and small duralumin parts, of human hands and faces. Most of the faces are young; and many who work here still wear sweaters bearing their high school numerals or insignia. Down one side of the building boys swarm about a line of enormous dural wings; at another point they

are bucking up rivets in fuselage and tail assemblies; at a third point they fit controls and instruments (the throttles one observes are marked "ouvert" and "ferme" instead of "open" and "shut") into bulbous dural noses. Only at the end of the building does the work integrate into finished product. There, standing in a far bay, is a line of sleek planes, landing gears in place, twin motors sheathed into wings, each unit representing 30,000 parts brought together into 16,500 pounds of fighting ship.

It seemed to be a vision of Brueghel crossed with Dickens.

Like others who had taken a sidewise journalistic glance at the company's founder, *Fortune's* reporters stated without elaboration that Glenn Martin was "tough to work for." They quickly mentioned "the Martin clothes, which always have a narrow ridge on cuff and lapel" and "the profound influence on Martin's life of his mother." Regarding his personality, the otherwise effusive magazine would only call him a "nonconformist" and venture cryptically that "many a man will simply shake his head and say, 'Well, Martin is different.' "

"One might suppose that the Martin Co., which has a history going back to 1908 [sic], might have obtained an unassailable niche in an industry where many a company is still adolescent," *Fortune* mused. "Such is by no means the case. As the Martin Co. has grown older, its experiences have become more harrowing."

The nonconformist head of a harrowing business nonetheless had a personal opportunity to influence Roosevelt's position on rearmament when he and his mother traveled through Germany, France, Holland, Switzerland, and England in the spring of 1938. After visiting German aircraft factories, Glenn Martin delivered a report to the American ambassador in Berlin, who used it to compose alarming messages to the President. FDR was soon on the path to proposing an increase in the strength of the Air Corps amounting to ten times the authorized limit of 2,320 planes.

The frenetic business activity of these months did not prevent Martin from attending to important personal matters. He found time, for example, to dictate a two-page reply to a salesgirl in a Fifth Avenue fur salon who was despondent about not selling his mother a new mink coat:

> Mother wishes me to inform you that she cannot possibly find any fault with the way you handled our visit to the salesroom of

Gunther's—as a matter of fact, she classes you above the average for sincerity and understanding of the needs of a customer who is in the market for merchandise.

I can concur in this opinion, as I have been unable to find the slightest failure on your part, as I review the steps taken in our interview.

We first purchased a good fur coat from Gunther's in the year 1917. The man selling us the coat promised correct fit, and, of course, complete satisfaction. The purchase was made at approximately 20 percent above competitive costs, because of our feeling that the house of Gunther's was a superior firm.

At the end of the second fitting, an important man in Gunther's talked us out of the desirability of the comfortable hang of a good coat and his superior knowledge and attitude bothered us a little at the time, but this was soon forgotten.

We again shopped at Gunther's for a sable wrap in the fall of 1922 and after making careful comparisons, purchased a sable wrap at another store, on the grounds that Gunther's should meet the competition.

The present sale was lost to Gunther's, I can assure you, from no lack of attention or effort on the part of the saleslady, but because we found a wild mink coat with better styling and quality for the price than we could find at Gunther's.

He also found time to have his male secretary convey his desire to the Maryland Commissioner of Motor Vehicles for special license tag numbers matching the company's telephone number. "Mr. Martin would deem it a great favor and privilege if you could comply," the commissioner was informed.

By contrast, Robert Gross was utterly absorbed in the whirling vortex of his business, though still wearing the blinders that had made sales to Japan seem like a good idea. "The unchallenged advance of the German march across Eastern Europe gives me furiously to think," he wrote at the end of March 1939. "I see our gallant little markets like Roumania, Jugoslavia and Poland falling before Hitler's steamroller every week, and it makes me realize how quickly the work of years cultivating on our part can be swept away over night. Not a very pleasant thought." Here was the destruction of a continent reduced to lost business opportunities. Gross's answer was to "do what we can to pad out in South America what we lose in Europe." An Air Corps solicitation for bids on building several hundred new interceptors might also help "take the place of some of Hitler's conquests," he hoped. Indeed, it would. In August, with

the Wehrmacht poised on Poland's western border, Lockheed won
a $4.8 million Army order for 210 P-38s.

JACK NORTHROP HAD become independently wealthy in 1937 when
Donald Douglas bought out his interest in the Inglewood subsidiary.
Within a year, forgetting his pledge never to enter the airplane
business again, he began a partnership with a forty-four-year-old
Wall Street figure, LaMotte T. Cohu, whose experience in the aviation industry traced back through the dark maze of directorships
and financial stakes in large holding companies of the past decade.
Before Congress passed legislation in 1934 that segregated manufacturing and transportation activities, Cohu had been a director of
United Aviation, Aviation Corporation, American Airways, North
American Aviation, and TWA. He was, among other things, still a
director of TWA, and thus long familiar with Northrop's design
work.

In New York, Northrop and Cohu's attempt to interest investors
in backing a new manufacturing company was frustrated by skepticism about whether demand could yet support a fresh venture.
Existing companies seemed able to absorb the upswing in business
so far, and there was no reason to risk supporting Jack Northrop—
recently purged by Douglas—when established firms were hungry
for expansion capital and returning healthy dividends.

In Los Angeles, however, they found a less conservative audience,
much as Robert Gross had discovered during Lockheed's restart, and
Donald Douglas long before that. Once again, Northrop was driven
by his fixation on building a flying wing. According to the investment banker who helped organize a syndicate of underwriters,
Northrop insisted that the stock not be narrowly controlled, because
he "wanted to be free to design an entirely new airplane—a flying
wing—and he felt that a larger number of shareholders with indi-
vidually less money at risk would enable him to do it." On the
condition that Cohu take charge of the business, the syndicate proceeded to put 400,000 shares up for sale at six dollars each. As a
final precaution, the investors met with Donald Douglas and Robert
Gross to obtain their assurance that Northrop could enter the local
market again with no interference.

"My aircraft manufacturing venture with Jack Northrop is pro-
gressing very well," Cohu wrote to Glenn Martin at the end of
February 1939. "We expect to raise between a million and a half

and two million dollars and build a small plant out here of about 150,000 square feet [Martin currently had 1.1 million square feet]. Confidentially, a substantial number of the key men of the [formerly Northrop] branch of Douglas will come with us, which will be our manufacturing setup." Public and private sales of stock actually brought in about $1.25 million.

In the summer of 1939, the new company leased seventy-two acres of farmland in Hawthorne, California. During construction of a $500,000 assembly hall, Northrop operated out of the Hawthorne Hotel, a former whorehouse known locally as the "Yellow Peril" because of its gaudy exterior. Cohu and Northrop took salaries of $15,000, with a provision for a lavish stock distribution after the company earned its first million dollars. But before it was able to garner any profit-making contracts at all, Northrop Aircraft Inc. started pouring time and money into Jack Northrop's private obsession, the flying wing.

ON AUGUST 31, 1939, Robert Gross sent a telegram to Lockheed's chief sales agent in Europe, a Jewish American named Nick Ebin. Three years earlier, Gross had delicately instructed his brother Courtlandt to "go into the problem of our European representation" by having a "good old fashioned heart-to-heart talk with Nick" about whether "'the unfortunate wave of anti-Semiticism sweeping Europe" was hampering his ability to sell the company's airplanes. Ebin had stayed on the job, moving his office from Warsaw to Amsterdam. Now, on the last day of peace before World War II, Robert Gross still had one urgent concern.

"What are your present plans if war comes. Are you prepared remain Holland. Would you feel secure there. The coming of war would undoubtedly strengthen our standing greatly and we shall have increased need for your help with commercial and military orders. Do not wish to unduly urge you to stay on job and desire you make decision but appreciate your advising your present attitude."

Ebin stayed until Blitzkrieg put Europe in Hitler's hands. By then, neither Lockheed nor any other airplane company needed a salesman. The world was at their doorstep, pounding down the doors for the sinews of war.

CHAPTER NINE

THE IRON
CORNUCOPIA

THE HISTORY OF World War II has been told in such multivious detail that its simplest lessons are easily obscured. Between Germany's invasion of Poland on September 1, 1939, and Japan's surrender on September 2, 1945, at least 50 million people were killed and 34 million wounded. Economies on both sides were wrecked for generations. Social and political changes were set in motion that are still determining the course of daily events. Under Nazism, Germany set a benchmark for depravity against which any and all means used to defeat it appeared good. This warping of values was, of course, yet another facet of the tragedy, a most sensitive one to discuss even fifty years later.

In order to win the war, the United States spawned a weapons industry of titanic scope. Its contribution to victory was profound, as is the ongoing burden of its creation. From 1939 to 1945, American manufacturers turned out 304,139 aircraft for the military services. Between 1940 and 1945, 812,615 aircraft engines were produced, along with 713,717 propellers (this number includes only the most sophisticated type, known as variable pitch propellers, which had different blade angles for take-off and cruising). In terms of airframe weight, a measurement that gives some tangible impression of the quantity of massive bombers and transports assembled, the Army alone procured 2,089,436,000 pounds. Between 1939 and 1944, the industry increased its total annual output, measured in pounds of airframe, by 13,500 percent.

In March 1944, the month of maximum output, the Army received 6,800 planes and the Navy took 9,113. Total production in the United States that year reached 96,318, compared to 39,807 in Germany, 28,180 in Japan, and 24,461 in Great Britain. This was the iron cornucopia that nourished what one mainstream American historian called, with a mixture of horror and awe, "the triumphs of technological fanaticism."

Of the $185 billion spent by the United States on armaments of all kinds between July 1940 and August 1945, about $46 billion went for Army and Navy air matériel. At peak production, prime contractors employed a labor force of more than 750,000 people—largely unskilled and untrained—in over 100 million square feet of factory space. Subcontractors added another 250,000 workers. (Desperate for bodies, Lockheed hired 4,000 high school children to work in shifts—four hours of school, four hours in the plant; or four weeks of school, four full-time weeks of work.) They assembled a typical fighter aircraft from at least 10,000 different parts, 10,000 feet of wiring, 3,000 feet of hydraulic tubing, and 36,000 rivets. A bomber might absorb 16,000 parts, 24,000 feet of wiring, and 200,000 rivets. All of this from companies that had never built more than a few score planes per order until 1939.

This was the "great business" that Glenn Martin had spoken of so wistfully in 1934. His company sold 8,983 planes to the Army and Navy during the war—more than ten times its total output from 1919 to 1939. Douglas sold 30,980 planes. Lockheed sold 19,077 planes. Even Northrop, starting from nothing in the summer of 1939, sold 1,098 planes. Federally mandated profit limitations merely prevented huge earnings from becoming mountainous. At the end of 1939, the leading dozen aviation companies had a net worth and total working capital comparable to the auto industry in 1911. Between 1939 and 1944, the aircraft industry rose from forty-first place in dollar value of output to first. It was not unusual, because of the sheer volume of production, for manufacturers to have backlogs valued at many times their entire invested capital.

In 1944, Lockheed's unfilled orders reached one hundred times its net worth. During the six years ending in December 1945, Lockheed realized a $35.9 million profit on sales of $2.39 billion—nine times the company's total profit during its first eight years of existence, yet only a 1.5 percent return on sales. During the same period, Douglas earned a profit of $60.8 million on sales of $3.5 billion—more than five times its cumulative profit from 1922 to 1939, but just 1.7 percent of sales.

The "great business" meant that great shareholders like Glenn Martin became rich men. As of July 1940, he owned 337,085 shares of Martin Company stock, which returned a dividend that year of two dollars a share. The dividend rose to three dollars in 1941, dipped to $1.50 in 1942, and returned to three dollars through 1946. His personal earnings from this source alone thus exceeded $5 million during the war years.

Robert Gross owned 33,763 shares of Lockheed, which had a market value of $540,208 in April 1944. From this and other stock holdings he received $69,776 in dividends that year, on top of his $125,000 Lockheed salary.

This wealth, coupled with the hyperbolic patriotism of wartime, turned military industrial leaders into pop icons. In November 1942, *The New Yorker* magazine published a two-part profile of Glenn Martin entitled "Hero for Business Reasons" that evoked all his favorite personal myths—building his first biplane under the glow of a kerosene lamp held by his mother, breaking the world record for over-water flight in 1912, "inventing" the first automatic parachute, creating the MB-1. None was exactly true, but heroes were not subjected to close scrutiny. He was barnstormer, engineer, business-man, and visionary genius all rolled into one. That he was also a friend of *New Yorker* owner and editor Harold Ross's wife, Ariane Allen, was not mentioned.

The magazine's gloss on the "merchants of death" movement of the 1930s showed how war was turning popular sentiment 180 degrees to the right. Its tone presaged the anti-intellectual vicious-ness that would soon infect the nation's political discourse. "The literati of America had convinced the country that wars were little games arranged by big industrialists in order to get orders for weap-ons," writer Alva Johnston said. "The intelligentsia had convinced the majority of the people and their congressmen that international trouble could be abolished by the simple device of hissing and de-nouncing manufacturers and capitalists. This feeling was summed up in Robert Sherwood's play 'Idiot's Delight,' in the line 'And who are the greater criminals—those who sell the instruments of death, or those who buy them and use them?' The current popular hero was Ferdinand the Bull, and millions believed that the way to baffle international menaces was to bask among the buttercups and daisies. Industrialists like the DuPonts got so tired of being called merchants of death that when war first broke out they would not accept orders from England for explosives. It became necessary for Washington to beg the merchants of death to become the saviors of civilization. With the fall of France, the new fashion came in of denouncing merchants of death for not making instruments of death fast enough." That the Neutrality Act was not repealed until November 1939 was not mentioned.

Martin flourished in this laudatory new atmosphere as never before. He bought thousands of acres of farmland near Chestertown on Maryland's bucolic Eastern Shore, using the duck and goose hunters' paradise to entertain General Hap Arnold, General George Marshall, and other military leaders who bought his company's airplanes. (In 1940 he disguised his contribution to the Wendell Willkie for President campaign by having it listed in the name of his farm manager.) Yet Glenn Martin was never part of the hail-fellow circle epitomized by Donald Douglas and Hap Arnold. Jerome Hunsacker later recalled that Admiral Jerry Land of the Navy "loved him" and made frequent visits to the Shore, but Martin was otherwise perceived as too peculiar for manly camaraderie. Like his interest in the Martin Company's employee baseball team (known as the "Bombers") and his habit of parading through the plant with chorus girls on his arm, the hunting trips seemed staged to convince everyone that Glenn Martin was a regular guy.

Elaborate, costly shooting blinds were built for VIP guests, one bizarre structure being made entirely of glass. In an attempt to guarantee successful outings, the Martin Company's engineering department was put to work designing and purchasing a system of recorders and loudspeakers to broadcast the feeding babble of geese from the roof of one of the farmhouses. The birds were also heavily baited with corn, an illegal tactic that local game wardens could never get higher authorities to condemn. The *Glenmar* became a cruise ship, ferrying guests across the Bay for weekend parties. A letter from June 1943, written by the daughter of an old Cleveland acquaintance, captured the scene:

> It took two hours to get to the Eastern Shore—and when we arrived I almost died. It is another world—with acres of rich ploughed fields and green trees. Pheasant and quail are wild and everything is so peaceful. To spend a week there would be better than medecine. Anyhow, I can't describe the charm and beauty of the land. And he

[Martin] has restored five houses—each is a young manor. I don't think we had a meal twice in the same place or house.

We slept on the yacht Saturday night and the others stayed at Green Oaks [one of the houses]. The station wagon came down for me at nine forty five and we went up to the house for breakfast. And what a breakfast! We had hot cakes and honey and butter and bacon and real ice cream and everything! I ate until I couldn't stand.

Then we went to this tiny church which Mr. Martin has restored. It was built in 1693 and has a tiny Scotch vicar. It was like a picture-book. Then we drove over the twenty-eight hundred acres and I felt like a Great Lady. I could have had a piglet and a puppy if I had just said "yes." And the MARS [a gigantic Martin/Navy flying boat that never found any important practical use] flew over the yacht and landed beside it. What a thrill! We are going to have breakfast aboard her—someday. Then to top it all off, we went to a movie in his own private theater. Ho hum—my deah!

Later that summer, when the Navy attempted to requisition the *Glenmar* for photographing fleet target practice off Norfolk, Glenn Martin sent letters to the Secretary of the Navy and the Vice Chief of Naval Operations. "This boat was purchased by the Company for use in connection with the development of its Naval aircraft," the Vice Chief was informed. "It is used as an observation platform from which to witness tests, experimental flights and current operations of PBM-3 naval patrol planes and of the XPB2M-1 (MARS) long range Navy patrol plane. The latter type of airplane is still under-going demonstration trials and will involve regular use of the Glen-mar for observation of these trials, not only by the operating personnel of this Company, but by representatives of the Navy Department and War Production Board."

The Secretary of the Navy was told that "the more important function of the yacht Glenmar is that in case of emergency or acci-dent, doctors, nurses and their equipment can be immediately rushed to the scene of accident out in the Bay, and would be the means of saving lives, as there is ample room on the Glenmar for many stretchers and the Glenmar is fully equipped with the services necessary for the emergency. For example, there is an abundance of instantaneous hot water always available and ready." That the hot water heaters were installed to supply showers in the vessel's three staterooms was not mentioned. Nonetheless, the Navy abandoned its attempt to get the *Glenmar*.

A year later, when the local branch of the Office of Price Administration queried the company about whether a day trip to Tolchester—an Eastern Shore amusement park and beach—was "for one of the purposes for which the oil ration for the boat was issued," Martin replied that the *Glenmar* was "not a private yacht" and that "obviously, emergency assignments cannot be fulfilled without occasional runs for the purpose of checking equipment and training the crew in their specialized duties."

In addition to big seaplanes, the war brought back to Martin's factory another of his lifelong ambitions—Army bombers. But the result was a mixed blessing that might have been disastrous in a peacetime business environment. In the frenzy of September 1939, the Army placed an order for 201 new high-speed medium bombers, called the B-26 "Marauder," directly off the company's drafting tables—an unprecedented leapfrog over the normal process of first testing a prototype. To meet the Army's specifications for speed, Martin engineers had conceived a twin-engine aircraft with radically small wings—a "hot" plane that also had an abnormally high landing velocity. It soon became known among pilots as the "Widow Maker" and the "Flying Prostitute," recalling the "Flying Coffin" DH-4 of World War I.

Accidents were so persistent that only one combat group was equipped with B-26s by December 1941. In order to obtain sufficient range, fuel had to be carried in the bomb bays, limiting the Marauder's bomb load to one ton. At the end of March 1942, General Arnold suspended production for two months until researchers recommended that the bomber's wings be enlarged. On several more occasions the Army considered scrapping it. As late as July 1943, congressional analysts deemed the B-26 "unsafe when operated by any pilots except those specially trained for its operation, because of unusual difficulties in landing and take-off." Intensified training of crews and a change in the type of mission assigned to the plane—from low-level to high-altitude attacks, then finally to tactical air support of infantry—redeemed its combat reputation, though high production and maintenance costs eventually led the Army to taper off the output of Martin's plants in favor of other aircraft.

The Marauder brought Glenn Martin under direct fire from a special investigative committee in Congress, chaired by Senator Harry Truman. The panel made its own engineering survey of the design, also finding that the wings were too small. Martin then testified that the program was too far along to do any major rework.

"I told Martin that if the lives of American boys depended upon the planes that were produced for the United States Army Air Force, the committee would see to it that no defective ships were purchased," Truman recalled in his memoirs. " 'Well,' Martin replied, 'if that's the way you feel about it, we'll change it.' "

Just as difficulties with the B-26 were becoming apparent in early 1941, Douglas offered a bomber called the A-26 "Invader," an improved version of the DB-7 that was close in physical dimensions to the Martin Marauder (its wings were even smaller), yet faster and with a better range, ceiling, and rate of climb. Three prototypes were ordered in June 1941, and by 1942 the Army was planning to replace the B-26 with the Invader. But production delays (General Arnold fumed that he wanted Douglas's A-26 "for use in this war and not for the next") kept deliveries low until mid-1944. Combat operations did not get under way until November of that year. After the war, however, the Douglas Invader became the Air Force's standard tactical bomber, taking away Martin's B-26 designation in 1948, when the Marauder was withdrawn from service, and serving through the Korean conflict. A few upgraded models even saw counterinsurgency action in Vietnam during the 1960s. This was just the sort of long, profitable contract life that the Martin Company would desperately need.

In late 1942 and early 1943, Glenn Martin was scrutinized by the Internal Security Division of Headquarters Third Service Command, a military investigative branch. Personal references, such as the president of Kansas Wesleyan University, were asked to attest to his character and loyalty. Whether this was launched by an overzealous bureaucrat or in retaliation for the *Glenmar* tussle, or because of concern about his lifestyle cannot be determined. He was one of the most well known men in American aviation, a contractor to the military for three decades, an acquaintance of every military and political figure of any consequence to his industry since the days of homemade biplanes. Nonetheless, his friends were asked to affirm that he was suitable "for employment in National Defense." Fortunately for him, it was not a management test.

DONALD DOUGLAS WAS the favorite airplane industry icon during the war, dubbed "the cornerstone of American air power" by aviation author Alexander P. de Seversky and, perhaps more accurately, "the passionate engineer" by the editors of *Time*. Unlike Glenn

Martin, Douglas had sex appeal, his tall, brown-eyed, tweedy good looks appearing regularly in the pages of slick magazines. With a salary of $120,000 a year in 1943 (up from $64,683 in 1939), a $150,000 Spanish-style home near Santa Monica, and a seventy-foot sailboat named *Endymion* (after the beautiful youth of Greek mythology, who was enthralled by the cold heart of the moon goddess Selene), he also fit the all-American bill of wealthy self-made businessman, though he now owned less than 1 percent of his company's 600,000 outstanding shares.

Just past the age of fifty, he was known to be a laconic individual of obsessively unvarying habits. He awoke every morning at seven-thirty without an alarm clock, ate a breakfast of one egg, one piece of white toast, and one cup of black coffee. He left his house at nine, driving his own black Lincoln Zephyr to the factory ten minutes away. He sat in a walnut-paneled office there behind a carved walnut desk adorned with his family coat of arms (motto: *Jamais arrière,* "never behind"). An old hand drill, brightly polished and mounted on a mahogany pedestal, was prominently displayed to remind him of what he had started out with in 1920. A pottery skunk stood for his battles against labor unions.

At exactly twelve-ten every day, he walked across Ocean Park Boulevard to a private dining room, where he had his invariable lunch of hamburger steak, black coffee, and a chocolate sundae. At five-thirty he left the plant and drove home, where dinner was always served at six-thirty—dessert always a chocolate sundae. He spent almost every evening alone in his library, usually reading books about yachting, while wife Charlotte listened to the radio upstairs. It seemed to be the life of a man coldly detached from his job and his family, who relied on strict routine to endure weekdays in order to escape on the *Endymion* every weekend. Aboard his beloved sailboat, a bit of color returned to his life. He cooked curried lobster for the crew of Douglas executives, cruised to Catalina to play bagpipes on the hillsides, and certainly spent time with Peggy Tucker. Her name never appeared in the magazines, which were still infinitely genteel in such fragile situations.

THE WAR WAS ONLY six months old when Robert Gross sent a telegram to the presidents of the major domestic airlines. "The expansion of major factories occasioned by these war orders is a splendid thing for the industry," he said, "and while I believe much of it can be preserved permanently, nevertheless a cessation of hostilities in Europe and a major disarmament program would present problems to the expanded aircraft industry. I am personally interested in seeing our company start in a campaign to provide means for sustaining this activity." Germany had not yet begun the Blitzkrieg, with which it would overrun Belgium, the Netherlands, and France in seven weeks. So Gross could not know that the lull of winter 1939–40 was false. But he was right about the detrimental effect of peace on the aircraft industry. His effort to start a campaign to sustain the level of wartime business if peace broke out was merely premature. On May 16, 1940, FDR asked Congress for 50,000 military aircraft.

By July, with Paris occupied and the Battle of Britain under way, Gross refined his prognosis for a Wall Street friend into four possibilities, each of which looked wonderful for business:

1. While the war goes on, we are going to sell all the airplanes we can build.
2. If England wins, they will readjust the map of Europe and a good many of the countries now under Hitler's thumb will have to be reorganized and will need airplanes, and furthermore the British will never allow themselves to be caught asleep again and will buy more than ever.
3. If Hitler wins, the United States will be so scared that they will buy their heads off for the next ten years.
4. Again if England wins, the United States now thoroughly aroused will still buy more planes than the industry can make, which orders, coupled with the enormous increase in commercial aviation which is coming all over the world, should make legitimate companies permanently prosperous.

Gross was also satisfied that wartime profit limitation statutes would not be a problem. Indeed, the issue of profiteering never arose as it had in World War I. "I believe that sentiment in the Army, the Navy, the Defense Committee [Congress], and in most sections of the Treasury is so sympathetic to seeing the aircraft companies prosper that they will greatly liberalize their interpretations as to what is cost and what is profit," he reasoned. "In other words, we may be able to write off more by agreement with the Treasury in the future than we ever have in the past." Anyway, only 5 percent of Lockheed's $167 million current backlog was subject to newly enacted profit limitations.

To Lockheed employees, on the other hand, Gross presented the face of sacrifice. "Getting around to patriotism, I want every employee to know that your company's heart is in the right place on this one," he wrote in the *Lockheed Star,* a weekly newsletter. "We are making financial sacrifices every day to get ready to do a big job for Uncle Sam and this is no time for anybody to start rocking the boat by questioning anybody's motives. . . . I wish somebody would tell me just what good it did a worker in a French aircraft factory to have gotten double time for everything over 38 hours a week, and all the other so-called social gains that he got during the Leon Blum government, when the Germans who worked 50 and 60 hours a week walked in and took him over. The Frenchman got his social gains all right, but he lost his country."

In an industry whose leaders had until now been remarkably apolitical, if only because they were so preoccupied with keeping their businesses afloat, Gross's thinly disguised comments about "so-called social gains" were a harbinger of polemics to come. Even Glenn Martin, who had never even registered to vote, drifted into hard-right rhetoric, exclaiming that "if our country can revert back to good old democracy and free enterprise and not become a social state or communist, I think everybody will have a job and can grow and prosper in the after-the-war period, but if by some chance the majority of the people should want to become communistic and fascist, there is no way to tell what the future will hold."

For the first six months of 1940, Lockheed turned a profit of over $2 million on sales just under $20 million. The extended Gross family was prospering as never before, with the wives meeting at the Savoy-Plaza in Manhattan for fall shopping sprees and Robert's daughter off to Sarah Lawrence College with a monthly pocket allowance of $100. "The signing of the treaty by Japan, Germany and Italy further intensifies the situation," Gross wrote to her in October about the Tripartite Pact between those countries, "and it may even be that we shall get into some trouble with Japan first. Basically I think the Japanese are afraid of us and won't start anything, but if their military party should get fresh and run off at the mouth somewhere we might have to pin their ears back." Back at home in Bel Air, the most pressing problem was that her two show horses, Egon and Miss Pat, had lost their good dispositions since she went away.

By the autumn of 1940, Lockheed was no longer a commercial aircraft company, having joined Martin, Douglas, and other major firms as an adjunct of the U.S. government (and, by extension, the British and French governments). Congress had already passed, at the end of June, legislation that granted military contracts priority over all commercial work. "We were shanghied," Donald Douglas groused about Washington's takeover of the business. In October, more new statutes made it highly profitable to convert to weapons production, permitting the complete tax write-off of new facilities within five years and thus taking some of the sting out of higher wartime tax rates. Ninety percent of Lockheed's $280 million backlog at the end of November consisted of military planes for Washington and Whitehall. Some $30 million in commercial orders were on the books, but only about $4 million worth was being put through the factory. Gross received a telegram from Lord Beaverbrook, the British procurement czar, thanking him for "your friendship and sympathy with us in this battle for freedom and liberty and the right to live our lives according to the great traditions of the Anglosaxon race."

Where Robert Gross and every other aircraft company president had once encountered stiff shirts when they sought financial assistance, now they found glad hands. "I think I should be frank about it and tell you that we cannot conscientiously encourage any more banks with hope of business from us," Gross told his father, who had recommended a friend's bank. "We have recently arranged bank credit of some $20 million. The credit was participated in by a good many of the leading institutions, such as the City, Chase, Morgan, First of Boston, etc. Banks are so loaded up with money and the airplane business, which is now backed up by Government contracts, looks so secure to them that the difficulty is to keep them all satisfied." Here was the engine of war, grinding away the last vestiges of the Great Depression.

Given the avalanche of war profits, Gross was careful about not looking piggish in public. Resisting pressure from one of the major investment houses to place a member on Lockheed's board, he explained to the firm's senior partner that Lockheed "started off its early years with a plethora of investment bankers, and I shall always feel that with no criticism of the men as individuals, and incidentally we like them all, the top-heavy character of our board worked against us in the eyes of the public. People took us for a promotion." Gross, of course, was himself a wily investment banker when he plucked Lockheed out of bankruptcy court. The "Wall Street element" did not sit well with the government or labor unions, he added. "We would be well advised not to appoint any banker, at least until the stage is reset."

(Gross touched on this matter again several years later, when a Merrill Lynch broker criticized the Lockheed annual report for lack of sales appeal. "My feeling during the war was that it was better taste not to be emphasizing the commercial side of our business and for that very reason we went to considerable pains to be very modest in our public utterances and statements," he explained. "I think you know that at the high pressure point in the war any thing that a company did to favor its stockholders or cater to the financial interests created a certain amount of unfavorable background in the Army and the Navy. This was extremely unfair and shortsighted, but nevertheless it was true and for this reason our reports were cut and dried.")

At the end of July 1941, Gross and his daughter left Los Angeles for ten days at Lake Tahoe. His wife, Mary—never of robust health— had been stricken by a brain hemorrhage four months earlier and was facing uncertain years of convalescence. He was thinking of buying Clark Gable's house in Encino, where she would have more room to rest outdoors. His father, who had recently moved to Bel Air and on whose advice—if not financial support—he continued to draw, was also in precarious health and would be laid to rest at Forest Lawn a year later. Thrust to the top of a thunderous industry at the age of forty-four, Robert Gross lived a grueling life. He was racked by chronic insomnia. These few days in the mountains were the last vacation he would take for a long time. In the fall, Leningrad was besieged by German forces.

"I don't think there is going to be a war with Japan," he wrote to his daughter on December 4. "I feel fair."

Three mornings later, the first of two waves of 353 Mitsubishi, Aichi, and Nakajima warplanes left the flight decks of the Japanese 1st Air Fleet two hundred miles north of Oahu. Within two hours, the vast American naval base at Pearl Harbor was in ruins. Nothing about Lockheed and the aircraft business would ever be the same again. The years from 1910 to 1940 would soon seem as distant and quaint as the antebellum South.

"It is almost unbelievable to realize that our peaceful and happy community here, which has been living in some comparative calm, has suddenly been transformed into a veritable war camp," Gross wrote to a friend in New York on Christmas Eve. "We are operating in an atmosphere which I can assure you is very realistic, and while we have been taking life seriously here for the past three years, the degree to which we are now geared to the task is very, very different indeed."

GIVEN THE SCOPE of the conflict, it is not surprising that the Army and the Navy became the nation's biggest consumers and investors. This was the meaning of "total war." Between January and June 1942, the two armed services signed more than $100 billion worth of new contracts.

Because the military possessed its own economic mobilization scenario and carried on its traditional mistrust of civilian agencies, these initial months were not as smooth as one might have expected after two decades of interwar planning. But a modus operandi was soon reached that brought military and corporate power together as allies—much as in World War I, but without gross scandal. This time around, such notorious instruments of corruption as cost-plus-percentage-of-cost contracts were avoided in favor of the more seemly cost-plus-fixed-fee deal. As historian Paul Koistinen wrote, the emergence of a "Big Military" was the most significant development of the World War II economy. "Big Government" and "Big Business" were already hand in hand, despite sporadic complaints about semi-socialist New Deal ideology, which was actually all but dead.

The more than two years of European war before Pearl Harbor gave these major players time to address, if not solve, many of the workaday problems—such as labor and material shortages, factory capacity, tools, and subcontracting—that follow massive shifts in a complex economy. To help prepare for American entry into the war, the old National Defense Advisory Committee (NDAC) had been revived after the fall of France in June 1940. Constrained by antiwar sentiments at home, President Roosevelt felt that NDAC was as far as he could go for the time being. Once again, industry leaders were given the opportunity to regulate their own fields of business, with the largest companies exerting the heaviest influence.

William "Big Bill" Knudsen, president of General Motors, served as commissioner of production (with ultimate authority over aircraft matters), and Edward Stettinius, board chairman of U.S. Steel, was in charge of raw materials. They were the tallest titans of American capitalism, comfortably situated to get what they wanted out of Washington. Beginning in October 1940, Knudsen made sure that the auto industry would play a substantial role in aircraft production, despite unsavory memories from World War I that lingered among senior Air Corps officers. When serious expansion of production facilities first got under way in the summer of 1940, thirteen established aviation firms received all the government funding for this purpose. Martin, for example, more than doubled its floor space at Middle River to accelerate the output of B-26 bombers, at a cost of $25 million. After Knudsen's initiatives on behalf of the automakers, however, Martin assembled additional B-26s at a government-built plant in Omaha, using parts supplied by Chrysler, Hudson, and Goodyear. Douglas operated a similar plant in Tulsa, initially planned for production of B-24 "Liberator" bombers with components supplied by Ford, but switched to other aircraft when Ford failed to meet its delivery promises.

In January 1942, the same month in which he called for production of 60,000 warplanes that year and 125,000 the next (actual output was 47,836 in 1942 and 85,898 in 1943), Roosevelt created the War Production Board, a superagency with totalitarian power to enforce industrial mobilization. The First War Powers Act of December 18, 1941, had given the President and his executive agents the right to enter contracts without regard to existing law whenever this would advance the war effort. The new board collided head-on with military planners in the War and Navy departments, but by the end of 1942 reached a compromise in which the armed services continued to procure their own weapons while the board allocated materials and determined production schedules. Operational authority came to rest in an aptly named Production Executive Committee—chaired by the president of General Electric—whose majority membership hailed from the War and Navy departments.

There was scattered political opposition to this system, because it naturally acted to obliterate weaker interest groups, but Roosevelt was successful in calming critics through the creation of committees for settling disputes without changing the basic distribution of power. Senator Truman, for example, endorsed a bill that would have set up an office of war mobilization staffed by civil servants. The bill was supported by organized labor, which usually found itself shut out of major decision-making, but it went nowhere.

Thus, by 1943, a mutually beneficial partnership between the armed forces and elite corporate interests was in position to distribute billions of dollars of federal contracts, always under a blanket of "military necessity." Unlike the World War I experience, production records were so miraculous that any vestigial concerns—about the deleterious effects of such concentrated power—raised by the "merchants of death" and other reform movements of the 1930s were completely neutralized. Victory was achieved, but only by granting large corporations and the military unprecedented authority over a democratic society. This would be the permanent legacy of two great foreign wars that never touched the nation's soil, yet still left a mark deeper than any bomb crater or trench.

THE SPECTER OF PROFITEERING, so noxious in the aftermath of World War I's aviation scandals, was not a hypothetical issue for weapons builders or buyers during World War II. The magnitude of federal expenditures needed to fight simultaneously in Europe and the Pacific was beyond comprehension, no more so than among airplane manufacturers. Congress did not legislate specifically against excessive profits until April 1942—after House investigators publicized the case of a newly organized Cleveland company that was charging the government $750 for aircraft engine starters that cost $292 to produce—but the topic was never a stranger to public discourse.

In June and July of 1940, Congress passed emergency laws that did away with competitive bidding but prohibited the military from signing cost-plus-percentage-of-cost contracts with manufacturers. They had been the most odious type of business deal during the Great War, encouraging firms to pad their costs in order to make ever larger profits. Congress also imposed a profit limit of 8 percent on all Army and Navy aircraft work. This curb, which did not apply to export orders, was removed entirely in October after fierce lobbying. In practice, aviation procurement officers tried to limit profits to between 9 and 10 percent on conventional fixed-price deals, where buyer and seller agreed on a total dollar figure from the outset.

To cover situations where cost could not be closely predicted before production was well under way—a common problem in aviation and other businesses at the forefront of technology—Congress authorized the use of cost-plus-fixed-fee contracts, where the "fee" or profit could not exceed 7 percent of the estimated cost of labor, materials, and myriad other certified items. The fee was reduced to 6 percent in the fall of 1940, 5 percent in 1942, and 4 percent in 1943. Between June 1940 and the end of 1944, despite the War

Department's official policy that cost-plus-fixed-fee contracts would be used only as a last resort, the Army Air Force placed $24.16 billion under this category, which amounted to 55.4 percent of all its contracts worth more than $10 million each.

None of these regulations caused any real pain beyond a crush of paperwork. Yet manufacturers sharply resisted the reduction of their fees on cost-plus work to 4 percent. Even at just 5 percent, companies with astronomical orders on the books were earning huge profits in relation to their net worth. In the fall of 1942, when the War Department decided on the 4 percent maximum, industry leader Douglas refused to accept it voluntarily. The Army then threatened to issue a mandatory command, which would have ripped apart the fabric of cooperation between "Big Business" and "Big Military." Douglas wisely retreated, along with every other major aircraft firm.

Besides fee percentages, the government had one other tool for keeping profits from reaching politically untenable heights. In April 1942, Congress passed a Renegotiation Act that gave military auditors the power to recoup excessive earnings. Like the first federal tax on income in 1913, it was a revolutionary concept in American business; but like the trade embargoes and neutrality laws of the 1930s, it contained enough loopholes to permit all but the most egregious activities. The definition of excessive, for example, was embedded in a dazzling tautology: "any amount of a contract or subcontract price which is found as a result of renegotiation to represent excessive profits." Contract officers who dealt with individual companies thus had broad latitude to interpret the legislation case by case. The baroque bookkeeping complexity of any major production order was, of course, a formidable barrier to these harried auditors, no matter how intent they might be on carrying out the spirit of the law. About $3.5 billion was nonetheless recovered by the Army Air Force through the end of 1944—an impressive figure, but no doubt only the tip of an iceberg that will always remain submerged in the pelagic darkness of the war.

Lockheed's earnings from P-38 contracts show how the government managed gradually to reduce profit margins somewhat. In June 1942, a cost-plus-fixed-fee order for 1,800 P-38s was figured on the basis of each fighter costing $82,418, with Lockheed earning $3,925 per plane. That summer, the House Committee on Military Affairs published a report that was highly critical of cost-plus contracting, urging an end to the "honeymoon at the expense of the taxpayers." In January 1943, an order for 800 more P-38s put the cost at $66,861 each, with a Lockheed profit of $2,572 apiece. And in June 1944, an additional 1,700 cost $67,589 each, bringing the company just $2,000 a plane.

"In a broad sense, we must recognize that we shall probably never again during this war be able to make the money that we made in the years 1942 and 1943," Robert Gross wrote to a discouraged stockholder in January 1945. "This is fundamentally due to the fact that we have felt in your protection we must take an increasingly large portion of our business on the cost-plus-a-fixed-fee basis [where the fee was now nominal]. However, you will be pleased to learn that we ended the year [1944] with our company in the best financial position that it has ever enjoyed. Our balance sheet will show a strong cash position, increased working capital, very substantial reserves, and a larger net worth."

This glowing image of what the war had done for Lockheed was not shown to Congress during the course of oversight hearings on procurement practices. In June 1943, for example, a spokesman for five West Coast airplane companies, including Lockheed and Douglas, painted quite the opposite picture for the House Naval Affairs Committee. "I say to you with the utmost sincerity," he intoned, "that our stockholders are risking every penny of their capital, up to the hilt, in the war effort. The general belief is that we have benefited by the immense volume of business which has been entrusted to us. This is far from being the case."

Ever the best friend of industry, the committee harbored only one member who was willing to stick out his neck and say he found this hard to believe. "It is going to be difficult to convince anyone," said Representative Michael Bradley of Pennsylvania, "that with the profits you have made that you can actually count, not that you are contemplating renegotiation on, that your company is going to be worse off because they participated in the wartime program than they would have been had they not. Now I just can't conceive that."

Nor could anyone else who bothered to examine the companies' balance sheets. But in the midst of a great patriotic war effort, such matters were understandably swept aside. The aircraft builders did the job they were called upon to do. Military aviation since 1917 had indeed been a school run by the government for businessmen, where they learned all the subtleties of federal contracting, as well as how to hurdle investigations of their work. By the 1940s, they had no use for the rough-shod graft of World War I.

"Surely a major deterrent to fraud must have been the relative ease with which ample profit margins could be acquired under war boom conditions," concluded the Army's official history of World War II aircraft procurement. There was plenty of cream for everyone, so few bothered to steal milk.

CHAPTER TEN

PERMANENT REARMAMENT

W HEN R OBERT G ROSS , Glenn Martin, Donald Douglas, and Jack Northrop walked out into the light of the first morning of peace after World War II, they faced a radically different universe. Much has been written about the tribulations of defense companies since 1945, but much can be explained by their transformation during the war into technological grotesqueries. They had accomplished a stupendous feat, one that left them now like whales stranded on a sandbar at ebb tide.

"As long as I live, I will not forget those short, appalling weeks," Robert Gross said. He was not talking about the last bitter days of war, but of the first days of peace after Japan surrendered.

Of the four industrial leaders, only Gross, who had entered the business as an investment banker with whale-sized ambitions to build an empire, would retire with his corporate power intact. The others—who had been seduced in their callow youth by a backyard engineering marvel, and then created a business just to stay near it—would be crushed by the new environment. Martin, Douglas, and Northrop would die as legendary pioneers, but only after their namesake companies had been torn from their grasp.

That the industry experienced a wrenching deceleration that burned enormous amounts of cash is unmistakable. In 1946, the first full year of peace, Douglas Aircraft had net sales of $106.7

million, nearly four times the figure for 1939. Yet its profit was only $2.18 million—$650,000 less than in 1939. Likewise, Lockheed had sales of $112.7 million in 1946, more than three times the 1939 level, but profits of just $3.06 million—$80,000 less than in 1939. This black ink was achieved only by applying a federal income tax credit of $11.2 million and transferring $13.8 million from a reserve fund set up during the war. Without such measures, Lockheed would have been $21.9 million in the red at the end of 1946.

During the war years, neither the industry nor the military had ignored the inevitable problems of demobilization, though near-panic production in 1944 made an orderly transition to peacetime unlikely. The grim economic lessons of World War I's aftermath, especially the effects of precipitous contract cancellations and the dumping of surplus government equipment onto commercial markets, were widely understood. But no one fully appreciated that the second global war had made the aviation companies so big that they could never return to the formulas of the pre-1939 era. The only business "great" enough to refloat the whales was government business—specifically military business, in a far deeper current than had ever been known in peacetime.

"V-J Day came with such suddenness that the Government was far off-base, as far as reconversion is concerned," Glenn Martin wrote to his old patron in Los Angeles, Frank Garbutt, near the end of September 1945. "We have found ourselves in the midst of a very exciting struggle to save some of the industrial debris floating around on the sea of uncertainty."

As long as the war was on, "we were not permitted to do any engineering or research work of any kind on post-war business or designs of any description, except war designs, as ordered from time to time by the Government," Martin explained. Practically all of the war contracts had to be cancelled, permitting only a few experimental projects and a few servicing jobs to continue. Thousands of employees had to be laid off, of course. "We are in for rather intensive planning to hold our productive strength until orderly business can be established," he said with a very stiff upper lip.

Under such circumstances, timing was critical to a company's re-entry into the commercial world. "If the war had ended six months ago, our development position would have been so favorable by comparison with anybody else's except Douglas that we would undoubtedly have been guaranteed a leading position in the market," Robert Gross stated in March 1945. Lockheed had bene-

fited during the war from military contracts for long-range transports that would be highly attractive in airline passenger service.
"Now, however, the war has dragged on and every month that it
lasts it gives other companies an opportunity to get development
work going, so that while I still consider our position to be among
the best, time works against ourselves and Douglas in favor of anyone who might have started advanced development later than we
did."

All of the major companies were banking on a sharp rise in commercial business as soon as peace came. The war's engineering miracles spread an intoxicating optimism about technology-based
businesses, especially aviation. "I honestly believe the industry is on
the threshold of a whole new realm of performance," Gross said in
a speech before the Southern California Council of the State Chamber of Commerce, probably referring to Lockheed's secret wartime
program to build a jet fighter. "Having made these new discoveries,
we must decide whether we will advance them as a means of security for our country, or abandon them, only to have other countries use them against us." He foresaw "extraordinary advances in
transport of passengers and mail all over the world" that would cut
costs and enable the "little fellow" to take two-week vacations
abroad. Chasing the same chimera that had teased Jack Northrop
and the Loughead brothers into bankruptcy after World War I, he
predicted "an enormous increase in private flying—in which the
helicopter figures—with a possible goal of an airbuggy for everyone."

Having lived through the humiliation of hitting rock bottom after
the Great War, Jack Northrop was less sanguine. "There will be
insufficient commercial business available to maintain even the minimum desired nucleus of aircraft engineering and manufacturing
talent in the postwar era," he told a congressional committee studying reconversion problems on August 24, 1945—ten days after Japan gave up. "Substantial quantities of military aircraft must be
designed and built" for the industry to remain solvent. Arthur Raymond, Douglas's vice president for engineering, whose experience
in the industry was almost as long as Northrop's, recommended to
the same panel "a policy of continuous replacement" of warplanes
in order to "maintain at all times a superior striking force instanta-
neously ready for use." No nation in history had ever kept such an
arsenal in peacetime.

Donald Douglas sent a pugnacious letter to the committee's chairman that essentially called upon Washington to save the aircraft companies as the companies had saved the nation in helping win the war. "The men and women who made victory their business during the war must be helped to make industry self-supporting," he wrote. "After telling industry to drop everything else and concentrate on war production, after dictating for several years how, where, and when to operate, Government should not, now that the war is over, say to industry: 'Shut down everything. Sell everything to the highest bidder. Fire everybody. You're on your own.' " Douglas remembered that yard full of rotten potatoes in Los Angeles in the lean spring of 1920.

Right after the war, the price of aviation manufacturing stocks on Wall Street reflected Gross's heady optimism, not Northrop's realism. Douglas traded at a high of 107⅞ in 1946, Lockheed at 45¼ and Martin at 45½. By the spring of 1947, however, they fell to 67, 15, and 26, respectively. All of the sweet forecasts were turning sour as the world convalesced from a six-year calamity.

Soon in a more sober mood, Gross confided to a friend that "we are rather bravely launching ourselves into the postwar world somewhat whistling to keep our courage up." He still believed that demand would rise, that "if we can swim through another two or three years of muddy water, we will find ourselves in an enormous industry—I only hope we can keep our own company right-side-up until we get there."

Perhaps because Lockheed looked in better financial shape than most of the other major aviation companies, perennially upbeat *Time* magazine featured Robert Gross on its cover in January 1946. Lockheed had just delivered the first of its postwar Constellation airliners to Pan American—aesthetically stunning four-engine transports whose development dated back to 1938. "Among planemaking tycoons, predominantly an inbred and individualistic group of onetime designers and pilots, Bob Gross is a sport," the flagship of popular American opinion said. "He is not a pilot. He knows little about aerodynamics. As a production man and administrator he is just so-so. Yet he has one talent which more than balances these apparent deficiencies. He has a seemingly intuitive salesman's sense of knowing what planes will be wanted a few years hence."

This was the kind of jolly simplification of a tortuous career that only a journal eager to please could make, but it conveyed the accurate impression that Lockheed had been fortunate in the marketplace since 1932. Gross embodied the flush of victory, however obtained, and the public was now happy to hear about a lifestyle that would have sounded "excessive"—to use Washington's euphemism for greedy—during the war: "His home in swank Bel Air, ten miles from Burbank, would be appropriate for a painter or musician. A roomy house of pink stucco (big enough to keep four servants busy), it has a large sweep of lawn in the rear and a badminton court. The decor, in subdued colors, was done by Hollywood's Dolena, the 'Greek modern' furniture designed by Manhattan's Robsjohn-Gibbings. Against this background hangs Gross's collection of modern and abstract art: Klee, Kandinsky, Braque. (A Modigliani blonde over the davenport is known to his friends as 'the happy housewife.') Gross began to collect pictures about the time he went into Lockheed, finds the same elements of composition in them that he finds in clean, functional plane design." His salary was listed as twice that of the company's longtime chief engineer, Hall Hibbard, who had been responsible for the creation of the Electra transport, the Hudson bomber, the P-38 fighter, and the Constellation airliner.

Time was better at describing the swankness of Bel Air than the truth about the aircraft industry, as related by Gross to an acquaintance in the Royal Air Force a month after the magazine story appeared. "The situation in the aircraft industry today is pretty grim," he said tersely. "The companies have no idea where any real business is coming from. The commercial market has proved to be far less than one would have thought, and everybody of any standing is fiercely competing to get what orders are offering . . . I think all of us are more tired, definitely older, and wondering what this torn world is going to do to us." Gross would not have to sell his Modigliani, but the next few years would be harder than he had foreseen in 1944 when Lockheed's backlog dwarfed its net worth.

His sense of beleaguerment could only have increased when the Constellation, which represented $75.5 million worth of orders at the end of 1945, began to experience serious mechanical failures that resulted in its being grounded by the Civil Aeronautics Administration in the summer of 1946. "The Constellation is tying up the entire net worth of our company," he said fearfully, "and if it were to falter, it would shake the whole foundation." Though fifty-seven Constellations would be delivered to airlines in 1946, a new jet fighter called the "Shooting Star" (400 deliveries in 1946) and a Navy patrol plane, the "Neptune," were the company's most "steady and solid" business.

Still, he clung to an idealistic notion that was long obsolete—that the business could somehow be sustained by commercial orders alone. "The war has left us with many problems and the uncertainty in our Government's foreign policy has made our military and naval progress, which has always been the foundation of aircraft development, of such reduced proportions that the art of design and production may lag for a few years. We have always had the ambition to make the business pay, just on the basis of commercial and private type airplanes, without having to be dependent upon military orders. As a matter of principle, the prospect of building weapons never appealed to me sufficiently to make me want to earn my living from such a source. . . . In spite of my personal feelings on the subject, the aircraft industry needs substantial Government support."

It did not take long for Gross's colleagues throughout the industry to reach this same conclusion, which had been valid at least since 1918. It was also quite obvious that some of the companies that had grown obese on wartime appropriations would have to die. To this end, Gross briefly pursued a takeover of one of the fattest wartime firms, Consolidated Aircraft, which had tripped badly as soon as peace broke out. But when he routinely submitted the proposal to the Justice Department's antitrust offices, it was opposed on the grounds that it would lessen competition and "tended toward monopoly." Gross was embittered by this stand, but admitted in private that the merger would have created a company under his control with "thirty or forty percent of the total aircraft backlog in America."

Gross summed up the industry's precarious condition in a letter to Edward Stettinius, the U.S. Steel chairman who had served as a key mobilization official during the war. "It might interest you to know that while the problems of the war were great and the pressure upon airplane manufacturers to produce was incessant, I feel that the hazards experienced then were never comparable to the ones we have had to face up to since. We had one underlying element of comfort and reassurance in the war—we knew we would get paid for whatever we built. Today we are almost entirely on our own." In other words, the airplane companies were now face to face again with free enterprise, and they did not like what they saw.

Oddly enough, Wall Street analysts were not so worried. An internal study by Smith, Barney found that earnings for 1946 would "not be impressive," due to the "disruption entailed in scaling down from a wartime basis." But "most managements are taking advantage of large excess profits tax credits to make a thoroughgoing reorganization of facilities and to write off as much new development expense as may be permitted." Many companies were in strong working capital condition, because a large part of their earnings during the war had been retained in the business and the government had permitted wartime plant costs to be quickly charged off. Smith, Barney concluded that selected issues, including Martin and Lockheed, were ripe for speculative appreciation, though not suitable for long-term investment—much the same evaluation as ten years earlier. What the Wall Street experts could not feel, apparently, was the psychology of collapse that had gripped the industry.

In this cutthroat atmosphere, where even the most venerable companies saw themselves on thin ice, it was natural that the industry's leaders might conspire for more favorable conditions. They had a long history of being coddled by Washington, and now was the time to renew old friendships. As a prelude, the Aircraft Industries Association published a long essay in its *Aviation Annual* for 1946 calling for preservation of wartime production capacity. "This war has shown how our air power is an essential part of our national life," the trade group effused. "Obviously there is an inescapable responsibility on the American people and their government to make certain that air power is not lost in the shuffle of reconversion of the aircraft manufacturing industry."

The association found no less of an authority than Harvard University to quote on its behalf. According to a recent study there, where Business School faculty were heavily funded by major aviation companies and Robert Gross was a trustee: "The principle of retaining military power as peace insurance so long as we must live in the midst of an armed world is now generally recognized. How to assure this power, especially in such a rapidly developing field as aviation, is an unanswered question we dare not neglect. Today America not only has the world's largest air force but, to supply the equipment which makes this force possible, has organized a gigantic industrial structure dwarfing even the peacetime automobile industry. These national assets will, however, become liabilities if inadequate planning leads to disorderly contraction." If there was to be a Pax Americana, there had to be a weapons industry of concomitant stature.

More than intellectual endorsement, what the industry needed was a lofty declaration like the Morrow Board report of 1926, that would unleash massive appropriations from Congress. Robert Gross had first urged the formation of such a board in a statement submitted to the special Senate Committee Investigating the National Defense Program in August 1945. ("It is a cause I would not be selfish enough to plead as a businessman," he wrote, "but it is my duty to plead for it as a citizen.") Now there was no time left to wait for a mandate to rise from the grass roots. In May 1947, executives from the primary aircraft manufacturers met in Williamsburg, Virginia, "in the hope of getting a five year building program for the country's air forces," as Gross succinctly put it. Their effort focused on Capitol Hill, where the Republican Party held a majority, and resulted by July in the Senate passing legislation, known as the "Brewster Bill," that called for a presidential panel exactly like the Morrow Board to consider the industry's plight. "This is the 'cause célèbre' which we have been fighting for the last few months," Gross exulted, "and I cannot overestimate the good effect the creation of such a board might have on our whole industry. It now goes to the House for conference and should become law before Congress adjourns at the end of this month. It provides for the creation of a board consisting of three members of the House, three Senate, the four Secretaries ex officia, War, State, Navy and Commerce, and two prominent civilians. That is the 1947 version of the 1926 Morrow Board, which laid the foundation for America's air power."

As is often the case in Washington, "things got a little bit confused" by the time Congress broke up to escape the August heat. "We had been working originally a year ago on the idea of having the President appoint a Board of private citizens, but we got nowhere," Gross explained to a Wall Street friend. "We then switched this spring to the tactic of working through Congress to have a law passed creating a Board, as opposed to a merely presidentially appointed one. This turned out to be the so-called Brewster Bill, which was a compromise between a bill introduced by [Senator Owen Brewster of Maine] and one introduced by Congressman Carl Hinshaw of Burbank.

"These bills were passed by both the House and the Senate, but while in conference committee the President, evidently sensing that the whole idea of an Air Policy Board was gaining momentum, jumped the gun and appointed his own commission somewhat along the lines that we had been advocating to no avail a year ago. For a few hours we were pretty shaky as to what course to pursue, but we finally decided to press for the legislation anyway and support the President's commission as well. As a result we now probably have two Air Policy Boards; one a legally constituted body consisting of congressional and senatorial members only and implemented by law, and the other a Policy Board created by an executive order of the President. At first blush this may seem to be unwieldy and perhaps controversial. Personally, however, I think we can work with both Boards and I am not chastened by the creation of two. It is ironic that for a while we could get none and now we have two."

Seldom has a corporate executive been so candid about the ability of his business to push the right buttons in Washington. In this case, the aviation weapons industry got more than it could ever have dreamed—in effect, a pronouncement that the manufacturers were so vital to national security that they should be freed from the normal pressures of supply and demand.

Of the two boards, the five-member Air Policy Commission appointed by President Truman on July 18, 1947, received the most attention. Called the "Finletter Commission" after its chairman, Thomas Finletter (a fifty-four-year-old Philadelphia lawyer and State Department adviser previously unknown to Gross and other industry leaders, but endorsed by key Wall Street figures), the panel published a report in December, ominously titled *Survival in the Air Age*, that laid the philosophical foundations for a Cold War armaments industry. Of the 150 witnesses called to testify—including Gross, Glenn Martin, Donald Douglas, Jack Northrop, and every other authority from major aircraft and airline companies (Charles Lindbergh, the former celebrity-saint of aviation whose noninterventionist speeches before Pearl Harbor had made him unwelcome in Washington, was notably absent)—none came from outside the circle of business, military, and government officials who had a direct stake in the expansion of air power. Hanson Baldwin, a reporter from the *New York Times,* was the only man (no women testified, there being no postwar analogue of Amelia Earhart) who could claim independence, yet he was well known as a conduit for pro-military views.

After dismissing disarmament and United Nations peacekeeping as utopian pipe dreams, the commission called for a permanent munitions economy of unprecedented scope: "This country, if it is to have even relative security, must be ready for war. Moreover, it must be ready for modern war. It must be ready not for World War II but for a possible World War III."

World War III, according to the panelists, had already begun. "The great contenders in a possible war of the future will first engage in the political and then the industrial phases of war. The political phase of the next war has been actively engaged in since VE-Day—and the industrial phase is clearly recognizable."

With an Orwellian flourish, the commission argued that constant diligence against defeat would require a whole new mindset, "nothing less than a reversal of our traditional attitudes toward armaments." A veritable turning point in history had been reached, when "our policy of relative security will compel us to maintain a force in being in peacetime greater than any self-governing people has ever kept." Since 1915, the Finletter panelists calculated, "about 85 percent of our total Federal budgets have been spent for war or preparation or payment for war." Nonetheless, "the evidence is overwhelming that even this amount is not enough." Therefore, the government should immediately "increase substantially" its expenditures for military aviation, since air power was now the pivotal element in national defense. "Self-preservation comes ahead of economy," the commissioners declared apocalyptically, making a vast aircraft manufacturing system a matter of life or death for all Americans.

To some extent, the Finletter report civilized the alarmist views about technological vulnerability that had been voiced by General Arnold in his final report on wartime Army Air Force activities, published in November 1945 and quickly sensationalized by *Life* magazine. Bombers with intercontinental range, guided missiles, and the atomic bomb would mean that security "in the visible future will rest on our ability to take immediate offensive action with overwhelming force," Arnold had warned. Such action would necessitate, of course, a military force maintained on a permanent basis, not one mobilized during an emergency.

The Finletter Commission made several specific recommendations for expanding the nation's various air forces, and thus the weapons industry. For example, the Air Force (newly independent from the Army) would rise from its then-current 10,800 active-status warplanes organized in fifty-five mission groups to 12,400 planes in seventy groups over the next five years. A reserve force of 8,100 new aircraft would also be required. The Navy's aviation programs would receive a similar boost. All in all, total defense spending would have to increase by at least 80 percent by 1952, from about $10 billion to $18 billion, to pay for self-preservation.

In 1950, the previously obscure Thomas Finletter became Secretary of the Air Force. His costly proposal for a seventy-group air armada would be tempered by Truman's budget restrictions until the Korean War broke out, but it established a floor for ensuing debate. The fact that the industry-inspired Finletter Commission preceded events usually cited as triggering the post–World War II buildup of U.S. military power—the Communist coup d'état in Czechoslovakia on February 24, 1948, for example, or the Berlin crisis of that summer—suggests a more mercantile context for the Cold War than is traditionally accepted.

The philosophical foundations for armed peace had already been started elsewhere—in a widely circulated 8,000-word telegram sent by the American chargé d'affaires in Moscow, George F. Kennan, to the State Department in February 1946, claiming that the Soviet Union was an aggressive threat; and in President Truman's March 1947 speech before a joint session of Congress, pronouncing the "Truman Doctrine" of military and financial support to anticommunist groups outside the hemisphere. Yet these policy expressions, though they signaled a radical departure from prewar concepts of America's role in world politics, did not pump up U.S. military appropriations at home in a way that would help the weapons industry. To the contrary, the first major spending they spawned was for reconstructing European economies via the Marshall Plan, outlined in nonconfrontational humanitarian terms by Secretary of State George C. Marshall in a June 1947 Harvard commencement address, followed by initial funding of $5.3 billion from Congress in March 1948.

This was the kind of money the aviation companies sorely needed, and eventually got many times over. Truman and Stalin launched a forty-year contest for power, but Robert Gross and his colleagues were maneuvering at home to justify big defense budgets long before Europe once again played into their hands.

GLENN MARTIN'S POSTWAR business strategy was the same as Robert Gross's—go out on a limb if necessary to corner part of the burgeoning commercial market and keep as much military work as possible to meet the payroll. It was a reasonable plan, except that the commercial market did not burgeon on time, military work sputtered for too long, and the cost of developing any new airplane became astronomical.

Drawing on an in-house survey of the airliner market, Martin decided to build with haste a twin-engine medium-range transport that could carry forty passengers and not have to compete against the bigger Lockheed Constellation or Douglas DC-6, which had both benefited from wartime development of military versions. The market for short-haul passenger planes was relatively small, but it was the only segment undominated by established manufacturers. "We did not utilize our engineering forces during the war in preparation for peacetime aircraft designs," Glenn Martin stated in January 1946. "So when V-J Day came we found ourselves suddenly without any designs on the shelf and no way to keep our people busy, and we have been driving pretty hard to keep our best people together and to be prepared for the markets knocking at our doors." The company had not built any type of civilian aircraft since the 1932-vintage China Clipper flying boats, on which it had lost heavily. There was no market banging on its doors.

Nonetheless, by July, Martin was touting tentative sales for 327 of the new transports, called the "2-0-2" (or "3-0-3" with pressurized cabin), from seventeen domestic and foreign airlines. This was supposedly close to the projected break-even level of production, where development costs of at least $15 million could be written off, assuming every order was covered by a signed contract. (The company inaugurated a costly system of computerized punch cards for tracking the myriad cosmetic differences between one airline's planes and another's. It was an avant-garde step that bogged down in piles of cards that workers found incomprehensible.) But there were already questions within the industry about whether Martin was wearing rose-colored glasses. As early as March, a San Francisco–based broker who specialized in the company's stocks had written to Glenn Martin's personal secretary asking for comment on a statement attributed to "a director of Douglas Aircraft" that "Martin had a very excellent two-engine plane" but "the company stood to lose money on the first two hundred to three hundred" because their price was too low. Martin was, indeed, playing the high-risk game of winning initial orders through loss-leader prices. He had used the tactic often before in military procurement, where the taxpayer always absorbed the risk sooner or later.

But this time Martin had only one major weapons contract in hand to help support the tremendous financial burden of developing a new commercial transport—a Navy torpedo plane called the "Mauler," which had been designed during the war. With the military services turning their attention to jet propulsion, the piston-powered Mauler was a futureless machine. Only 151 were built between 1946 and 1949, a significant number by prewar standards, perhaps, but a mere trickle in the new environment. (When the Navy began to buy quantities of new aircraft again after 1948, it concentrated on carrier-based jets, not the seaplanes advocated by Martin.) The company had an experimental contract to build two prototype jet bombers, designated B-48, but the project was a mismanaged failure that cost the Air Force at least $9 million.

Wall Street kept pressing for information about the 2-0-2 program, which was regarded by the financial community as a bellwether. The mentality of Cold War was only in its fetal stages, with informed people outside of military-industrial circles still skeptical about the "Soviet threat" that would become a catch-all justification for rearmament. The following telephone conversation took place in August 1946 between Glenn Martin (whose office recorded it) and the West Coast broker. Martin, as usual, was prone to carnival-barker exaggerations.

> BROKER: Everybody figures that if they can't do as you can, you must be losing plenty on it [the 2-0-2]. It's a natural feeling for anybody. Incidentally, are you getting much in the way of war business?
>
> GLM: Yes, we are. Our war contracts are increasing—our War Department business is going right along. As a matter of fact, we're talking about more business now than we originally planned we would get. It runs into a good many millions of dollars. We've got about eighty million or more Government work that we're negotiating but haven't signed yet.
>
> BROKER: Is that right? Tell me, does this Russian situation look serious or not?
>
> GLM: I think it does. You see, we've got this world court, but every question is voted on six to fifteen. Russia and her satellites six votes and the opponents are fifteen votes. That's the way it goes every time, so there's a definite cleavage of purpose. The Russians know that they can never get their way in the world court, and I'm positive that they plan that they may have to fight if they can find it advantageous. I know—I can't tell you how I know—that they are working night and day in increasing their armament and their design of new equipment.
>
> BROKER: The Russians are?

GLM: The Russians are working on missiles and all of the latest war weapons—are going forward with great pressure.

BROKER: Well, are we doing the same?

GLM: Not too vigorously.

BROKER: In other words, we're lapsing back to prewar status.

GLM: That's right. We're getting ready to throw away our guns.

BROKER: While the Russians at the moment are probably not able to fight, they are making great strides in that direction.

GLM: They are increasing their science every way they know how in war equipment.

BROKER: They aren't in a dangerous position yet, do you think?

GLM: I don't think so yet.

BROKER: Now, getting back to one other thing and then I won't bother you anymore. The dividend doesn't look so good, does it?

GLM: Oh, yes. You see, we voted a dividend in August for payment September 23.

Here were the primary elements of postwar business coming gradually into focus—weapons, the Russians, and dividends. Wall Street had not quite figured out how they were interrelated, but Finletter would soon explain it all.

At the end of 1946, when the 2-0-2 took its maiden flight and his company paid its last wartime-level dividend of three dollars per share, Glenn Martin was still in a position to give away some of his personal income. He could no longer afford huge gifts (which also eased his considerable wartime income tax burden), such as $2.5 million given to the University of Maryland in 1944 and 1945 or the $500,000 "Minta Martin Aeronautical Endowment" to the Institute of Aeronautical Sciences in 1942. So he sent $25,000 to the League of Maryland Sportsmen and $10,000 to Ducks Unlimited. He was less generous with the Republican National Committee, giving them just $250. Though his opinions about labor and international relations were sliding toward the far right, he had never been actively political (he turned down an offer in 1945 from the Maryland GOP to run for a U.S. Senate seat), preferring to aim his philanthropy at hunting organizations, churches, and colleges. He also continued to gamble on rather shady investments—oil exploration in Canada, diamond mining in Arkansas—that lost hundreds of thousands of dollars. Unlike the sophisticated Robert Gross, Glenn Martin was a rube with his own money, a mark for unscrupulous promoters who were like characters from his carney youth.

By the spring of 1947, Martin was already dropping projections of
that year's business from $80 million to $69 million, due mostly to
cancellations on 2-0-2 orders, which were down to about 250
planes. Wall Street was getting jittery about the company's health,
about whether "we have permitted ourselves to become oversold on
the Martin situation," in the words of one major shareholder. Alarm
bells rang at the end of May, when Martin approached the Reconstruction Finance Corporation for a $25 million loan—after having
been turned down by Guaranty Trust in New York, which had a
director on Martin's board—of which $16.6 million was approved
on condition that the company's choice of a new treasurer be satisfactory to the RFC and that no new experimental projects be
started without the agency's permission. Dividends were restricted
until the loan was paid off, depressing the market value of Martin
stock. It seemed like a replay of 1934. Still, Glenn Martin insisted to
anyone who asked that the company would make a profit and pay
dividends in 1948. "All next year will be a good year," he told one
broker.

In August, the 2-0-2 became the first postwar twin-engine airliner
to pass its CAA certification tests—a considerable achievement under the troubled circumstances of the time. By October, however,
Martin was under fire from Northwest Airlines, Lockheed's old nem-
esis and now Martin's premier commercial customer, because of
manufacturing deficiencies. Northwest's vice president for engineering submitted a list of eighty-six "squawks," or defects, that "can be
laid directly to poor workmanship." Among them were fuel leaks,
loose or broken rivets, and sloppy controls. Martin was not the only
airplane builder discovering that commercial jobs could not be
rushed through the plant at the same pace as wartime military
work, but the company was far more vulnerable to doubts about its
new product than Douglas or Lockheed. Martin's books would show
a crushing loss in 1947 of $20.9 million.

When the Finletter report was published at the end of the year,
Glenn Martin, who had testified before the commission, felt that it
was not "enough of a report to be the stem of any action." The "real
work" of rooting out the industry's problems was being done by the
congressional panel. "I don't know whether it was planned that the
Finletter committee would only go so deep and hurry up a preliminary report," he said, "but I do know that the Congressional committee is really in earnest and their questions are intelligent."

The Congressional Aviation Policy Board, which did not call its own witnesses, issued a report in March 1948 that was more critical of the defense establishment than Finletter, though its basic conclusions were the same. The primary problem, the board found, was "one of providing well-balanced military and naval air forces rather than one of finding means to maintain an aircraft industry." If the Joint Chiefs of Staff would just come up with a "unified plan of action" with precise calculations of aircraft requirements, the industry's health would be assured. This was the sort of advice that Washington perpetually ignored.

In mid-January 1948, Glenn Martin appeared before the Senate Banking and Currency Committee, which was investigating irregularities in RFC loan awards. The committee was concerned about the fact that Martin had used some of its RFC loan to complete a $6 million polyvinyl chloride plant in Ohio—part of an effort to diversify after the war—instead of protecting its ability to carry out Army and Navy contracts critical to national defense. The chemical plant also seemed to have taken precedence over refunding to the government some $3.7 million from wartime profits that were determined by military auditors to have been excessive (later increased to $7.9 million). Martin skirted these issues, but his testimony was nonetheless a stunning confession of disaster in the ambitious 2-0-2 program. It merits quoting here at length, because it provides a measure not just of Martin's errors, but of the magnitude of problems that struck the whole industry after the war and have reappeared ever since:

> Immediately after V-J Day, we departed from our previous tradition of concentrating exclusively on military business and began the development of the 2-0-2. For a substantial time, our judgement in embarking on this commercial program appeared to be abundantly confirmed by the enthusiasm for the airplanes expressed by our potential customers among the airlines. We received contracts and expressions of intent to buy covering 152 2-0-2's and 159 3-0-3's; and at the peak, our backlog of commercial business appeared to be as high as $83 million.
>
> In fact, this enthusiasm on the part of the airlines caused us to make a substantially greater investment in our commercial program than would otherwise have been necessary. Our customers were pressing us for early deliveries; and we were anxious to reach the market ahead of competition. Under this compulsion, we followed the policy of tooling for production concurrently with the engineering and refinement of our prototypes. At the same time we intensified our engineer-

ing research to achieve compliance with the new safety regulations being promulgated by the C.A.A. As well we knew it would, this program proved much more expensive than the alternative but slower method of deferring our tooling and production until the prototypes had been perfected. Even now, we believe we were right in adopting this policy, because among other things the alternative would have delayed for many months the availability of modern twin-engine equipment to the airlines.

As a result, by March 31, 1947, we had invested $25,514,522 in the 2-0-2 and $8,121,338 in the 3-0-3 in design, engineering, tooling and manufacturing, plus about $8.5 million spent for materials and parts.

Here was the boggling new arithmetic of postwar aircraft production. The original DC-1 prototype had cost Douglas just $307,000 to design, build, and test. The company took a loss on more than half of this figure when delivery was made to TWA in 1933. But other business, mostly military, was more than enough to carry it safely through. A dozen years later, Martin proceeded to spend over $42 million on an admittedly larger, though operationally analogous plane. ("We thought we had to have a ship that would go anywhere that the DC-3 would go," Martin later said.) Even accounting for inflation, the industry had clearly passed into new technological and financial territory of unfathomable dimensions. This was part of the "turning point in history" that Finletter had argued would require radically different thinking.

Under the era's tax laws, which had been crafted early in the war with liberal concessions to private manufacturers in order to coax them into government work, Martin's attorneys figured that between $15 million and $20 million of the 2-0-2 loss could be eliminated through refunds on taxes paid in 1945 (the Internal Revenue Service stood by the lower figure, then compromised at $18.3 million). Glenn Martin thus believed—and told the senators—that the combination of RFC loan and tax refund would be enough to "tide us over" until cash started to flow in from fresh sales of 2-0-2s.

He was very wrong. On April 1, 1948, the Soviet commander in Germany announced that his troops would begin to inspect Allied trains and trucks entering Berlin. The resulting crisis helped jump-start military appropriations on the kind of upward spiral called for by Finletter. But it also stymied commercial purchases of new passenger planes like the 2-0-2 while the airlines waited to see whether the government would build a new fleet of transports and then hand them back to the airlines at bargain rates. Ultimately, Washington depended on World War II surplus for the Berlin airlift, especially military versions of the Douglas DC-3. Thus, the 2-0-2 suffered the double blow of lagging commercial orders and no sales to the Air Force.

In May, Glenn Martin received a confidential report from a stock-broker who had recently met with Robert Gross that the Lockheed president believed "Martin is in the toughest spot of all—they're apparently going to get less than anybody else in the country" from new military appropriations. Martin replied crossly that Gross was "opposed to anything that would help Martin—he's been that way, I don't know why, but he's prophesied that we would not continue in business." This sort of ad hominem sniping was rare in the aviation business's clubby upper circles, but the Martin Company was definitely in dire straits. Gross's crystal ball was evidently much clearer than Martin's.

In mid-June, Glenn Martin had a melancholy telephone conversation with Larry Bell, his oldest friend still in the business. Bell, who harked back to the "Battle of the Clouds" at Pomona Speedway in the faraway spring of 1914, had been president of his own aircraft company in Buffalo, New York, since 1935, and had hoped to join with Martin in converting wartime B-29 bombers (the plane that dropped the atom bomb on Hiroshima, assembled at Martin's Omaha plant) for use as tankers. When the deal fell through, Bell felt that his appetite for military work was almost gone. Their commiserations would be familiar to anyone in the postwar aerospace business.

BELL: From my standpoint, I either want to have a very substantial amount of commercial business in our plant along with military, or I don't want to be on the military. It's too rugged.

GLM: It's pretty rugged.

BELL: You've seen it a lot longer than I have, but I've seen it long enough. It's a business of peaks and valleys, which is about the worst kind of business you can have. There's nothing under the sun that I know we can do to military procurement to level it out. It's either haste or famine. It seems to me that if we had fifty percent of our work in commercial, whether it be aircraft or non-aircraft, and carry forty percent or fifty percent of our work in military, we would not be over a barrel all the time.

GLM: A little bit of diversification, if you can get it, is a wonderful thing to have. It will cost a little gamble sometimes, but then you're not at the mercy of the peaks and valleys.

BELL: I suppose, though, that the Army will resent anybody trying to build a commercial business.

GLM: Yes, I believe that's the feeling.

BELL: How does the [2-0-2] market look?

GLM: Very, very paralyzed, because everybody is waiting for a subsidy from somewhere.

BELL: That's the trouble. There's something the matter with the whole business.

On August 29, 1948, two Northwest Airlines 2-0-2s flew into a violent storm near Fountain City, Wisconsin. One crashed after its wings sheared off, killing thirty-seven passengers. The other flew out with a cracked wing spar. Martin claimed that Northwest was negligent in flying the planes that day. Northwest claimed that the 2-0-2s harbored a design flaw. All 2-0-2s were grounded, and subsequent tests suggested that Martin was indeed responsible for metal fatigue failure in the wing structure. The program obviously could not endure such intimations of technical scandal.

Glenn Martin seemed increasingly out of touch with the company's plight, and perhaps with reality in general. He was convinced that there would be a "shooting war" with the Soviet Union within twelve months, that the Russians would foment revolution in America by ordering the labor unions to strike. "With a five year rearmament program if there is peace, and a big expansion of business in case of war, I cannot understand the low market on Martin stock," he wrote to a family friend one month after the Northwest crash. "We have about $120 million worth of unfilled government orders, all of which will show a reasonable profit. We do have too much money invested in 2-0-2 inventory and it will take some time to work out of that problem, but I find nothing in the picture that interferes with our sound long term." The company was, in fact, sinking rapidly beneath his feet, and he seemed powerless to stop it.

In October, perhaps to escape the darkening shadows of his business life, certainly to dramatize his faith in what he had wrought, he and his eighty-four-year-old mother took a sentimental cross-country journey aboard one of the company's unsold 2-0-2s. They landed in Salina, Kansas, and toured the prairie wheat towns of Glenn's childhood, just like visiting royalty, which in many ways is what they were. Photographs in the local newspapers showed them posing together in their expensive coats, an elderly woman and her aging son, next to the plain clapboard houses that had once been home.

They went on to Santa Ana, California, for more testimonial dinners with those few neighbors who remembered the days of Martin's Garage and the silly contraption that emerged one morning long ago from the Methodist church. Glenn Martin was, after all, a showman at heart, a carnival actor who knew that to win the most cynical audiences over he must put his own neck on the line, season after season. It did not matter if they called him the "Flying Dude," a mollycoddle in leather britches and thick glasses who blanched when the local cantaloupe queen or Hollywood starlet planted a sticky kiss on his cheek. The whole idea was to sell flying machines, to convince the bumpkins and the bankers that they could all be birdmen someday, tripping to the Orient or chasing the bandits way below. Here they were, ladies and gentlemen, Glenn Martin and his wonderful mother, riding their latest aeronautical wonder, ready to perform the rise and the dip and the circle. It was, actually, their farewell performance.

In January 1949, after the company recorded a loss of $15.5 million for the year just ended, the military services took a hard look at whether the Martin company's financial condition had deteriorated to a point where it could no longer handle federal contracts. The 2-0-2 had just lost a competition for navigational trainers because of "lower performance and higher price." In all, only twenty-five 2-0-2s were sold and delivered to Northwest, plus six to two airlines in South America. Aggregate losses on the 2-0-2 and 3-0-3 programs, the latter of which was abandoned before reaching production at a cost of $17.9 million, totaled $37 million after various tax refunds were applied.

During the war period from December 1941 to June 1945, the auditors found, Martin did $1.148 billion worth of government business. So far in the postwar years, there had been $14.4 million with the Air Force and $70 million with the Navy. (Demonstrating the incoherent technical and economic problems of the time, more than half of the Air Force business was accounted for by just two prototype jet bombers that were never put into production. Martin was not alone in such profligacy, however—two four-engine piston-powered "Constitution" prototype transports built for the Navy by Lockheed cost $15 million each.) "This company has not been successful in competition on types required by the Air Force," the auditors concluded, because "they have been comparatively high cost producers." It was the same charge that had haunted Glenn Martin for a quarter century.

Two days after the report was released, he put a large section of his beloved Eastern Shore farmland up for sale. The company was still carrying the *Glenmar* and her eight-man crew on the books at a monthly expense of more than $2,000 for salaries alone. This was small change, of course, but symbolic of how out of touch the chief executive had become.

In September 1949, when it became apparent that the Glenn L. Martin Company would no longer be able to find a federal or commercial bank to lend it money without a drastic change at the top, its founder surrendered. He was replaced as president by a forty-two-year-old vice president of Curtiss-Wright and former general manager of a Douglas war plant in Oklahoma City, Chester Pearson. Martin was then elected to the newly created position of board chairman, at the same $60,260 annual salary that Pearson received. From this position one of his first actions was to order the Mishawaka Rubber and Woolen Company of Mishawaka, Indiana, to manufacture elevator shoes for the Martin "Bomberettes" ladies basketball team.

JACK NORTHROP'S FATE, for better or worse, had been tied since the days of the Lockheed Vega to his dream of building flying wings. As soon as he had established a new company in March 1939 with financier LaMotte Cohu, he began hiring engineers to construct an experimental prototype, which he called the "N-1M" (for Northrop Model 1 Mockup). While other managers drummed up production orders to pay the infant company's bills—twenty-four seaplanes for the Norwegian Navy, and a subcontract for empennages (tail assemblies) on a Consolidated flying boat—Jack Northrop nursed his N-1M at company expense, more or less oblivious to the harsh realities of running a for-profit business.

"The U.S. military were not very interested in Northrop's ideas for a flying wing," recalled the company's vice president for sales. "They considered it futuristic and commented privately that Northrop did not seem to understand their priorities." Likewise, the Northrop board of directors was dominated by men from the investment firms that had underwritten the company's initial stock offering, and they were hardly eager to pour precious capital into a blue-sky project.

Though the company had tried not to ruffle the feathers of other airplane manufacturers long established in the Los Angeles area, Northrop quickly enraged the most prominent one of all, Donald Douglas. When the Navy invited Northrop to bid against Douglas for production of 200 "Dauntless" dive bombers—which had been derived from a plane designed under Jack Northrop's aegis while he was still operating as a Douglas subsidiary—Northrop snared the business with a suicidally low bid. Donald Douglas, who was already miffed by how Northrop had raided his plants for employees (some 80 percent of Northrop's new work force had walked away from Douglas), went directly to Washington and applied the full weight of his reputation to convince the Navy that he, not Jack Northrop, should be the prime contractor. The Navy complied with his wishes and then offered Northrop a subcontract, but the petulant Jack Northrop refused to play second fiddle, thus ruining a deal that his new company desperately needed.

Like Lockheed and Martin, Northrop was saved by the ravenous export market for American weapons. When the British Purchasing Commission proffered a $17 million order in the summer of 1940, Jack Northrop did not like the fact that it was for co-production of another company's product—200 Vultee "Vengeance" dive bombers—but he was in no position to scotch another contract. Northrop's net worth was represented by plant and equipment, deferred charges for development, and an operating deficit. Working capital was severely depleted. The Consolidated flying-boat subcontract was in the red. "There was serious doubt as to whether the company could have continued," Northrop later admitted. But by the end of 1940, by which time the U.S. Army had ordered the first of more than 700 Northrop-designed "Black Widow" night fighters, the company appeared to be on firm financial ground with $20.6 million of unfilled orders from domestic ($1.5 million) and foreign ($19.1 million) military customers.

"Had it not been for the British government, there probably would have been no Northrop Aircraft of sufficient size to be of material value to the war effort," Jack Northrop told a congressional panel after the war. Face-to-face with the Luftwaffe at home, the British were very generous with American warplane makers, paying to double the Northrop plant's capacity, as well as bolster the company's shaky credit so that working capital could be secured from hesitant banks.

As for the Black Widow, a coal-black oddity with a twin tail that resembled Lockheed's smaller P-38, Northrop suffered through a painfully long gestation period that would have been ruinous under normal economic circumstances. Its prototype first flew in May 1942, but was not delivered to the Army until July 1943. Far behind schedule, production models did not reach combat theaters until the summer of 1944. Nonetheless, Jack Northrop, who was never much interested in manufacturing problems, felt free again to indulge his imagination on the flying-wing project without going bankrupt. The N-1M made its first flight in July 1940, followed by a successful test series that convinced him of the design's promise.

"Of course, the desire for the flying wing still persisted," Northrop said many years later. "I insisted on doing enough development so that it wouldn't die on the vine." Without the war, however, the whole vine would have dried up in short order.

Northrop, whose technical education had been purely empirical, was abetted in his research by Theodor von Karman, a Hungarian-born "promoter, globe-trotter, and scientific jack-of-all-trades" who had joined the California Institute of Technology in 1926. Von Karman's forte was aerodynamics, and he had worked for aviation companies in Germany, Japan, and the United States since the 1920s. To Jack Northrop, he represented the intellectual prowess that would always be beyond reach. He was also something of a scientific guru to General Hap Arnold, who pumped Army funds into Cal Tech's research on jet propulsion and rocketry. One of Von Karman's students, William Sears, became Northrop's chief of aerodynamics during the war.

In January 1941, the Army's Air War Plans Division began to consider the development of bombers with intercontinental range, spurred by the possibility of losing Britain as a base of operations against Germany. In May, Jack Northrop wrote to the Army Air Corps Matériel Command, saying that the N-1M experiment could be the basis for a large bomber. At the end of that month, the Army asked him for a formal design proposal that would meet definite specifications. The initial success of Hitler's invasion of Russia that summer accelerated the Plans Division studies, drawing four companies into bomber design work—Douglas, Boeing, and Consolidated on conventional planes, and Northrop on a flying wing. By September, Northrop was able to submit an offer to build one bomber for $2.87 million and a second for $1.55 million, the first to be delivered within two years of signing a contract.

Though senior Air Corps officers believed that none of the companies could complete a prototype in less than two and a half years, the project was given highest priority. Consolidated won a contract to build two mammoth experimental bombers, weighing 265,000 pounds and powered by six 3,000-horsepower engines, designated XB-36. On November 1, Northrop was awarded $2.91 million for one XB-35 flying-wing prototype, plus a wooden mock-up. Shortly after Pearl Harbor, the Army ordered a second XB-35. The flying wing was considerably smaller, at 155,000 pounds, than the Consolidated entry, and was powered by four of the 3,000-horsepower engines instead of six. Many different claims were made by each company for their aircraft's range and payload, but these remained on paper as each project encountered long delays.

The Army also encouraged Northrop and Von Karman to dream up other applications of the all-wing concept, resulting in some of the most bizarre inventions of the war. There was the "Flying Ram," a rocket-propelled wing with stainless steel leading edges, which the pilot was expected to steer directly into enemy aircraft. "It was designed as a projectile," Northrop said without irony, "with the thought that it could be used to intercept and knock wings or tails off other airplanes." Fortunately, none ever reached combat status, though a Northrop pilot was killed in a test flight. There was also a "Rocket Wing," a pilotless "Jet Bomb," and a manned "Silver Bullet" fighter shaped like a beer bottle. These absurd, hazardous creations were the first offspring of a budding relationship between certain academic scientists and the military that would be one of the war's most problematic legacies.

The Army approved Northrop's XB-35 mock-up in early July 1942. It was, by any standard, a stunning aircraft—a completely tailless boomerang-shaped span of 172 feet that seemed to have leaped from the pages of science fiction. It was revolutionary, not evolutionary, and as such occupied the highest category of risk for wartime weapons development. Not surprisingly, engineering difficulties delayed the first test flight of the all-wing bomber for four years, until June 25, 1946.

As early as November 1942, the Army realized that Northrop simply could not handle the full load of tasks associated with assembling the aircraft, and that mass production at the relatively small Hawthorne plant was impossible. The Air Corps therefore decided to transfer all production engineering to the Martin Company in Baltimore. Even though he was also offered a deal to build

200 B-35s, Glenn Martin was not enthused, finding the project so full of unknowns that in August 1943 he asked for his company to be released. That fall, test of a one-third scale model indicated that the XB-35's range and speed would fall below expectations. General Arnold questioned the merits of continuing toward mass production.

Glenn Martin flaunted his disdain for the project—he had been left out of the original competition to build a bomber of his own company's design—by subcontracting work to the Otis Elevator Company. Finally, with schedules slipping farther into the future, the 200-plane production order was cancelled in June 1944. When engineering on the two prototypes reverted back to Northrop, Martin quipped, "The headaches are all theirs now."

Jack Northrop got back full control of his obsession, but he lost several of the top managers who had helped him start the company in 1939, including LaMotte Cohu. Largely because Northrop's proportion of research work to production had been the highest of any major aircraft firm, the company's postwar prospects were among the bleakest in the industry, with no promising commercial projects on hand and only one innovative military contract—for a new multiseat jet fighter—that was years away from quantity production. Jack Northrop's diamond-in-the-rough genius had been squandered on aeronautical white elephants, leaving little that might keep the business going through the first years of peace.

The piston-powered XB-35 itself fell victim to the era's most unforgiving technological advance—jet propulsion. Though its flight test program was plagued by faulty propeller mechanisms and nagging questions about stability, its maximum unrefueled range (with no bomb load) of 10,000 miles was impressive for the era. But when jet engines were rigged onto the airframe in early 1947 (with a redesignation to YB-49) in an attempt to boost its speed, fuel consumption increased so dramatically that the plane's range plummeted to 3,000 miles or less. Carrying bombs, it would have been much less.

"What happened in the 1940s was obviously a mistake by Northrop," recalled Joseph Foa, an engineer at the Cornell Aeronautical Laboratory in Buffalo, New York, where mathematical analyses showed flying wings to be the worst possible aerodynamic configuration for jet-propelled long-range bombers. William Sears was part of the postwar exodus from Northrop, becoming chairman of Cornell's Graduate School of Aeronautical Engineering in Ithaca.

When Foa brought his findings to the attention of Sears in April 1947, suggesting that the laboratory submit a proposal to the Air Force for further research—"especially in view of the large sums of money that are now being spent to maintain leadership in a race which is apparently running on the wrong track [i.e., the YB-49]" —Sears quashed the move.

"Sears responded by stating, in effect, that what I was claiming was absurd, that he and an assistant had rigorously proven in a Northrop report (of which he could not provide me a copy) that the optimum configuration for range in the case of the YB-49 was indeed a flying wing," Foa remembered. Three months later, he was allowed to see the report, which turned out to be an arcane appendix to a secret 1945 Air Force study directed by Theodor von Karman.

In the study—titled "Toward New Horizons," a survey of fertile military technologies undertaken in the immediate aftermath of World War II by a team of prominent scientists and engineers, then delivered to General Arnold and circulated among top Air Force officials—Sears had made copious reference to the promise of all-wing aircraft. Only two technical appendices were attached to the general discussion, one that analyzed some of the still-novel flight characteristics of rockets and another that claimed to prove mathematically that, for best range, an airplane's volume should be contained almost entirely in the wing—just like the XB-35.

When Foa examined the appendix written by Sears and his colleague at Northrop, however, he discovered that an embarrassingly simple error had been made in the calculations. Instead of maximal range, the all-wing shape would result in minimal range, just as Foa's independent research had predicted.

In mid-July 1947, Foa sent a letter to Sears pointing out the mistake. Sears quickly replied, admitting that it certainly appeared that the optimum configuration was not a flying wing. Nonetheless, he ended his letter by restating his opinion that the laboratory should not undertake further studies of the problem for the Air Force. "I suspect you will find that they have enough studies from airplane manufacturers so that they wouldn't be particularly excited about the proposal," he wrote.

Foa found this response to be "shocking." In deference to Sears, however, he asked the laboratory's director to let it be known that if Sears or his co-author would make some public admission of error, he would remain silent. The reaction to this ultimatum was not exactly a public confession (the Air Force report would remain tightly classified for thirty years) but a new paper on the subject that was published in the February 1948 issue of the authoritative *Journal of the Aeronautical Sciences*. Far more abstruse than the original appendix, it seemed to show that under certain conditions, the all-wing shape gave best range. Foa's interpretation of the text indicated, however, that the wing structure would have to be implausibly thick to attain this goal.

Meanwhile, flight testing of the YB-49 had been deeply scarred by a fatal crash that killed all five crewmen, including Captain Glen Edwards, for whom Edwards Air Force Base was later named. The disaster added to fears that the plane was sometimes uncontrollable. On January 11, 1949, the Air Force cancelled all further activity on a contract that had grown to $88 million. The official reason was budget limitations, but many years later Jack Northrop would broach a far different possibility.

In an interview aired by the Public Broadcasting System in December 1980, Northrop—then a frail eighty-five and only two months from death—insisted that the contract had been cancelled punitively by Secretary of the Air Force Stuart Symington, who had demanded that the Northrop company merge with Consolidated Aircraft. In mid-July 1948, Northrop and his board chairman, Richard W. Millar (a former investment banker who had urged Donald Douglas to acquire Northrop from United Aircraft in 1931), met with Symington at the Los Angeles home of John McCone, a member of President Truman's Air Policy Commission and then Symington's assistant. Also in attendance was the owner of 18 percent of Consolidated's stock, Floyd Odlum (chairman of the Atlas Corporation, a large aviation holding company, who had arranged a crucial purchase of 50,000 shares of Northrop stock in 1939), and General Joseph McNarney, head of the Air Force Matériel Command (who later became president of Consolidated). It was an ingrown elite circle, the kind where weighty business could be conducted off the record. Consolidated, moreover, was then in the middle of a costly, technically controversial bomber program of its own—the B-36.

When he balked at the merger, Northrop said, Symington retaliated by killing the YB-49—almost literally, since in this extraordinary case the Air Force ordered that eleven unfinished airframes and all production tools be destroyed for scrap. In both a personal and corporate sense, Jack Northrop was never the same man again. In the PBS interview, long after his eventual retirement, he confessed that he had perjured himself before the House Armed Services Committee in 1949, when he had twice stated that there was no political corruption behind the flying-wing cancellation. Symington, who went on to win a Senate seat and campaign for the Democratic presidential nomination in 1960, always denied the accusations.

The YB-49 achieved a range of 3,155 miles with a 16,000-pound bomb load. In Theodor von Karman's introduction to the secret 1945 Air Force study, he had stated the range goal as 3,500 miles with 20,000 pounds of bombs. Later in 1949, the Air Force told the House Armed Services Committee that "the YB-49 showed considerable promise in speed and altitude, but had inadequate range."

This was the official requiem for Jack Northrop's fondest ambition, the pinnacle of his career since the days when he had watched seagulls dive along the Santa Barbara coast. He spent the rest of his life a broken man.

As THE DEAN OF the American aviation industry, Donald Douglas had every right to expect a postwar career as luminous as his past. He was the only major aircraft builder whose civilian products—the DC line of passenger planes—were more famous than his weaponry. He had suffered none of the financial embarrassments of his contemporaries in the business. Through the marriage of his daughter to Hap Arnold's son, he and the Air Force were "family." The future must have looked limitless on V-J Day.

Despite these enviable advantages, Douglas Aircraft was not ultimately better at adapting to the new universe than any other company. The dizzying drop in sales from $744.68 million in 1945 to $106.7 million in 1946 defied management palliatives. The $2.18 million profit shown for 1946 was due to substantial tax credits, and a $2.14 million loss reported in 1947 was the first in the company's history. Still, because of ultraconservative provisions during the war to set aside some $15.4 million in cash for general contingencies, the company was able to pay dividends in each of the postwar years.

In part, Douglas had to compete against its own past success. By the time the company delivered its last DC-3 transport in 1945, more than 10,000 planes of this model had been produced. Most were still flying—about 500 for domestic and foreign airlines and the rest in military service. For two years after the war, despite vivid memories within the industry of the disastrous effect of surplus dumping after World War I, the U.S. government sold 1,800 DC-3s at about one-third cost and leased large numbers at bargain rates. By the end of 1946, an estimated 5,000 DC-3s were in commercial service around the world, most from surplus disposal.

But the major airlines were still eager for new high-speed long-range airplanes above the DC-3 category. Douglas's answer to the Lockheed Constellation was the DC-6, a fifty-two-passenger transport developed from the prewar DC-4. Though the DC-6, like the Constellation, had been supported by military development work, initial deliveries to the airlines were made at a loss to the company because of the establishment of low "upset prices" in wartime contracts. Rather than lose customers to Lockheed, Douglas sold the first DC-6s at a ceiling price of $600,000 that rising production costs had greatly exceeded (new Constellations were selling at about $1 million apiece). More money was lost in 1947, when two crashes led to a grounding of all DC-6s and forced the company to foot the bill for modifications.

Like Lockheed, its principal post-1945 competitor, Douglas survived the uncertain few years that preceded Cold War rearmament by judicious use of vast contingency funds accumulated during World War II, successful marketing of a passenger transport developed with the help of government funds, and a profitable though relatively lean series of military orders. Oddly enough, the problems that eventually led to the company's demise as an independent business entity were encountered during the huge weapons buildup triggered by war in Korea.

The traditional explanations for Douglas's gradual lethal decline during the 1950s, leading to an overdue acquisition by the McDonnell Company in 1967, center on missteps made while management struggled to understand the dawning market for jet-powered airliners. These errors—foremost, an initial strategy to develop propeller-turbine transports rather than turbojets—stand out clearly as mistakes only in retrospect, however. No one gossiped about Martin-style blunders. Since moving to California in 1920, Donald Douglas had never been an imprudent risk taker for nongovernment business. The original DC-1, progenitor of his commercial fortune, had been placed in his lap as a deal he could not refuse. Turbojets were still an exotic technology, and their commercial potential, while universally acknowledged, was largely a matter of guesswork.

When a seminal jet airliner developed in Great Britain, known as the "Comet," was knocked out of contention in 1954 after a series of crashes, Douglas appeared to have chosen the wisest path. But the first flight of the Boeing 707 that same year, which would start the Seattle company's campaign to capture Douglas's reputation as the world's pre-eminent manufacturer of passenger aircraft, threw Douglas into a catch-up race of extraordinary cost.

"Our hand had been forced," Donald Douglas later admitted. Unlike the Boeing effort, which benefited from jet bomber and tanker projects for the Air Force, Douglas had to develop a jet airliner without tangential government support. The resultant DC-8, whose initial orders for a while overtook the 707, was a "stupendous task for any private company to finance," a Douglas vice president said. "It almost broke the company." DC-8 development costs exceeded $200 million, demonstrating both the financial strength of the company and the Pyrrhic battle it fought.

Yet the bruising contest with Boeing was insufficient to account for Douglas's downfall. Lockheed, for example, did not even enter the fray, waiting until 1970 to field its first jetliner. ("The strain of trying to figure out the most intelligent and logical airplane to build for the commercial field for four or five years hence is burdensome and I am hard put to solve this problem," Robert Gross wrote in 1949.) But Lockheed profited from an unbroken string of major military contracts, Douglas's traditional mainstay, whereas Douglas began to lose ground. September 1952 marked the last Douglas order leading to quantity production of new aircraft for the Navy, and the last Air Force order passed in May 1954. Given the company's long history as a premier aviation weapons manufacturer, the evident collapse of its military business was far more startling than its commercial miscalculations.

To fully explain the slide, one must enter the subjective, unquantifiable realm of corporate culture. By the late 1940s, a very sensitive aspect of Donald Douglas's personal life had begun to unravel his authority. Peggy Tucker, the mistress he had supported since 1931 and hired into the company as his personal driver ten years later, rose to a position of influence in his business life. She took charge of all transportation and food supply contracts, a powerful role in a company that employed 117,000 people in 1945. Inevitable questions about the handling of these accounts even reached the ears of Donald Douglas's wife, Charlotte. "My mother would ask 'do you know about these contracts—what's going on?' " his daughter remembered. "My father would say 'nothing.' " At the end of the decade, Peggy appeared in organizational charts on a parallel line with Donald Douglas, the president and founder. As his personal assistant, she controlled access to his office, virtually isolating him from managers who did not court her favor.

"People who had been loyal to Donald Douglas left—they couldn't stand it," Barbara Douglas Arnold recalled. "They were afraid of her and quit the company. Everybody in the industry knew about her, though I don't believe Hap Arnold knew. He wouldn't have tolerated it. [General Arnold retired from the military in 1946 and died in 1949.] I think even the banks may not have liked it, because at one crucial point they refused to give my father an extra sixty days' credit. My father didn't know how to get out of the problem. It was like drowning."

Sometime in 1951, Charlotte Douglas learned about her husband's two-decade love affair with Peggy Tucker. They were divorced in 1953. Donald Douglas then married his mistress, but did not see his embittered daughter again for twenty years. In 1957, his son, Donald Douglas Jr., was named president of the company. A few years later, however, it became another family's concern.

In June 1954, on the occasion of the thirtieth anniversary of the first round-the-world flight in Douglas biplanes, Donald Douglas gave a speech before the Institute of Aeronautical Sciences. His company, unlike his family, had not yet been torn apart. He was an elder statesman, sixty-two years old, full of wise-sounding words about the industry he had helped found. "The aeronautical sciences have been consistently employed in furthering the twin causes of peace and war," he said, "and it is very difficult at times to distinguish between what is being done for the benefit of mankind and what is being done for the purpose of obliterating man and his works.

"We must all have asked ourselves many times whether we are agents of death and destruction when we have seen the uses to which our products have been put. The answer, I think, is quite simply this: We know in our hearts that we as individuals, and the great body of mankind as well, are far from being warmongers.

"It has been written that 'as a man thinketh, so he is.' Ours has been the task and duty of designing and building certain fundamentals into our products. As aircraft people, we know there can be no compromise with integrity in design or with honesty of purpose and performance.

"Through the years we have been mainly supported by our government, and I consider it significant and heart-warming that I know of no leader in aviation who hasn't always through the years devotedly done his part to see that every dollar was spent to the fullest advantage.

"Let us never forget this, or permit this dedication to principles, physical and financial, to change."

These were noble sentiments, and there is no doubt that Donald Douglas believed what he said. But if his words had any connection to reality—if, that is, there really were no militarists at heart and the men at the top saw to it that the taxpayers' dollars were spent in the best way—then he had become a curious relic from a brief bygone age. Perhaps when he had first read about the Wright brothers' flights on a summer day at the Connecticut shore, aviation was still in some sense pure. Or when he had watched Glenn Curtiss jink the *Gold Bug* around a racetrack in the Bronx, flying was uncorrupted. Those must have been the moments he had in mind.

But he knew very well that the century turned out to have a barbaric touch, one that bloodied almost every aspect of science, technology, and business. Donald Douglas and the other founders neither sought to capitalize on this when they began their careers, nor ever relished it, yet they went along—rarely with regret, always sustained by platitudes about the "twin causes" of peace and war. In this regard they were emblematic of the entire nation, which convinced itself rather easily that transforming much of its great wealth into weapons was an act of integrity.

EPILOGUE

FATES

WITH ACTIVE MANAGEMENT of his company in another man's hands after September 1949, Glenn Martin occupied himself with the vestiges of power. He corresponded with J. Edgar Hoover about the competition between the Martin "Bomberettes" women's basketball squad and the FBI team. He fended off an attempt by a local U.S. Fish and Wildlife Service agent to prosecute him for baiting his duck blinds on the Eastern Shore. And he presided over board meetings of the Glenn L. Martin Company, which could not seem to find its legs in the postwar market. Perhaps of greatest concern to him was his mother, who was bedridden with arteriosclerosis, tended at home by three nurses around the clock.

Like other major companies, Martin was in the black on military orders, with heavy backlogs. At the end of 1951, it held some $425 million worth of weapons contracts. Since the outbreak of the Korean War in June 1950, employment had jumped from 7,300 to 20,000. The company had reentered the quicksand commercial transport market—after a two-year hiatus following the financial debacle of developing the 2-0-2 airliner—with a forty-passenger plane called the 4-0-4. Echoing the mistaken optimism of the 2-0-2 program, Martin trumpeted the initial sale of 101 4-0-4s to Eastern and TWA.

Once again, however, the commercial work brought disaster, causing a precipitous $22 million loss that destroyed the Defense Department's confidence in the new management. Although in its public statements the company took the position that the 4-0-4 was derived from the earlier 2-0-2, so that large portions of the 2-0-2 loss could be carried over into the new program to create substantial tax savings, the 4-0-4's tooling was actually 80 percent different from the 2-0-2's. It was a new airplane, with all the attendant risks. Korean War expansion of the work force compounded engineering problems, as average length of employment dropped from twelve years to six months and annual turnover soared from 12 percent to 70 percent.

Under Glenn Martin, management had ignored its own team of production-cost estimators, selling 2-0-2s at a competitive price rather than on a profitable basis. The new managers, led by Chet Pearson, gave their estimating group full attention, but sold the 4-0-4 at a fixed price calculated on pre-Korea labor and material costs. This figure turned out to be too low by at least 44 percent. Even with $18 million in loans from the RFC and Mellon Bank, plus another $18 million in advances from the airlines, cash flow spiraled downward at an astonishing rate.

Once again, the federal government saved the company by injecting $32 million in new loans, largely guaranteed by the Navy. As part of the bailout, Eastern and TWA agreed to pay $25,000 more for each 4-0-4 than the $475,000 they had contracted for, and Martin submitted to yet another management shake-up. Chet Pearson, forty-five, was replaced by George M. Bunker, forty-four, an industrial engineer with no aviation experience, who had enjoyed the reputation of a boy wonder while president of Trailmobile, the truck-trailer manufacturer. Bunker was reportedly nominated by Mellon banking interests, which not only handled a large portion of the government-guaranteed loans, but were acquainted with him through financial ties to the Pullman Company, Trailmobile's parent.

Glenn Martin was completely severed from influence. He was named Honorary Chairman of the Board, a pathetic sinecure, and forced to move his office away from the Middle River plant to downtown Baltimore. Though he still owned 26 percent of the company's stock, by far the largest block, he became merely the minority figure in a voting trust dominated by representatives from the Department of Defense and a group of investors organized by Smith, Barney & Co. His holdings were soon cut to about 13 percent by a stock offering that specifically excluded him. Laurence Rockefeller, a major Eastern Airlines stockholder, was reported to be among the providers of new capital.

Thus Glenn Martin saw the enterprise he had lifted from the dust of California fairgrounds in 1910 wrenched finally from his grip. The company would never again attempt another major nonmilitary venture, and by the end of the decade would phase out its production of airplanes altogether, in favor of missiles and electronics. In July 1952, the symbol of his forty-year reign as the industry's odd sultan, the fabulous yacht *Glenmar*, was removed from the company's books and sold.

On the morning of March 14, 1953, he suffered the most grievous blow of all, when Mother Minta died at their home in Baltimore. She was eighty-nine years old, he was sixty-seven, and they had lived together continuously except for the few months of his ill-fated association with Wright-Martin in 1917. He placed her next to his father at Fairhaven Mausoleum in Santa Ana, covering her coffin with a blanket of white orchids and gardenias. Tiny Broadwick, the "Doll Girl" of 1913, sent a telegram of condolence. He lingered in Santa Ana for a week, visited Mary Pickford in Beverly Hills, then flew to Salina, Kansas, for the solace of his oldest family friends.

After losing Minta, his world soon contracted to occasional speeches at aviation ceremonies, where he played the only role left to him, that of a peculiar relic from an era few listeners could possibly comprehend. For the first time in his life, he sent gifts—a box of chocolates, a wool stole—to his sister Della, who still resided in a California sanitarium. In early November 1955, he entered University Hospital in Baltimore with a severe upper respiratory infection. He recovered enough to convalesce at his farmhouse near Chestertown on the Eastern Shore, where, on the morning of December 4, he was found stricken by a cerebral hemorrhage, alone in a room that looked out over barren Chesapeake marshland.

His body was taken back to Santa Ana and placed next to his mother's. Honorary pallbearers included Donald Douglas, Lawrence Bell, William Boeing, Robert Gross, Grover Loening, and the comedian Joe E. Brown, whom he had befriended on the carnival circuit in 1911. Roy Beall, the mechanic who had helped build his first flying machine, helped push his casket into the mausoleum wall.

His knack for publicity served him well at the end, with a front page obituary in the *New York Times* repeating all the myths about his career. Even in death, he maintained a showman's disdain for reality. "Mr. Glenn L. Martin was a great man in many respects," wrote his longtime personal secretary, William Ruttig, after the funeral. "It was too bad that a lot of people did not understand him."

The last ten years of Martin's life were tragic, Ruttig reflected honestly. "He should have retired in the calm and serenity of his past achievements. But this was not to be."

The Martin estate was valued at $15 million. He left $3.635 million to various charities, including $2 million to the Minta Martin Aeronautics Research Foundation at the University of Maryland. The will was contested by four obscure cousins.

In 1957, THE year he turned sixty-five, Donald Douglas passed his corporate mantle to his oldest son, Donald Jr. Riding the crest of Korean War and Cold War rearmament, plus healthy sales of propeller-driven airliners in fierce competition with Lockheed, the company's annual net income soared past World War II heights to more than $30 million. But there were time bombs ticking under the glittery surface—no new quantity production contracts from the Navy or Air Force, and lethal hits from Boeing in the nascent jet transport field.

In 1958, net income fell by half. In 1959 and 1960, the company suffered tremendous back-to-back losses of $33.8 million and $9.4 million. Profits began a slow climb in the early 1960s, but a $27.5 million loss in 1966, caused by double failures in the military and commercial markets, attracted the Wall Street scavengers.

Lazard Freres stipulated a merger that would install new management as well as fresh capital. The McDonnell Company of St. Louis, which had been founded just before World War II by James Smith McDonnell Jr., a former engineer at the Glenn L. Martin Company, offered $68.7 million and generous stock exchanges. The inimitable Douglas name thus drew second billing in a new corporate entity that would become one of the dominant forces in the worldwide aerospace-weapons business.

Though he stayed on as a member of the McDonnell Douglas board, Donald Douglas attended few meetings and withdrew from the industry. He designed a new yacht for himself, a motor-sailer named the *Ladyfair*, but led an otherwise modest existence. In the early 1970s, he began to see his daughter, Barbara Douglas Arnold, again after a twenty-year estrangement. His first wife, Charlotte, died in 1976. His second, Peggy, lived until 1989. Prostate and bone cancer claimed the one-time cornerstone of American air power on February 1, 1981, at Desert Hospital in Palm Springs.

Donald Douglas Jr. left the McDonnell Douglas board in 1989, at the age of seventy. The second-eldest son, William, had already retired from the company's astronautics division in Huntington Beach, California. In June 1990, a third son, fifty-nine-year-old James S. Douglas, and twenty-nine-year-old grandson, James Jr., were laid off from their jobs at the troubled Douglas Aircraft unit of McDonnell Douglas in Long Beach, California, ending the family's involvement with the company.

"We were just peons," James Jr. said. "It is tough to be tossed out when your name is on the door. It is the McDonnell corporation now, as far as I am concerned. I am very sad that it is over."

JACK NORTHROP QUIT the company that bore his name at the end of 1952. Board chairman Oliver Echols, a retired general who had been chief of Air Force procurement, hastened the departure by vetoing Northrop's choice of a successor as chief engineer. This humiliation and the brutal destruction of his beloved flying wings in the late 1940s broke his mental and physical health. In 1950, after divorcing Inez, his wife since 1918, he married his secretary of twelve years, Margaret Bateman. They lived together quietly in Santa Barbara until her death in 1977. His health then deteriorated rapidly, and strokes ruined his ability to speak. In 1980 he was taken in a wheelchair to see drawings of the Air Force's secret new "Stealth" bomber, the Northrop B-2, which strongly resembled the Northrop B-35 flying wing.

He died peacefully in a Glendale hospital on February 18, 1981, believing that his technical vision had been vindicated.

IN 1947, when Lockheed recorded a loss of nearly $2.5 million, its first since 1934, Robert Gross still ordered custom-made suits from Lesley & Roberts in London. He and his wife, Mary, continued to live in Bel Air at 649 Stone Canyon Road. The company was quite big enough by now that when trouble came, it did not reach the Gross family.

Like many prominent American businessmen who had never in-dulged in especially strong political rhetoric, he was adopting the slogans of hardline anti-Communism. "The inner ring of Commu-nists are merely using it as a tool to forcefully and relentlessly take over our lives and way of living," he wrote to a family friend in August 1948, echoing the paranoia that would soon infect the entire nation. In March 1949, he told a gathering of the Advertising and Sales Executives Club in Kansas City that "we can spend ourselves into communism or we can two-bit ourselves into communism by not spending enough to prevent the Commies from taking over." Manufacturers "want steadiness in government appropriations so there will be steadiness in the air force," he added. Coincidentally, Lockheed's business was overwhelmingly military. Between 1948 and 1954, annual military sales accounted for between 78 percent and 93 percent of total sales.

In June 1949, his twenty-six-year-old daughter, Palmer, and thirty-five-year-old Charles Emil Ducommun shocked the Los Angeles society columnists by getting married in a small ceremony at St. Thomas Episcopal Church in Manhattan. "Duke" Ducommun was vice president of a family business, Ducommun Metals and Supply Company of Los Angeles, a supplier of aluminum and other materials to the aircraft industry. He had spent the war as a personal aide to Admiral Ernest King, Chief of Staff of the U.S. Fleet. He would later become an important fund-raiser for the Republican Party in Southern California. Palmer and Duke honeymooned for several months, first on safari in Kenya and then on a tour through Turkey, Greece, Italy, and France. When they returned to California, her parents feted them with a Christmas gala for 600 people.

"Seldom since the 'roaring twenties' has such a large and magnificent party been given in this vicinity as the one hosted by Mr. and Mrs. Robert Ellsworth Gross at their palatial Bel-Air home," gossip writer Cholly Angeleno gushed in the *Los Angeles Examiner*. Among the guests noted as dining and dancing under a circus tent that covered the lawn were Mrs. Donald Douglas and son, Donald Jr., her husband apparently elsewhere.

In December 1950, Robert was awarded the French Legion of Honor. "We had a little ceremony here at the factory and the French Government sent a general out from Washington who kissed me on both cheeks and pinned a ribbon on me," he wrote. "It was very nice."

Lockheed thrived throughout the 1950s, buoyed by Korean War rearmament. Robert drove an Aston-Martin, kept a beach cottage at Carpinteria. In 1956, he became chief executive officer and chairman of Lockheed, handing active management of the company over to his brother, Courtlandt, who was named president. They both had their shirts made by T. Hodgkinson Limited in London, and sent them back there regularly for laundering. In 1959, the actor Adolphe Menjou named Robert one of the thirteen best-dressed men in the world.

In May 1961, Robert entered Good Samaritan hospital in Los Angeles for stomach surgery. He read the latest novel of one of his favorite authors—*Thunderball* by Ian Fleming. On June 1, apparently recovered, he wrote to the Hotel Du Cap D'Antibes on the French Riviera, asking the manager to reserve rooms in July for him, his wife, and Blanche Satchel Yeager, a friend from New York and Beverly Hills. A large corner suite with terrace was made available, containing a single room that Blanche had also occupied when the Grosses stayed at the hotel the previous summer.

During the third week of June, Robert developed "a little touch of hepatitis," he wrote to a Lockheed director in France, attributed to needle injections while in hospital. On June 30, his doctors prohibited all travel. "The world looks far, far away," he wrote in disappointment over having to cancel the trip to Antibes.

In July, he was confined to bed on a restricted diet, suffering terribly from nausea, severe headaches, and high fever. He described his wife as "very tired and at times distraught." His thoughts took a philosophical turn.

"Although I am a Republican and likely to remain one, I have been quite impressed with President Kennedy personally since he took office," he wrote. "I feel that most of the problems that he has are ones that he inherited from my Republican administration. The Cuban thing, for example, had its roots 25 years ago when the rich Cuban families dug their toes in and flatly refused to give the peons anything more than a bag of beans a day, and when neither Truman nor Eisenhower went down there and tried to beat some sense into Batista, of course the Cuban scene was wide open for a revolution. Any broken-down Joe McGee could have taken the place over and Castro just did. Same thing in Laos. The froggies messed the whole thing up years ago by not facing facts but President Kennedy gets the unpaid bills."

As for his personal life, he admitted regret. "My life has been so ill-spent from an entertainment and social point of view," he wrote. "For 30 years we have entertained, but 98.2 percent, I am sure, of all the entertaining we have done has been in one way or another connected with the Gargantua Behemoth, the aircraft and now the space and rocket business. Don't ever, ever, let yourself get in for this kind of life. As I lie here looking at the ceiling going over my sins, this is one that stands out. I have really treated Mary abominably and if I ever get out of this mess I intend to put things right."

At 4:45 A.M. on the morning of September 1, 1961, Robert Gross died of pancreatic cancer while murmuring "I won't give in." Dwight Eisenhower wrote to Mary that "he was one of the distinguished leaders of American industry—and an individual with whom I had the most pleasant, but far too few, contacts." Richard Nixon wrote to her remembering him as "a man whose courage and skill as a corporation executive I admired greatly, and whose friendship and support I valued highly." Barry Goldwater extended his "heartfelt sympathies in this hour of great loss to you and America," prompting Mary to reply: "You, too, are striving for the same purpose. May you succeed with God's help!"

Under the terms of Gross's 1934 will, Mary received the bulk of his estate. The probate petition simply stated that his fortune was worth over $10,000. A 1961 codicil contained a provision that left $250,000 to Blanche Yeager.

In July 1982, Courtlandt Gross and his wife, Alexandra, were shot to death in the kitchen of their secluded mansion in Villanova, Pennsylvania, on Philadelphia's Main Line. Courtlandt had been retired from Lockheed since 1967, when he stepped down as chairman of the board. A twenty-two-year-old "drifter," Roger Peter Buehl, was later convicted of the murder. Buehl was the son of Uwe Buehl, owner of a business in Pottstown, Pennsylvania, that sold airplanes and helicopters.

In 1930, a year after quitting the wildly successful Lockheed Aircraft Company when it was taken over by Detroit Aircraft Corporation, Allan Loughead tried to reenter the business by organizing Loughead Brothers Aircraft in Glendale, California. With shallow backing from various smalltime promoters, he managed to build one airplane, a Vega-like machine with unusual side-by-side twin engines on its nose, called the "Olympic Duo." Its performance was unsatisfactory, and it was destroyed in a crash, wiping the company out in 1934.

In 1937, with aviation again popular among speculators, he formed the Alcor Aircraft Corporation in Oakland. In 1938 the firm completed one six-passenger plane, a modern low-wing transport with retractable landing gears that kept the Olympic Duo's peculiar side-by-side engine placement. It was lost that same year in a test flight over San Francisco Bay, where the Lougheads had first tasted fame and fortune in 1915.

Allan went back to selling real estate in Palm Springs, then followed his second wife and their son (his first wife had died in 1922; he remarried in 1926, and was divorced in 1936) to Tucson, where he lived out the rest of his life with no further involvement in the aviation business. Just before he died, in May 1969, the mighty Lockheed Aircraft Corporation acknowledged its roots by hiring him as a part-time consultant.

Malcolm Loughead, who sold his profitable hydraulic-brake company in 1929 for several hundred thousand dollars, lent his name to Allan's 1930 venture, but sank most of his fortune into a Calaveras County gold mine. His biggest strike was worth only about $20,000. In 1920 he had married his brother's wife's half sister, but divorced her in Detroit in 1928 on grounds of infidelity. He remarried in 1930. For many years he operated mining machinery and drove logging trucks to make ends meet. He died, probably of Alzheimer's disease, in 1959.

NOTES

Prologue: Fantasies

pages 19–24: Didier Masson's mission in Mexico is culled from reports in the *New York Times*, 5/8/13, 5/9/13, 5/20/13, 5/31/13, 6/22/13; *Los Angeles Tribune*, 5/13/13; *Santa Ana Register*, 5/28/13; and *Tucson Citizen*, 2/2/14. Also from Masson's personal letters in the collection of the National Air and Space Museum library, Washington, D.C., dated 8/5/35, 10/7/35, 8/13/36, 2/24/37, and 7/27/37; and from a letter written by his son, Didier Jr., dated 10/23/50. The *Los Angeles Tribune* reported that Glenn Martin was called to testify before a federal grand jury investigating a conspiracy to smuggle an airplane—bought from Martin by Van M. Griffith, secretary of the Aero Club of Southern California—to Mexican Constitutionalists. Martin claimed ignorance of the plan. The *Santa Ana Register* reported that Masson, his mechanic, and six Mexicans were then indicted. The *Tucson Citizen* reported that Martin appeared in U.S. District Court as a government witness against two Mexicans in the case. "Masson had been trying to buy an aeroplane for six months," Martin told the paper. "He couldn't find the money, so he came back from San Francisco to work in my shops awhile. When he did arrange for payment, I had no idea that the machine was for the Constitutionalists. Masson told me his first stopping place would be Tucson. The Government's detective asked me about the transaction and I told him that the machine had been shipped to Tucson. It was then in the railroad yards, I suppose." Further reports in the journal *Aeronautics*, 2/28/14 and 5/15/14. Air warfare during the Mexican Revolution of 1913 is discussed by James R. Hinds in "Bombs over Mexico," *Aerospace Historian*, September 1984. The wider context of the conflict is treated in *The Mexican Revolution*, volume 2, by Alan Knight, Cambridge University Press, 1986; and *The Course of Mexican History*, by Michael C. Meyer and William L. Sherman, Oxford University Press, 1987.

Chapter 1: Invention Before Industry

page 28: The AAAS meeting is described in *A Dream of Wings*, by Tom Crouch, Norton, 1981.

pages 29–34: Langley's career is analyzed by Wallace R. Beardsley in *Samuel Pierpont Langley: His Early Academic Years at the Western University of Pennsylvania*, an unpublished Ph.D. dissertation in the collection of the

National Air and Space Museum library. His aeronautical experiments are described by Crouch, *Dream*. The Wright brothers' regard for him is discussed in *Wilbur and Orville*, by Fred Howard, Knopf, 1987.

pages 35–36: Wright quotations here and elsewhere, unless otherwise noted, are from *The Papers of Wilbur and Orville Wright: Including the Chanute-Wright Letters and Other Papers of Octave Chanute*, edited by Marvin W. McFarland, McGraw-Hill, 1953.

pages 38–40: Glenn Curtiss's work is described by C. R. Roseberry in *Glenn Curtiss: Pioneer of Flight*, Doubleday, 1972. His relationship with the Wrights is treated in Howard, *Wilbur and Orville*.

page 41: War Department aviation expenditure cited in *Ideas and Weapons*, by I. B. Holley, Archon, 1971, p. 29.

page 42: Arnold quotation is from his autobiography, *Global Mission*, Harper & Row, 1949, p. 2.

page 42: Aviation expenditures by foreign governments from Holley, *Ideas*. Appeal from Secretary of War appeared in the War Department's *Annual Report of 1908*, also cited in Holley, p. 27.

Chapter 2: *Money-Minded Eccentrics*

page 44: Glenn Martin's life has not previously been the subject of an authoritative biography. The primary source for this volume is the extensive collection of his papers in the Library of Congress (hereafter noted as GLM papers), which includes personal and business correspondence, company records, scrapbooks, and memorabilia. Unless otherwise noted, details of his private life, as well as the financial course of his company, have been culled from documents in this collection. The National Air and Space Museum library contains a minor file of secondary biographical material (hereafter noted as NASM file).

page 47: First flight test mishap described in *Southern California Industrial News*, 8/1/66, NASM file.

page 47: "You were so interested," letter from Ella M. Perry, 1/22/38, GLM papers.

page 48: Quotation from *Scientific American*, 10/28/11, p. 382.

pages 51–52: Martin's offer to pay a fee to the Wrights described in the journal *U.S. Air Services*, December 1955, p. 7.

page 52: "frightening them out of their wits," *Pomona Progress*, 9/13/11.

pages 52–53: Della Martin's patient history described in letter from Patton State Hospital superintendent, 8/23/50, GLM papers, box 31.

page 53: Lecture to Santa Ana businessmen, *Santa Ana Blade*, 11/23/11.

page 56: "This show and others like it," from CBS *We the People* radio broadcast, 1/27/42, transcript in GLM papers.

page 57: "Sentiments seemingly opposed to each other," in *The Rise of American Air Power*, by Michael S. Sherry, Yale, 1987, p. 6.

page 59: Incorporation of Glenn L. Martin Company reported in *Santa Ana Register*, 6/21/12, 8/16/12.

page 59: Chicago earnings reported in *Chicago Tribune*, 9/24/12; rescue attempt reported in *Los Angeles Times*, 12/16/12; *Los Angeles Tribune*, 12/16/12; and *Los Angeles Examiner*, 12/17/12. In a letter dated 2/12/42, Garbutt recalled Martin's "undismayed smile when you lost practically everything you had on our search for Lawrence and Kearney and you said 'I'll play the game with the cards I have left.' " Garbutt wrote that Martin's determination was one of the "outstanding things" that "made me resolve that Zimmerman & Co. would never get the best of you."

page 61: "The United States is far behind," quoted in *Los Angeles Express*, 12/5/12.

page 62: Brashear quotation in *Los Angeles Tribune*, 12/11/13.

page 64: Bell's description of the "Battle of the Clouds" is from interview transcript in the collection of the Oral History Research Office, Columbia University.

Chapter 3: Southern California

page 68: Northrop quotations, unless otherwise noted, are from 1972 oral history transcript in the collection of the University of California at Santa Barbara library; copy provided by Northrop Corporation historian Ira E. Chart.

page 70: Biographical details of the Loughead brothers' early careers are drawn from a multipart series written by Denham Scott in the weekly *Southern California Industrial News*, which was published between 5/1/67 and 5/5/69. According to Allan Loughead's son, John Lockheed, who provided a complete copy of the series, Scott "reviewed each section with my father" before publication—"I believe it to be quite accurate." Further biographical material is in the Lockheed corporate journal *Lockheed Horizons*, issue no. 12, 1983; and *Revolution in the Sky*, by Richard Sanders Allen, Orion, 1988.

pages 71–72: "the safest means," quoted by John Lockheed during the July 1986 enshrinement of his father into the National Aviation Hall of Fame; videotape provided by Mr. Lockheed.

pages 75–76: For the formation of World War I era military-industrial bureaucracy, see *The Military-Industrial Complex*, by Paul Koistinen, Praeger, 1980; and *The War Industries Board*, by Robert D. Cuff, Johns Hopkins, 1973. Coffin quotation is from his essay "New Deal Performance" in *Truth About the New Deal*, edited by Earl Reeves; Longmans, Green, 1936.

page 76: "Twentieth century warfare," from testimony before the Senate Committee on Military Affairs investigation of the War Department, 1917, quoted in Koistinen, p. 26; "the boogaboo," from a letter to Josephus Daniels, 3/9/16, quoted in Cuff, p. 23.

page 77: "opens up a new," quoted in *New York Times*, 10/12/16, p. 10.

page 77–78: Army war expenditures cited by Koistinen, *Military-Industrial*, p. 31.

page 78: "clubbable," from "Background and Incorporation of the Institute of the Aeronautical Sciences," by Jerome Hunsaker and Lester Gardner, March 1952, in GLM papers, box 75.

page 78: Estimate of 1916 aircraft production from *Aerospace Facts and Figures*, 1962," Aircraft Industries Association; "obvious," cited in Holley, *Ideas*, p. 35.

page 79: Secrecy measures and Signal Corps ignorance described by Holley, *Ideas*, p. 36.

Chapter 4: The Great Proposition

page 81: "I don't think my mother would like it," from a letter dated 2/16/54, GLM papers, box 75.

page 82: Douglas quotations here and elsewhere, unless otherwise noted, are from 1959 oral history transcript in the collection of the Oral History Research Office, Columbia University.

page 82: "a bitch," from an interview by the author with Barbara Douglas Arnold, 2/9/90. Donald Douglas's childhood literary efforts described in *Saturday Evening Post* profile, 11/27/43.

page 84: Hunsaker's opinion of Douglas's financial condition is from oral history transcript in the Columbia University collection.

page 86: Description of Ogg as domineering is from Barbara Douglas Arnold interview.

page 87: Caleb Smith Bragg biographical information is drawn from a file in the collection of the National Air and Space Museum library and from his obituary in the *New York Times*, 10/26/43.

pages 88–89: The Wright-Martin merger is described by contemporaneous newspaper articles preserved in scrapbooks of the GLM papers, and briefly in *The Aviation Business*, by Elsbeth E. Freudenthal, Vanguard Press, 1940.

page 89: Martin's technical critique of Curtiss airplanes, specifically their use of ailerons, is in letter to Orville Wright dated 3/27/15, Wright papers, Library of Congress, box 39. NACA report of 1917 extracted in hearings of the Select Committee on Expenditures in the War Department, U.S. House of Representatives, 1919–20, serial 2, vol. 1, p. 411.

pages 90–91: The Wright-Martin patent gambit is described in the NACA report of 1917; the cross-license agreement, dated July 24, 1917, was published in the *Congressional Record* of May 9, 1918, and reprinted by the House Select Committee, op. cit., p. 403.

page 92: "was to some extent possibly forced," House Select Committee, op. cit., p. 422; "in all probability," quoted in *Birth of an Industry*, by

Howard Mingos, W. B. Conkey Co., 1930. Mingos, a public relations agent for various aircraft industry groups, dismissed allegations of an "aircraft trust" as a "hymn of hate" against the patent pool.

page 93: Letter from Gregory to Baker dated 10/6/17, published by Senate Military Affairs Committee in *Senate Report 555*, 1918.

page 93: Royalty figure cited in *House Report 1653*, 1925, p. 3180.

pages 94–95: Mismanagement at Simplex was examined in the report to President Wilson by Charles Evans Hughes, extracted by the House Select Committee, op. cit.; also by Senate Military Affairs Committee.

page 95: British and German warplane evolution described in *Strike from the Sky*, by Richard P. Hallion, Smithsonian, 1989.

page 96: "At that time," from a statement on company history dated 4/17/34, GLM papers.

page 97: Holley quotation from *Ideas and Weapons*, p. 50.

page 97: "sheer waste of time," the judgment of General P. C. March, Army Chief of Staff, in his book *The Nation at War*, Doubleday, 1932, p. 47.

page 99: SPAD program described in Hughes report and discussed by Holley, *Ideas*, p. 61.

page 100: "they are going to," Orville Wright to Glenn Martin, 5/19/17.

page 101: Attorney General's findings re Deeds extracted in House Select Committee, op. cit., p. 58.

page 101: "I have viewed," Wright to Martin, 5/19/17.

pages 103–104: Liberty engine development history described in House Select Committee hearings, serial 2, vol. 3.

page 104: "not a suitable motor," from memorandum for Director of Air Service, 5/2/19, Mitchell papers, Library of Congress, box 7.

page 105: *New York Times* account of Baker speech, 12/13/17.

page 106: "winter weather," *New York Times*, 3/21/18.

page 106: The MB-1 contract was described in an opinion by the U.S. Comptroller General dated 1/13/25, published in the hearings of the Select Committee of Inquiry into Operations of the United States Air Services, U.S. House of Representatives, 1925, pp. 1441–51.

pages 108–109: "being pestered" and "Have you shown" quoted in *Southern California Industrial News*.

page 109: The Lougheads' encounter with Spencer is recounted in a letter dated 7/13/57, Robert Gross papers, Library of Congress, box 12.

page 111: "This will introduce," *Southern California Industrial News*.

Chapter 5: The Rat Race

page 113: Arnold quotation from *Global Mission*, p. 64.

page 113: Rickenbacker quoted in *House Report 637*, 1920, p. 9.

pages 114–15: "War means waste," ibid., p. 3. Casualty figures cited in *A*

Short History of World War I, by James L. Stokesbury, Morrow, 1981, p. 310.

page 115: Freudenthal quotation from *Aviation Business,* p. 61.

page 116: Loening quotation from *Our Wings Grow Faster,* Doubleday, 1935, p. 100. Ninety percent liquidation estimated in "Memorandum for Colonel O'Brien," 9/30/19, Mitchell papers, box 7.

page 116: MB-1 contract details in 1925 Comptroller General report.

page 118: Bane to Mitchell, 8/19/19, Mitchell papers, box 7; "up to any," from notes titled "Airplanes," 10/28/19, Mitchell papers, box 7.

pages 118–19: Transcontinental test described in letter from Mitchell to Director of Air Service, 10/21/19, Mitchell papers, box 7.

page 119: Mitchell to Bane, 2/10/20, Mitchell papers, box 8. Martin capitalization cited in 1925 House Select Committee hearings, p. 1128; "a direct result," from U.S. Senate Banking and Currency Committee hearings, 1933–34, *Senate Report 1455,* p. 295.

page 120: Bane to Mitchell, 2/25/20, Mitchell papers, box 7. Hunsaker comments from Columbia oral history transcript, pp. 163–64. Bane to Mitchell, 3/3/20, Mitchell to Bane, 3/20/20, Bane to Mitchell, 12/3/20, Mitchell papers, box 7.

page 121: "remarkable," 1925 Comptroller General report.

page 121: Arnold quotation from *Global Mission,* p. 96. Hunsaker quotations from Columbia oral history transcript. Arnold also remembered the flamboyant Mitchell at headquarters in Souilly, France, wearing "that blouse with the outsize patch-pockets, and the famous pink breeches," an outfit that would surely have appealed to Glenn Martin.

page 122: Wilson quotation is from oral history transcript in the Columbia University collection.

page 123: "no better way," quoted in *Saturday Evening Post* profile, 12/4/43.

page 124: "Many people argue," from "The Airplane as a Commercial Possibility," in *Transactions,* Society of Automotive Engineers, 1919, pp. 444–62.

page 125: NC transatlantic flight described in *History of United States Naval Aviation,* by Archibald D. Turnbull and Clifford L. Lord, Yale, 1949.

pages 132–33: Chandler's business dealings also described in *Saturday Evening Post* profile, 12/16/39.

page 135: "to take necessary actions," quoted in *McDonnell Douglas Aircraft,* vol. 1, by René J. Francillon, Naval Institute Press, 1988, p. 56.

page 136: The round-the-world flight was described in *The First World Flight,* by Lowell Thomas, Houghton Mifflin, 1925.

page 136: Douglas production figures compiled in *McDonnell Douglas Aircraft;* in 1938 Douglas Aircraft Company annual report; and in *Douglas*

Aircraft Company and the Aircraft Manufacturing Industry, Kuhn, Loeb & Co., 1954, NASM file.

page 137: MB-2 contract detailed in 1925 Comptroller General report.

pages 142–43: The Washington Conference is discussed in *The United States and the Washington Conference*, by Thomas H. Buckley, University of Tennessee, 1970. "I want to warn you," quoted in *Cleveland Plain Dealer*, 12/20/21.

page 143: Figures for federal aviation expenditures are from *House Report 1653*.

page 146: Rae discussed this period in *Climb to Greatness*, MIT, 1968. The five-year military aviation expansion programs are detailed in *The Politics of Military Aviation Procurement, 1926–1934*, by Edwin H. Rutkowski, Ohio State, 1966.

page 147: Coolidge quotation is from *1928 Budget Message of the President*, cited in Rutkowski, p. 32.

pages 151–54: The literature about Lindbergh is extensive, with his own writings still the most insightful. *Charles A. Lindbergh: An American Life*, edited by Tom Crouch, Smithsonian, 1977, contains a careful bibliography as well as essays on various aspects of his career. For his antiwar views, see especially *Charles A. Lindbergh and the Battle Against American Intervention in World War II*, by Wayne S. Cole, Harcourt Brace Jovanovich, 1974.

page 155: Wilkins quotation from his book *Flying the Arctic*, Putnam, 1928.

pages 156–57: The peak flotation of $1 billion against annual earnings of $9 million was proffered by Loening, *Our Wings*, pp. 193–94.

page 157: Douglas brochure in NASM file.

page 158: Testimony of J. H. Kindelberger, Douglas vice president, 2/21/34, before aeronautics subcommittee of House Naval Affairs Committee, p. 835 of hearing transcript.

page 158: The Deeds stock deal is described in Freudenthal, *Aviation Business*, pp. 95–96; "an aeronautical education," *Fortune*, May 1935, p. 178. "After we had to hire," related by Barbara Douglas Arnold.

page 160: "who even remotely," from "Aircraft Production Facts," by Colonel George Mixter and Lieutenant H. H. Emmons, January 1919, cited in Freudenthal, p. 56.

pages 160–61: Martin production figures from testimony before aeronautics subcommittee of House Naval Affairs Committee, 2/16/34, pp. 732–33, and from *The Glenn L. Martin Company, Supplementary Data*, November 1929, GLM papers. "Sometimes a man," quoted in *Cleveland Press*, 1/9/28.

page 161: "Glenn L. Martin was driven out," quoted in *Cleveland News*, 2/5/28.

page 162: "People think," quoted in *Baltimore Evening Sun*, 7/30/28.

pages 162–63: "Ignoring questions," quoted in *Baltimore Sun*, 8/4/28.

Chapter 6: Depression

pages 166–67: The War Department's 1930 *Industrial Mobilization Plan* is described by Koistinen, *Military-Industrial,* pp. 51–52. Market share and cost increases of military aircraft cited in Holley, *Buying Aircraft: Matériel Procurement for the Army Air Forces,* Department of the Army, 1964, p. 20.

page 167: Lockheed production figures cited by Allen, *Revolution,* p. 243.

page 168: Production numbers for military and civilian aircraft are from *Aerospace Facts and Figures,* various years.

page 168: The participation of Colonel Boots in the XP-900 program was recounted by Robert Gross in a letter to Randolph Walker dated 3/5/35, Gross papers, box 5. No further light was shed on Boots's identity. A Colonel Norman Jay Boots, apparently the proprietor of the Boots Aircraft Nut Corporation in New York, corresponded with Lester Gardner at the Institute of the Aeronautical Sciences during World War II—letter dated 3/9/44, papers of the American Institute of Aeronautics and Astronautics (AIAA), Library of Congress, box 88.

page 169: Lockheed deficit figures from "Analysis of Deficit," Gross papers, box 3. Unless otherwise noted, post-1929 Lockheed financial data are from documents in the Gross collection.

page 171: Like Glenn Martin, Robert Gross has not previously been the subject of a serious biography. The primary source for this volume is the collection of his papers in the Library of Congress, especially his extensive personal and business correspondence, which provides a detailed account of Lockheed's development after 1930. Gross was a careful writer. The letters are quoted liberally here to convey his eloquence and acumen. Aspects of his private life have been culled from this source as well, unless otherwise noted.

page 176: "The world was right flat," quoted in *Biographical Notes,* an undated post–World War II typescript, stamped by Lockheed publicity department but likely written by Gross himself, Gross papers, box 1.

page 177: "It's yours," quoted in *Southern California Industrial News,* 5/5/69.

page 178: The formation of United is described by Freudenthal, *Aviation Business,* pp. 103–105.

page 179: United's stock payments to Northrop and Jay described in *Wall Street Journal,* 8/9/29. United's share of aircraft markets compiled by House Naval Affairs aeronautics subcommittee, 1934, pp. 502–503 of hearing transcript.

page 180: Douglas profit levels compiled by House Naval Affairs subcommittee, p. 499; 1930–32 dividend and profit figures from 1938 Douglas annual report to stockholders.

page 181: The 1932 stock deal was described in the testimony of J. H. Kindelberger, pp. 835–37 of hearing transcript. At the time of these hearings, the third-largest holder of Douglas shares, behind North American Aviation and the Douglas family, was the New York firm of E. A. Pierce, which traded heavily in aviation stocks. In July 1929, Douglas had granted Pierce options to buy 150,000 shares of the company's stock at $10 a share. Pierce waited until June 1933 (six months before the option would expire), when the stock was selling on the New York Stock Exchange for about $17, to exercise its rights on 125,000 of these shares, thereby realizing a profit of $875,000.

page 183: Douglas's 24 percent profit level was noted in House Naval Affairs subcommittee transcript, p. 499.

pages 184–85: The account of Douglas's meeting and subsequent relationship with Tucker is according to his daughter, Barbara Douglas Arnold, in interview. In a letter to the author dated 7/26/90, J. Leland Atwood, one of the designers of the DC transports and a longtime Douglas senior executive, said that the romance was "reasonably well known to most of his associates at a fairly early stage."

pages 185–86: Sale of Martin seaplanes to Brazil described in *Baltimore Sun*, 10/24/30 and various issues, 11/30; sale to Soviet Union described in various issues, 6/30.

page 191: "It was suggested," from Martin's testimony before the U.S. Special Committee on Army Air Corps, Department of War, 5/9/34.

page 193: Douglas's absolute control over the Northrop subsidiary was described in an office memorandum dated 12/12/34, AIAA papers, box 145. Douglas actually held 100 percent of Northrop's common stock in escrow until all preferred stock (which Douglas received in return for cash advances) was retired.

Chapter 7: Merchants of Death

page 194: The PWA grants were examined in a special investigation by the House Military Affairs Committee, *House Report 1506*, 5/7/34. Book review appeared in the *New York Times*, 4/29/34, but on the same day the paper editorialized against the "merchants of death" thesis.

page 195: Vinson's actions reported in *New York Times*, 2/1/34.

page 196: Martin testimony dated 2/16/34, pp. 728–42 of hearing transcript. Perhaps the most useful parts of his testimony were two tables showing net profits since 1920 and the financial disposition of major contracts for the U.S. Navy.

pages 197–98: Kindelberger testimony dated 2/21/34. Included was a list of principal stockholders.

page 200: "the activities," from *Senate Resolution 206*, cited in *In Search of Peace*, by John E. Wiltz, Louisiana State University, 1963, p. 36. FDR statement dated 5/18/34, also cited by Wiltz, p. 40. FDR asked Congress to prohibit arms sales to Bolivia and Paraguay, where a regional conflict was being fed by munitions from abroad.

page 201: DC royalty noted in *Wall Street Journal,* 2/27/34.

pages 201–205: Testimony and documentation relevant to Lockheed and Douglas is from *Hearings Before the Special Committee Investigating the Munitions Industry,* part 39, 2/20/36, and from the Nye committee's final report, *Senate Report 944,* part 3, 1936.

page 202: Regarding Lockheed sales to Japan, the committee noted in its report that "in 1935, Lockheed sold two planes to Japan via Okura & Co. A. J. Miranda, whose firm of Miranda Bros., as Lockheed's general export agents, sold the planes to Okura & Co., of Japan, testified that it was his understanding that Okura & Co. sold the planes to the Japanese Navy." This was typical of how American manufacturers kept one or two steps away from controversial transactions. The U.S. government was not ignorant of such practices. In 1933, three Japanese aviation experts representing Okura were barred from visiting the Martin Company's Middle River plant, where B-10 bombers were under construction (*Washington Herald,* 2/25/33).

page 204: Starting in 1932, Nakajima developed the B5N2 carrier-borne torpedo bomber that devastated the U.S. Pacific Fleet at Pearl Harbor and elsewhere during the first year of war. Like the splendid Mitsubishi Zero fighter, these aircraft shocked Allied military experts who had considered Japan capable of producing only shoddy imitations of foreign planes.

page 205: The Fokker incident is culled from reports in the *New York Times,* 9/4/35, and *New York Herald Tribune,* 9/5/35. The Elliott Roosevelt incident is recounted by Wiltz, *Search,* pp. 67–68.

page 206: Truman quotation from his memoir *Year of Decisions,* Doubleday, 1955, p. 190.

pages 206–207: Martin financial data from an FTC registration statement dated March 1934, GLM papers, box 78.

page 207: Charles R. Crane, then seventy-five years old, was a member of a wealthy Chicago family involved in the manufacture of valves and fittings. He had been vice chairman for finance of the Wilson presidential campaign and served on various diplomatic missions to Russia, Turkey, and China during and after World War I. He was thus in a position to be closely familiar with some of Martin's prime export customers.

pages 210–11: Lockheed figures from "Statement of Cash Receipts and Disbursements," Gross papers, box 4.

page 218: Elliott Roosevelt meeting recounted in letter dated 6/20/33, Gross papers, box 4.

page 226: The scissored letters were Courtlandt to Robert, 8/8/34, Gross papers, box 4, and Robert to Courtlandt, 9/13/34, box 3. Courtlandt made specific reference to Carter Tiffany, Anthony Fokker's representative in New York, as a participant.

Chapter 8: Exports

page 232: American notions about aerial bombardment have been examined with wide scope by Michael Sherry, *Rise of American Air Power.*

page 233: Mitchell quotation cited in Sherry, p. 54. Martin quotation from the transcript of his testimony before the U.S. Special Committee.

page 234: Munitions definition cited by Freudenthal, *Aviation Business,* pp. 265–66. Aviation export figures compiled by Freudenthal from U.S. government statistics.

page 235: *Inter Avia,* 10/11/34. Export figures compiled by Freudenthal, pp. 295–96, and in Holley, *Buying Aircraft,* p. 19.

page 236: Observations about Glenn Martin's clothes in *Santa Ana Journal,* 9/24/35; *Saturday Evening Post,* 8/14/37.

page 237: B-10 losses in Spain reported by *Baltimore Evening Sun,* 2/9/38; "from photos," *New York Times,* 5/6/39.

page 239: *Wall Street Journal,* 4/26/37; *Financial World,* 7/14/37.

page 241: Lindbergh's attendance at the Lilienthal meeting is discussed by Cole, *Against Intervention,* pp. 36 and 39. Glenn Martin's membership in the society was deleted from otherwise copious lists of honors and affiliations distributed by the Martin Company publicity office after World War II.

pages 244–45: The background of labor strife in the aircraft industry is discussed in *The CIO Challenge to the AFL,* by Walter Galenson, Harvard, 1960, pp. 184–87 and 506–509; also in *Turbulent Years,* by Irving Bernstein, Houghton Mifflin, 1970. Koistinen, *Military-Industrial,* mentions the exclusion of unions from mobilization planning.

pages 245–48: The account of the Douglas sit-down strike is culled from reports in the *New York Times,* 2/24/37 through 3/3/37.

pages 249–50: Douglas actions against Northrop culled from reports in the *New York Times,* 9/8/37, 10/22/37, 2/2/38; the *Wall Street Journal,* 9/9/37, 9/22/37, 9/23/37, 10/21/37, 4/9/38; the *New York Herald Tribune,* 1/6/38; the journal *Aviation,* November 1937; memoranda titled "West Coast Report to National Aviation Corp." dated 9/3/37, 9/17/37, 10/6/37, AIAA papers, box 161; and statements distributed by the Douglas Aircraft Company, AIAA papers, box 144.

page 251: Douglas figures from 1938 annual report.

page 251: "turned it into the Zero," from Eugene Wilson oral history transcript in the Columbia University collection. (The Japanese began development of the Zero's direct antecedent in 1934, with first flight trials in February 1935. The prototype Northrop 3A was lost in July 1935 and the design was sold to Chance Vought, which in turn sold its version to Mitsubishi. In May 1937, Mitsubishi received specifications for what would become the Zero from the Imperial Navy, completing a prototype by March 1939.)

pages 251–52: DC-4 sale to Japan described in the journal *American Aviation*, 9/1/39, and the *New York Herald Tribune*, 8/15/39. DC-3 work for Japan noted in memorandum titled "Douglas Aircraft Company" dated 9/3/38, AIAA papers, box 147.

page 254: Airmail subsidies cited in Freudenthal, *Aviation Business*, p. 310.

page 258: Lockheed sold thirty Super Electras to Japan between March and September 1938. They were flown as airliners on routes to China. Military versions subsequently produced by Tachikawa and Kawasaki were used for Army transport.

page 262: *New York Journal and American*, 5/28/38.

page 263: Courtlandt Gross was conscious of what had brought him to Lockheed. "I'm just the dumb kid brother," he told the *Saturday Evening Post* in April 1946. "I wouldn't even be the office boy if it weren't for Bob."

page 264: DB-7 incident described in *New York Herald Tribune*, 1/27/39, and *New York Times*, 1/29/39.

page 265: Douglas testimony before Special Senate Committee Investigating the National Defense Program, 8/22/41; "thoroughly angry," *Fortune*, December 1939.

pages 265–66: *Fortune*, ibid.

pages 268–69: "wanted to be free," in *Jack Northrop and the Flying Wing*, by Ted Coleman, Paragon House, 1988, p. 54. Northrop Company financing described in *Wall Street Journal*, 5/13/39, 6/3/39, 6/28/39, and *New York Herald Tribune*, 5/14/39; also in memoranda dated 3/4/39, April 1939, and 6/19/39, AIAA papers, box 161.

Chapter 9: The Iron Cornucopia

pages 270–71: War production figures from Holley, *Buying Aircraft*, chapter 21.

page 271: "the triumphs of technological fanaticism," from Sherry, *The Rise*, who argued that "the leaders and technicians of the American air force were driven by technological fanaticism—a pursuit of destructive ends expressed, sanctioned, and disguised by the organization and application of technological means." Lockheed hiring of students described in a company publication titled *Of Men and Stars*, chapter 6.

pages 272–73: *The New Yorker*, 11/28/42.

page 273: Glenn Martin started a trend by entertaining military officers on the Eastern Shore. By the 1960s and 1970s, major aerospace companies maintained retreats there for Washington VIPs, charging off the expense as overhead on federal contracts.

page 275: The B-26's "Flying Prostitute" sobriquet derived from the droll observation that it had no visible means of support. Congressional comments in report of Senate Special Committee to Investigate the National Defense Program, part 10, 7/13/43, p. 349.

page 276: It took some twenty-eight months after the A-26's maiden flight for combat operations with the aircraft to begin. Production was plagued by design changes and slow delivery of parts. The company was criticized for appearing to be more interested in transport work, which would bring greater payoffs after the war. General Arnold was hard on Donald Douglas in public comments about the program, but their friendship endured.

page 276: Douglas was the subject of a worshipful *Time* cover story, 11/22/43. Looking ahead to peacetime, the magazine stated that "many airplane makers are convinced that they can survive only if the Federal Government steps in and underwrites the industry."

page 280: "We were shanghied," quoted in *Time*.

pages 282–84: The military-corporate bureaucracy is described by Koistinen, *Military-Industrial*, and in Holley, *Buying Aircraft*. These two important works tend to balance one another, Koistinen being rather suspicious of the growth of a military-industrial complex, while the official Army history refrains from criticism.

pages 284–86: Wartime contract types and legislation against excessive profits are discussed exhaustively in *Buying Aircraft* and *The Army and Economic Mobilization*, also in the official history series.

page 286: Testimony of Francis A. Callery, vice president of Consolidated Aircraft Corporation, dated 6/22/43, in report titled *Investigate Progress of the War Effort*, vol. 2, pp. 777–99.

Chapter 10: Permanent Rearmament

page 290: Northrop testimony published as *Exhibit 1531* in record of Special Committee to Investigate the National Defense Program, part 31, pp. 15606–13; Raymond testimony in same record, pp. 15401–19; Douglas letter, dated 8/21/45, published as exhibit 1528.

page 291: *Time*, 1/14/46.

page 293: Smith, Barney study dated 10/4/46, GLM papers, box 23.

page 295: "I find it very difficult to talk about the airplane as a weapon of war," Gross told the committee. "Not perhaps for the reason you think. It may seem strange to you that one who has made many thousands of planes for this war finds a reluctance to talk about war planes, but the prospect of an airplane maker pleading the cause of air security is somewhat tragic."

pages 296–97: *Survival in the Air Age*, reprinted by Arno Press, 1979.

page 297: Arnold quotation cited by John Lewis Gaddis in *The Long Peace*, Oxford University, 1987, p. 25. See also Gaddis's *Strategies of Contain-*

ment, Oxford, 1982, for background on postwar diplomatic and military policies.

page 301: Martin's investment failures produced significant tax write-offs, of course—a benefit noted by a local Arkansas newspaper.

page 308: "The U.S. military were not very interested," in Coleman, *Flying Wing,* p. 62.

page 309: Northrop company organization described in AIAA memoranda and in Northrop testimony before Senate Special Committee; 1940 Northrop financial figures from annual report.

page 310: The quoted description of Von Karman is from Sherry, *American Air Power,* p. 200.

pages 310–11: For policy background of bomber contracts, see *The Army Air Forces in World War II,* U.S. Air Force Historical Division, vol. 6, pp. 243–46. Coleman, *Flying Wing,* describes the XB-35 program from a company perspective.

pages 312–14: For investigation of Joseph Foa's contentions about the Northrop flying wing program, see "Skeleton Alleged in the Stealth Bomber's Closet," by Wayne Biddle, *Science,* 5/12/89, pp. 650–51, and "B-2 Comes Up Short," 10/20/89, p. 322.

page 317: "stupendous task," statement by Arthur Raymond, quoted in Rae, *Climb to Greatness,* p. 208.

pages 317–18: Appearance of Tucker in Douglas company organization charts recalled by J. L. Atwood in telephone interview with author, 5/23/90.

pages 318–19: Copy of Douglas speech in GLM papers, box 39.

Epilogue: Fates

pages 320–21: In October 1951, the accounting firm of Touche, Niven, Bailey & Smart presented a summary of the Martin Company's financial plight to the Mellon, National City, and Chase banks, noting that the 4-0-4 had been priced and sold in February 1950 on a basis of 2.94 direct factory labor hours per pound of airframe. Hours per pound had risen to 4.23, an increase of 44 percent above the sales price estimate. Each extra tenth of a labor hour per pound boosted costs by $750,000. In order to achieve the 2.94 ratio, Martin needed to "produce much more effectively than for any of its projects in the past and, in fact, much more effectively than any manufacturer in the industry," the Touche report concluded. Copy of report in GLM papers, box 83. In a letter dated 1/12/52, Secretary of the Navy Dan Kimball told Defense Mobilization Director Charles Wilson that "the confidence of both the Navy and the Air Force procurement authorities in the effectiveness of the present top management has been virtually destroyed." Copy of letter in GLM papers, box 83.

page 321: George Bunker's nomination by Mellon reported in *Aviation*

Week, 2/11/52. Laurence Rockefeller participation reported in *Aviation Week,* 1/21/52. Stuart Symington, then head of the RFC, preferred that the Martin Company raise new cash by selling stock, which he thought would be attractive to investors, given the huge backlog of military contracts. He believed that bankers would reap extraordinary profits under the bailout plan when Glenn Martin's equity position was diluted, since they would get the right to buy shares at a price considerably below then-current market value. He was overruled by Charles Wilson. *Philadelphia Inquirer,* 3/9/52.

page 322: *New York Times* obituary, 12/5/55.

page 324: "We were just peons," quoted in *Los Angeles Times,* 6/15/90.

page 325: "we can spend ourselves," quoted in *Kansas City Star,* 3/7/49.

page 327: Robert Gross will reported in *Los Angeles Examiner,* 9/13/61. Murder of Courtlandt Gross reported on front page of *New York Times,* 7/17/82, and further in issues of 7/18/82, 7/19/82, 9/9/82, 9/19/82, 10/10/82, 1/11/83, 1/19/83.

pages 327–28: Biographical details provided in telephone interview by John Lockheed, 1/30/91.

SELECT BIBLIOGRAPHY

Books

Allen, Richard Sanders. *Revolution in the Sky: The Lockheeds of Aviation's Golden Age*. Orion Books, 1988.

―――. *The Northrop Story, 1929–1939*. Orion Books, 1990.

Arnold, Henry H. *Airmen and Aircraft*. Ronald Press, 1926.

―――. *Global Mission*. Tab Books Military Classics, 1989.

Beardsley, Wallace R. *Samuel Pierpont Langley: His Early Academic Years at the Western University of Pennsylvania;* unpublished Ph.D. dissertation, University of Pittsburgh, 1978; collection of the National Air and Space Museum library, Washington, D.C.

Bennett, Richard R. *Aviation: Its Commercial and Its Financial Aspects*. Ronald Press, 1929.

Bernstein, Irving. *Turbulent Years*. Houghton Mifflin, 1970.

Bilstein, Roger E. *Flight Patterns: Trends of Aeronautical Development in the United States, 1918–1929*. University of Georgia Press, 1983.

―――. *Flight in America: From the Wrights to the Astronauts*. Johns Hopkins University Press, 1984.

Buckley, Thomas H. *The United States and the Washington Conference, 1921–1922*. University of Tennessee Press, 1970.

Calvocoressi, Peter, et al. *Total War: The Causes and Courses of the Second World War*. Pantheon Books, 1989.

Cole, Wayne S. *Charles A. Lindbergh and the Battle Against American Intervention in World War II*. Harcourt Brace Jovanovich, 1974.

Coleman, Ted. *Jack Northrop and the Flying Wing*. Paragon House, 1988.

Corn, Joseph J. *The Winged Gospel: America's Romance with Aviation, 1900–1950*. Oxford University Press, 1983.

Craven, Wesley Frank, and James Lea Cate, eds. *The Army Air Forces in World War II*, vol. 6, *Men and Planes*. University of Chicago Press, 1955.

Crouch, Tom D. *A Dream of Wings: Americans and the Airplane, 1875–1905*. Norton, 1981.

―――, ed. *Charles A. Lindbergh: An American Life*. Smithsonian Institution Press, 1977.

Cuff, Robert D. *The War Industries Board*. Johns Hopkins University Press, 1973.

Foulois, Benjamin D. *From the Wright Brothers to the Astronauts*. McGraw-Hill, 1968.

Francillon, René J. *Lockheed Aircraft Since 1913*. Naval Institute Press, 1987.

———. *McDonnell Douglas Aircraft Since 1920*, vol. 1. Naval Institute Press, 1988.

———. *Japanese Aircraft of the Pacific War*. Naval Institute Press, 1990.

Freudenthal, Elsbeth E. *The Aviation Business: From Kitty Hawk to Wall Street*. Vanguard Press, 1940.

Futrell, Robert F. *Ideas, Concepts, Doctrine: 1907–1964*. Arno Press, 1980.

Gaddis, John Lewis. *The Long Peace: Inquiries into the History of the Cold War*. Oxford University Press, 1987.

———. *Strategies of Containment: A Critical Appraisal of Postwar American National Security Policy*. Oxford University Press, 1982.

Galenson, Walter. *The CIO Challenge to the AFL*. Harvard University Press, 1960.

Goldberg, Alfred. *A History of the United States Air Force, 1907–1957*. Arno Press, 1972.

Hallion, Richard P. *Strike from the Sky: The History of Battlefield Air Attack, 1911–1945*. Smithsonian Institution Press, 1989.

Hammond, Paul Y. *Organizing for Defense: The American Military Establishment in the Twentieth Century*. Princeton University Press, 1961.

Hanle, Paul A. *Bringing Aerodynamics to America*. MIT Press, 1982.

Heinemann, Edward H. *Ed Heinemann, Combat Aircraft Designer*. Naval Institute Press, 1980.

Higham, Robin D. S. *Air Power: A Concise History*. St. Martin's Press, 1972.

Holley, Irving Brinton, Jr. *Ideas and Weapons: Exploitation of the Aerial Weapon by the United States During World War I*. Archon Books, 1971.

———. *Buying Aircraft: Matériel Procurement for the Army Air Forces*. Department of the Army, 1964.

Howard, Fred. *Wilbur and Orville: A Biography of the Wright Brothers*. Alfred A. Knopf, 1987.

Hurley, Alfred F. *Billy Mitchell: Crusader for Air Power*. Indiana University Press, 1975.

Kennedy, David M. *Over Here: The First World War and American Society*. Oxford University Press, 1980.

Knight, Alan. *The Mexican Revolution*, vol. 2, *Counter-Revolution and Reconstruction*. Cambridge University Press, 1986.

Koistinen, Paul A. C. *The Military-Industrial Complex: A Historical Perspective*. Praeger, 1980.

Kolko, Gabriel. *The Triumph of Conservatism: A Reinterpretation of American History, 1900–1916*. Free Press of Glencoe, 1963.

Levine, Isaac Don. *Mitchell: Pioneer of Air Power*. Arno Press, 1972.

Leoning, Grover Cleveland. *Our Wings Grow Faster.* Doubleday, 1935.
————. *Takeoff into Greatness.* Putnam, 1968.
Lougheed, Victor. *Vehicles of the Air.* Arno Press, 1972.
MacCloskey, Monro. *The United States Air Force.* Praeger, 1967.
Marcosson, Isaac F. *Colonel Deeds: Industrial Builder.* Dodd, Mead, 1974.
Marshall, S. L. A. *World War I.* Houghton Mifflin, 1987.
McFarland, Marvin W., ed. *The Papers of Wilbur and Orville Wright.* McGraw-Hill, 1953.
Meyer, Michael C., and William L. Sherman. *The Course of Mexican History.* Oxford University Press, 1987.
Miller, Ronald E. and David Sawers. *The Technical Development of Modern Aviation.* Praeger, 1970.
Mingos, Howard. *The Birth of an Industry.* W. B. Conkey, 1930.
Obendorf, Donald Leroy. *Samuel P. Langley: Solar Scientist, 1867–1891;* unpublished Ph.D. dissertation, University of California at Berkeley, 1969; collection of the National Museum of American History library, Washington, D.C.
Ogburn, William F. *The Social Effects of Aviation.* Houghton Mifflin, 1946.
Overy, R. J. *The Origins of the Second World War.* Longmans, 1987.
Pearson, Henry Greenleaf. *A Business Man in Uniform: A Biography of Raynal C. Bolling.* Duffield, 1923.
Rae, John B. *Climb to Greatness: The American Aircraft Industry, 1920–1960.* MIT Press, 1968.
Reeves, Earl, ed. *Truth About the New Deal.* Longmans, Green, 1936.
Roseberry, Cecil R. *Glenn Curtiss: Pioneer of Flight.* Doubleday, 1972.
Rutkowski, Edwin H. *The Politics of Military Aviation Procurement, 1926–1934.* Ohio State University Press, 1966.
Sherry, Michael S. *The Rise of American Air Power.* Yale University Press, 1987.
Simonson, Gene R., ed. *The History of the American Aircraft Industry.* MIT Press, 1968.
Smith, R. Elberton. *The Army and Economic Mobilization.* Department of the Army, 1959.
Stekler, Herman O. *The Structure and Performance of the Aerospace Industry.* University of California Press, 1965.
Stokesbury, James L. *A Short History of World War I.* William Morrow, 1981.
Swanborough, Gordon, and Peter M. Bowers. *United States Military Aircraft Since 1909.* Smithsonian Institution Press, 1989.
————. *United States Navy Aircraft Since 1911.* Naval Institute Press, 1990.
Truman, Harry S. *Memoirs,* vol. 1, *Year of Decisions.* Doubleday, 1955.
Turnbull, Archibald D., and Clifford L. Lord. *History of United States Naval Aviation.* Yale University Press, 1949.
Vecsey, George, and George C. Dade. *The Pioneers of Aviation Speak for Themselves.* Dutton, 1979.

Wiebe, Robert H. *Businessmen and Reform: A Study of the Progressive Movement.* Harvard University Press, 1962.
Wilkins, George Hubert. *Flying the Arctic.* G. P. Putnam's Sons, 1928.
Wiltz, John E. *In Search of Peace: The Senate Munitions Inquiry, 1934–36.* Louisiana State University Press, 1963.

Government Publications, listed chronologically

Expenditures in the War Department—Aviation, U.S. House of Representatives, Report No. 637, 66th Congress, 2nd session, 1920; and associated Hearings, Select Committee on Expenditures in War Department, 66th Congress, 1st and 2nd sessions, especially Serial 2, vols. 1 and 3.
U.S. House Report No. 1653 ("Lampert Report"), 68th Congress, 2nd session, 1925; and associated Hearings, Select Committee of Inquiry into Operations of the United States Air Services, 68th Congress, 1925.
U.S. President's Aircraft Board ("Morrow Board") *Report,* November 1925; and associated Hearings, 1925.
U.S. Senate Report No. 1455, 73rd Congress, 2nd session, 1934; and associated Hearings, Banking and Currency Committee, 72nd Congress, 1933–34.
U.S. House Report No. 1506, Committee on Military Affairs, 73rd Congress, 2nd session, 1934.
Hearings before Committee on Naval Affairs of the House of Representatives, Sundry Legislation Affecting the Naval Establishment, 73rd Congress, 1933–34: No. 18, Hearings before the subcommittee on aeronautics making an investigation into certain phases of the manufacture of aircraft and aeronautical accessories as they refer to the Navy Department ("Delaney Hearings"), 1934.
U.S. Special Committee on Army Air Corps and Air Mail ("Baker Board"), Department of War, Hearings and Final Report, 1934.
U.S. Senate Report 944 ("Nye Report"), Special Committee to Investigate the Munitions Industry, 74th Congress, 2nd session, parts 3–7, 1936; and associated Hearings, part 39.
Investigation of the National Defense Program, Hearings before a Special Committee Investigating the National Defense Program, U.S. Senate, 77th Congress, 1st session: parts 6–8 (July–October 1941), *Third Annual Report* (March 1944), and part 31 (July–August 1945).
Investigation of the Progress of the War Effort, Hearings before the Committee on Naval Affairs, U.S. House of Representatives, 78th Congress, 1st session: vol. 1 (March–April 1943), vol. 2 (June 1943), vol. 4 (October–December 1943), and *Report No. 733* (October 1943).
U.S. President's Air Policy Commission ("Finletter Commission"), *Survival in the Air Age,* Government Printing Office, 1948.

INDEX